T0100232

Machine Learning and AI for Healthcare

Big Data for Improved Health Outcomes

Second Edition

Arjun Panesar

Apress®

Machine Learning and AI for Healthcare

Arjun Panesar
Coventry, UK

ISBN-13 (pbk): 978-1-4842-6536-9 ISBN-13 (electronic): 978-1-4842-6537-6
https://doi.org/10.1007/978-1-4842-6537-6

Copyright © 2021 by Arjun Panesar

This work is subject to copyright. All rights are reserved by the Publisher, whether the whole or part of the material is concerned, specifically the rights of translation, reprinting, reuse of illustrations, recitation, broadcasting, reproduction on microfilms or in any other physical way, and transmission or information storage and retrieval, electronic adaptation, computer software, or by similar or dissimilar methodology now known or hereafter developed.

Trademarked names, logos, and images may appear in this book. Rather than use a trademark symbol with every occurrence of a trademarked name, logo, or image we use the names, logos, and images only in an editorial fashion and to the benefit of the trademark owner, with no intention of infringement of the trademark.

The use in this publication of trade names, trademarks, service marks, and similar terms, even if they are not identified as such, is not to be taken as an expression of opinion as to whether or not they are subject to proprietary rights.

While the advice and information in this book are believed to be true and accurate at the date of publication, neither the authors nor the editors nor the publisher can accept any legal responsibility for any errors or omissions that may be made. The publisher makes no warranty, express or implied, with respect to the material contained herein.

Managing Director, Apress Media LLC: Welmoed Spahr
Acquisitions Editor: Celestin Suresh John
Development Editor: Laura Berendson
Coordinating Editor: Divya Modi

Cover designed by eStudioCalamar

Cover image designed by Freepik (www.freepik.com)

Images by Krystal Sidwell and Giverny Ison

Distributed to the book trade worldwide by Springer Science+Business Media New York, 1 New York Plaza, Suite 4600, New York, NY 10004-1562, USA. Phone 1-800-SPRINGER, fax (201) 348-4505, e-mail orders-ny@springer-sbm.com, or visit www.springeronline.com. Apress Media, LLC is a California LLC and the sole member (owner) is Springer Science + Business Media Finance Inc (SSBM Finance Inc). SSBM Finance Inc is a **Delaware** corporation.

For information on translations, please e-mail booktranslations@springernature.com; for reprint, paperback, or audio rights, please e-mail bookpermissions@springernature.com.

Apress titles may be purchased in bulk for academic, corporate, or promotional use. eBook versions and licenses are also available for most titles. For more information, reference our Print and eBook Bulk Sales web page at http://www.apress.com/bulk-sales.

Any source code or other supplementary material referenced by the author in this book is available to readers on GitHub via the book's product page, located at www.apress.com/978-1-4842-6536-9. For more detailed information, please visit http://www.apress.com/source-code.

Printed on acid-free paper

Dedicated to the infinite majesty of Kirpa, Ananta, and Charlotte.

Table of Contents

About the Author

Arjun Panesar is the founder of Diabetes Digital Media (DDM), which provides clinically validated digital health solutions to over 1.8 million people. Arjun holds a first-class honors degree (MEng) in computing and artificial intelligence (AI) from Imperial College London. He has a decade of experience in big data and affecting user outcomes and leads the development of intelligent, evidence-based digital health interventions that harness the power of big data and machine learning to provide precision patient care to patients, health agencies, and governments worldwide.

Arjun's work has received international recognition and was featured by the BBC, *Forbes*, *New Scientist*, and *The Times*. DDM has received innovation, business, and technology awards, including the award for the best app for prevention of type 2 diabetes.

Arjun is an advisor to the Information School at the University of Sheffield, Fellow to the NHS Innovation Accelerator, and was recognized by Imperial College as an emerging leader in 2020 for his contribution and impact to society.

About the Technical Reviewer

Ashish Soni is an experienced AIML consultant and solution architect. He has worked and solved business problems related to computer vision, natural language processing, machine learning, artificial intelligence, data science, statistical analysis, data mining, and cloud computing. Ashish holds a BTech degree in chemical engineering from the Indian Institute of Technology, Bombay, India; a master's degree in economics; and a postgraduate diploma in applied statistics. He has worked across different industry areas such as finance, healthcare, education, sports, human resource domain, retail, and logistics automation. He currently works with a technology services company based out of Bangalore and also maintains a blog (Phinolytics.com) that focuses on data science and artificial intelligence application in the field of finance.

Acknowledgments

Much to my grandfather's dismay, I am not a doctor. Having said that, it was his diagnosis of type 2 diabetes during the first year of my undergraduate master's degree in computing and AI at Imperial College that has defined the last two decades of my life. What started off as a hobby to enquire about what other people ate to manage their condition has turned into a passion to empower people through data-driven digital health innovation, and evidence it.

I have been heavily influenced by my grandparents, firstly, by their sacrifices, having migrated to foreign lands carrying hope and endured unimaginable hardships to give us the world, and, secondly, by their experiences. Had it not been for my grandfather's quadruple heart bypass and subsequent diagnosis of type 2 diabetes, it is highly unlikely that I would be writing this.

Growing up, I observed healthcare holding the hands of my paternal grandmother. I recall she would be able to explain any of her medical symptoms to the local Sikh doctor in Punjabi but couldn't explain them sufficiently in English and often heard racist and condescending remarks as a result—unless said doctor was present. (It originally inspired me to be a doctor, but I soon realized I was not made for that life after repeatedly fainting after seeing blood.) Back then, this helped me to understand bias and prejudice.

Today, it inspires me to ensure underrepresented communities are represented and engaged in digital health. Quite simply, if the data is not there and we do not provide said data, bias becomes intrinsic, and that is a problem for everyone. Having traveled extensively around the world to implement digital health innovation, it is clear data and AI require an ethical and moral framework which it is our duty to protect and uphold.

A lot of this book refers to case studies or research generated from my own experience working with global stakeholders to engage people with digital health innovations—not just people with diabetes, but people who are lean; obese; diagnosed with mental health conditions, epilepsy, heart disease, and high blood pressure; or just looking to engage in their general health and well-being.

Thank you to the clinical professionals who have shared our mission of empowering others, particularly Professor Jon Little, Professor Jeremy Dale, Dr. Emma Scott, Dr. Kesar Sadhra, Dr. Jason Fung, Drs. Jen and David Unwin, and Dr. Peter Foley.

From conception to production, this book has been in preparation for well over 2 years. This book would not have been possible without the patience, tolerance, and encouragement of many people.

A huge thank you to the Diabetes Digital Media (DDM) team—particularly Amar, Harkrishan, and Dom—with whom many discussions on the future of artificial intelligence (AI) were had. Thank you to Krystal and Giverny for creating beautiful imagery to complement the book.

Thank you to my partner, Charlotte, for quite literally wandering around the world with me listening to endless conversations about the future of healthcare. You continue to inspire me.

Thank you to Ashish Soni and Girisha Garg for the meticulous detail and technical rigor they brought to this book and the fantastic team at Apress—Divya and Celestin—for their encouragement, support, and motivation.

Finally, thank you to time: what a funny old thing.

Introduction

The world is changing. There are more phones than people in the world, and it is increasingly connected. People use virtual assistants, self-driving cars, find partners through digital apps, and search the Web for any symptom of ill-health. Each digital event leaves a digital exhaust that is datafying life as we know it. The success of many of the world's most loved services, from Google to Uber, Alexa to Netflix, is grounded in big data, artificial intelligence (AI), and optimization.

Although medicine has been receptive to the benefits of big data and AI, it has been slow to adopt the rapidly evolving technology, particularly when compared to sectors such as finance, entertainment, and transport—that is, until now.

Recent digital disruption has catalyzed healthcare's adoption of big data and AI. Data of all sorts, shapes, and sizes are used to train AI technologies that facilitate machines to learn, adapt, and improve on their learning. Academic institutions and start-ups alike are developing rapid prototype technologies with increasingly robust health and engagement claims. The blending of technology and medicine has been expedited by smartphones and the Internet of Things (IoT), which is facilitating a wealth of innovation that continues to improve lives. With the arrival of health technology, people can monitor their health without the assistance of a healthcare professional (HCP); healthcare is now mobile and no longer in the waiting room.

At the same time, the world's population is living longer, and unhealthier, than ever — and in a financial crisis. Healthcare services are turning to value-based and incentivized care as non-communicable diseases such as obesity become global pandemics.

And that is before COVID-19, an infectious disease, transformed the digital universe. On January 30, 2020, the World Health Organization declared the coronavirus disease 2019 (COVID-19) outbreak a public health emergency, and six weeks later, it was categorized as a pandemic [1]. Digital health has always had an important role in healthcare. Still, the need to keep people connected and well at home during a period of living history has paved the way for mainly preexisting technologies to measure, monitor, and influence the wellness of people. Whether patients, public workers, healthcare workers, or otherwise, remote monitoring and self-management have never been more critical, particularly for those advised to self-isolate, quarantine, or shield.

Coronavirus disrupted healthcare in that it expedited the adoption of digital technologies by at least a decade. Digital health was one of the most critical weapons in the response against the pandemic and maintaining healthcare delivery during lockdowns [2].

AI was indeed used to weather the COVID storm. Adoption of novel technologies like health IoT was expedited, but it has been simple things like data sharing that lay the foundations of the rapid response. Sharing of unprecedented amounts of population health data enabled healthcare providers to consume vast amounts of data and information and segment it in the attempt to deliver more precise medicines and treatments [3]. In some areas, doctors were (even) able to email patients; and that has quite literally changed the fabric of public health as we know it.

The cybersecurity challenges of a distributed health ecosystem are complicated. With the pandemic came a rise in cyberattacks, particularly targeting national healthcare sectors and research. From setting up fraudulent Internet domains to organizing fraudulent COVID charities and false promise of delivery of masks and protective equipment, cybercriminals are sadly very creative [4]. Personal health information is extremely valuable on the black market, with a price tag in the thousands of dollars. In comparison, credit card numbers can be worth as little as a few cents.

Questions about misinformation, privacy, and credibility are more prominent now than ever, and rightly so [5].

COVID has shown that we need data. It just needs to be on *our* terms.

Data-Driven Decision-Making

Data-driven decision-making (also known as DDDM) enables consistency, assurance, and growth. It enables observers to spot patterns and opportunities, confirm hypotheses, predict future trends, optimize operational efforts, and produce actionable insights.

Data is an invaluable commodity. All sorts of data are available for scrutiny, and when it comes to health, information is empowering.

Data science, the science of data, its analysis, and intelligent programming layers have now become a pillar to achieve traction and success in healthcare. Digital health can democratize and personalize healthcare—and data is the golden key. This comes at the same time as a growing appetite to measure and quantify more aspects of human life. People want to live well—and data enables that.

Data is the new oil. It's valuable, but unrefined it is seldom usable. Data usually needs cleansing before it becomes a valuable entity.

Machine learning sits on top of robust data. Machine learning provides the ability to process big datasets which are outside the scope of human capabilities. It can then be used to reliably transform the analysis datasets into clinical insights to plan and provide care. Machine learning models can be developed to spot patterns, identify abnormalities, and highlight areas that need attention, thereby increasing the accuracy of all these processes. Ultimately, this leads to better outcomes, incentivizes success, and improves patient and healthcare professional stakeholder satisfaction.

Long term, machine learning will be beneficial for everyone, whether that is family physicians or intern anesthetists at the bedside. Machine learning can offer objective opinions to improve efficiency, reliability, and accuracy.

Data insight and real-world evidence are facilitating rapid technology innovation that regulators are struggling to keep up with. The consequences of digital health democratization have not only a health impact but also ethical implications. Big data and machine learning enable stakeholders to uncover hidden connections and patterns, including predictions of the future. The effects of understanding this data have moral and legal consequences that require appropriate governance to mitigate risk and harm.

Healthcare providers, individuals, and organizations all house a fountain of data that can be used for machine learning. Many people have a fair idea of what they would like to learn from data but are unaware as to how much data is required and what can be achieved before more technical aspects of uncovering hidden patterns, trends, and biases are found.

Bias itself can tamper even the most promising machine learning project. Bias can manifest itself in many forms (even before data handling) and can exclude, penalize, and demonize populations based on underrepresentative or poor-quality data. Now more than ever, humanity has an obligation to be fair and reasonable.

This book takes a practical, hands-on approach to big data, AI, and machine learning, and the ethical implications therein. We cover the theory and practical applications of AI in healthcare—starting with data, the cornerstone of any AI solution, and preparing it for your project. We then move on to the essentials of AI and machine learning, where and how you can apply machine learning algorithms and techniques, how to validate and evaluate model accuracy and performance, and, notably, AI ethics.

Digital technologies, regardless of industry, must be fully understood to ensure their behaviors are not violating our own human ethical and moral compass.

The book concludes with a series of case studies from leading health organizations that utilize AI and big data in novel and innovative ways.

Audience

This book is a radically different alternative to books on AI. It introduces healthcare professionals (nurses, physicians, practitioners, innovation officers), medical students, computing students, and data scientists to the use of big data and AI in healthcare and the critical considerations in its application without the need for endless code and mathematics.

CHAPTER 1

What Is Artificial Intelligence?

> *Knowledge on its own is nothing, but the application of useful knowledge? That's powerful.*
>
> —Osho

Artificial intelligence (AI) is considered, once again, to be one of the most exciting advances of our time. Virtual assistants can determine our music tastes with remarkable accuracy, cars are now able to drive themselves, and mobile apps can help reverse diseases once considered to be chronic and progressive.

Many people are surprised to learn that AI is nothing new. AI technologies have existed for decades. It is, in fact, going through a resurgence—driven by the availability of data and exponentially cheaper computing.

A Multifaceted Discipline

AI is a subset of computer science that has origins in mathematics, logic, philosophy, psychology, cognitive science, and biology, among others (Figure 1-1).

© Arjun Panesar 2021
A. Panesar, *Machine Learning and AI for Healthcare*, https://doi.org/10.1007/978-1-4842-6537-6_1

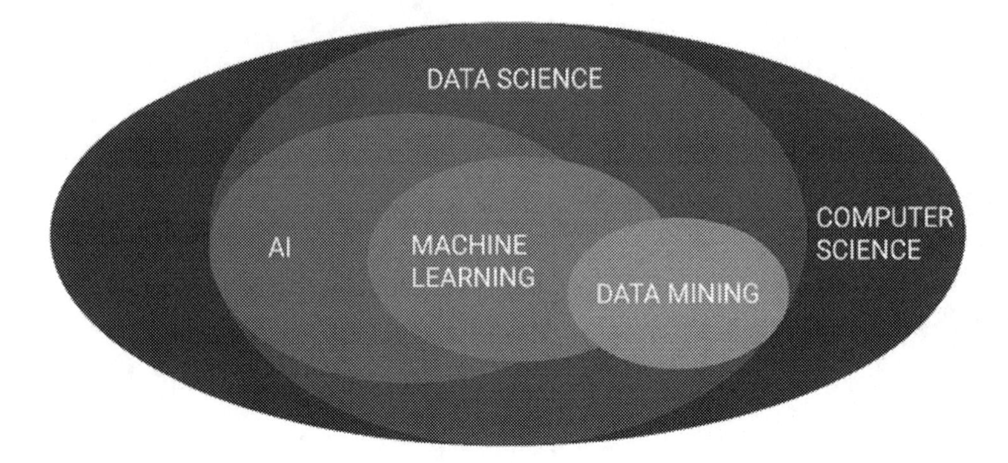

Figure 1-1. *AI, machine learning, and their place in computer science*

The earliest research into AI was inspired by a constellation of thought that began in the late 1930s and culminated in 1950 when British pioneer Alan Turing published "Computing Machinery and Intelligence" in which he asked, "Can machines think?" The Turing Test proposed a test of a machine's ability to demonstrate "artificial" intelligence, evaluating whether the behavior of a machine is indistinguishable from that of a human. Turing proposed that a computer could be considered to be able to think if a human evaluator could have a natural language conversation with both a computer and a human and not distinguish between either (i.e., an agent or system that is successfully mimicking human behavior).

The term AI was first coined in 1956 by Professor John McCarthy of Dartmouth College. Professor McCarthy proposed a summer research project based on the idea that "every aspect of learning or any other feature of intelligence can in principle be so precisely described that a machine can be made to simulate it" [6].

The truth is that AI, at its core, is merely programming. As depicted in Figure 1-1, AI can be understood as an abstraction of computer science. The surge in its popularity, and so too its ability, has much to do with the explosion of data through mobile devices, smartwatches, wearables, and the ability to access computer power cheaper than ever before. It was estimated by IBM that 90% of global data had been created in the preceding 2 years [7].

It is estimated that there will be 150 billion networked measuring sensors in the next decade—which is 20 times the global population [8]. This exponential data generated is enabling everything to become smart. From smartphones to smart washing cars, smart homes, cities, and communities await.

With this data comes a plethora of learning opportunities, and hence, the focus has now shifted to learning from available data and the development of intelligent systems. The more data a system is given, the more it is capable of learning, which allows it to become more accurate.

The use and application of AI and machine learning in enterprise are still relatively new, and even more so in health. The Gartner "Hype Cycle for Emerging Technologies" placed machine learning in the peak of inflated expectations, with 5–10 years before plateau [9].

As a result, the applications of machine learning within the healthcare setting are fresh, exciting, and innovative. With more mobile devices than people today, the future of health is wrought with data from the patient, environment, and physician [31]. As a result, the opportunity for optimizing health with AI and machine learning is ripening.

The realization of AI and machine learning in health could not be more welcome in the current ecosystem, as healthcare costs are increasing globally, and governmental and private bill payers are placing increasing pressures on services to become more cost-effective. Costs must typically be managed without negatively impacting patient access, patient care, and health outcomes.

But how can AI and machine learning be applied in an everyday healthcare setting? This book is intended for those who seek to understand what AI and machine learning are and how intelligent systems can be developed, evaluated, and deployed within their health ecosystem. Real-life case studies in health intelligence are included, with examples of how AI and machine learning are improving patient health and population health and facilitating significant cost savings and efficiencies.

By the end of the book, readers should have a confident grasp of key topics within AI and machine learning. Readers will be able to describe the machine learning approach and limitations, fundamental algorithms, the usefulness and requirements of data, the ethics and governance of learning, and how to evaluate the success of such systems. Most importantly, readers will learn how to plan a real-world machine learning project—from preparing data and choosing a model to validating accuracy and evaluating performance.

Rather than focus on overwhelming statistics and algebra, theory and practical applications of AI and machine learning in healthcare are explored—with methods and tips on how to evaluate the efficacy, suitability, and success of AI and machine learning applications.

Examining Artificial Intelligence

At its heart, AI can be defined as the simulation of intelligent behavior in agents (computers) in a manner that we, as humans, would consider to be smart or humanlike [10]. The core concepts of AI include agents developing traits including knowledge, reasoning, problem-solving, perception, learning, planning, and the ability to manipulate and move.

In particular, AI could be considered to comprise the following:

- Getting a system to reason rationally: Techniques include automated reasoning, proof planning, constraint solving, and case-based reasoning.
- Getting a program to learn, discover, and predict: Techniques include machine learning, data mining (search), and scientific knowledge discovery.
- Getting a program to play games: Techniques include minimax search and alpha-beta pruning.
- Getting a program to communicate with humans: Techniques include natural language processing (NLP).
- Getting a program to exhibit signs of life: Techniques include genetic algorithms (GA).
- Enabling machines to navigate intelligently in the world: This involves robotic techniques such as planning and vision.

There are many misconceptions of AI, primarily as it's still quite a young discipline. Indeed, there are also many views as to how it will develop. Interesting expert opinions include those of Kevin Warwick, who is of the opinion robots will take over the earth. Roger Penrose reasons that computers can never truly be intelligent. Meanwhile, Mark Jeffery goes as far as to suggest that computers will evolve to be human. Whether AI will take over the earth in the next generation is unlikely, but AI and its applications are here to stay.

In the past, the intelligence aspect of AI has been stunted due to limited datasets, representative samples of data, and the inability to both store and subsequently index and analyze considerable volumes of data. Today, data comes in real time, fueled by exponential growth in mobile phone usage, digital devices, increasingly digitized systems, wearables, and the Internet of Things (IoT).

Not only is data now streaming in real time but it also comes in at a rapid pace, from a variety of sources, and with the demand that it must be available for analysis, and fundamentally interpretable, to make better decisions.

There are four distinctive categories of AI.

Reactive Machines

This is the most basic AI. Reactive systems respond in a current scenario, relying on taught or recalled data to make decisions in their current state. Reactive machines perform the tasks they are designed for well, but they can do nothing else. This is because these systems are not able to use past experiences to affect future decisions. This does not mean reactive machines are useless.

Deep Blue, the chess-playing IBM supercomputer, was a reactive machine, able to make predictions based on the chessboard at that point in time. Deep Blue beat world champion chess player Garry Kasparov in 1996. A little-known fact is that Kasparov won three of the remaining five games and defeated Deep Blue by four games to two [11]. More recently, Google's AlphaGo triumphed over the world's leading human Go player [73].

Limited Memory: Systems That Think and Act Rationally

This is AI that works off the principle of limited memory and uses both preprogrammed knowledge and subsequent observations carried out over time. During observations, the system looks at items within its environment and detects how they change and then makes necessary adjustments. This technology is used in autonomous cars. Ubiquitous Internet access and IoT is providing an infinite source of knowledge for limited memory systems.

Theory of Mind: Systems That Think Like Humans

Theory of mind AI represents systems that interpret their worlds and the actors, or people, in them. This kind of AI requires an understanding that the people and things within an environment can also alter their feelings and behaviors. Although such AI is presently limited, it could be used in caregiving roles such as assisting elderly or disabled people with everyday tasks.

As such, a robot that is working with a theory of mind AI would be able to gauge things within its worlds and recognize that the people within the environments have their own minds, unique emotions, learned experiences, and so on. Theory of mind AI can attempt to understand people's intentions and predict how they may behave.

Self-Aware AI: Systems That Are Humans

This most advanced type of AI involves machines that have consciousness and recognize the world beyond humans. This AI does not exist yet, but software has been demonstrated with desires for certain things and recognition of its own internal feelings. Researchers at the Rensselaer Polytechnic Institute gave an updated version of the wise men puzzle, an induction self-awareness test, to three robots—and one passed. The test requires AI to listen and understand unstructured text as well as being able to recognize its own voice and its distinction from other robots [12].

Technology is now agile enough to access huge datasets in real time and learn on the go. Ultimately, AI is only as good as the data that's used to create it—and with robust, high-volume data, we can be more confident about our decisions.

Healthcare has been slow to adopt the benefits of big data and AI, especially when compared to transport, energy, and finance. Although there are many reasons for this, the rigidity of the medical health sector has been duly grounded in the fact that people's lives are at risk. Medical services are more of a necessity than a consumer choice; so historically, the medical industry has had little to no threat that usually drives other industries to seek innovation.

That has expedited a gap in what healthcare institutes can provide and what patients want—which subsequently has led to variances in care and in health outcomes and a globally recognized medication-first approach to disease.

The explosion of data has propelled AI through enabling a data-led approach to intelligence. The last 5 years has been particularly disruptive in healthcare, with applications of data-led AI helping intelligent systems not only to predict, diagnose, and manage disease but to actively reverse and prevent it—and the realization of digital therapeutics.

Recent advances in image recognition and classification are beginning to filter into industry too, with deep neural networks (DNNs) achieving remarkable success in visual recognition tasks, often matching or exceeding human performance (Figure 1-2).

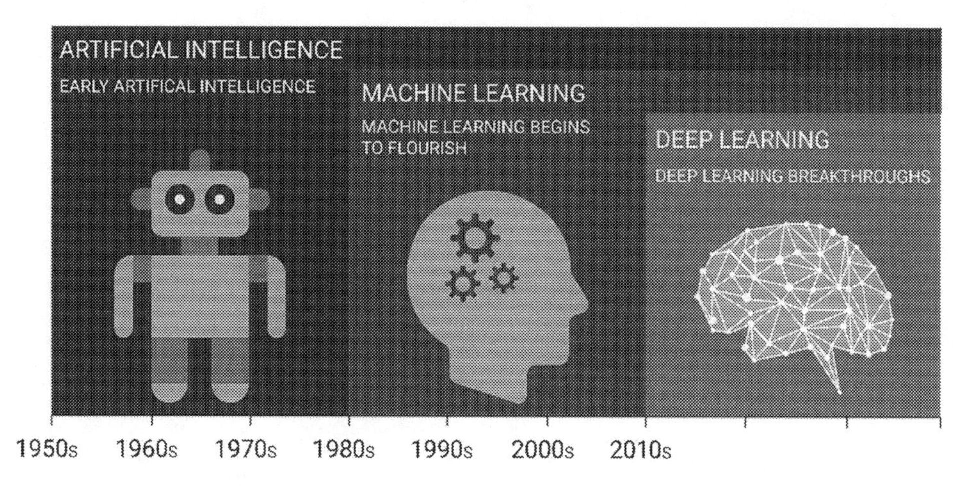

Figure 1-2. *AI and its development*

Types of AI can also be explained by the tasks that it can perform and classified into weak AI or strong AI.

Weak AI

Weak or narrow AI refers to AI that performs one (narrow) task. Weak AI systems are not humanlike although if trained correctly will seem intelligent. An example is a chess game where all rules and moves are computed by the AI and every possible scenario determined.

Strong AI

Strong AI is able to think and perform tasks like a human being. There are no standout examples of strong AI. Weak AI contributes to the building of strong AI systems.

What Is Machine Learning?

Machine learning is a term credited to Arthur Samuel of IBM, who in 1959 proposed that it may be possible to teach computers to learn everything they need to know about the world and how to carry out tasks for themselves. Machine learning can be understood as a form of AI.

Machine learning was born from pattern recognition and the theory that computers can learn without being programmed to perform specific tasks. This includes vtechniques such as Bayesian methods, neural networks, inductive logic programming, explanation-based natural language processing, decision tree, and reinforcement learning.

Systems that have hard-coded knowledge bases will typically experience difficulties in new environments. Certain difficulties can be overcome by a system that can acquire its own knowledge. This capability is known as machine learning. This requires knowledge acquisition, inference, updating and refining the knowledge base, acquisition of heuristics, and so forth.

Machine learning is covered in greater depth in Chapter 3.

What Is Data Science?

All AI tasks will use some form of data. Data science is a growing discipline that encompasses anything related to data cleansing, extraction, preparation, and analysis. Data science is a general term for the range of techniques used when trying to extract insights (i.e., trying to understand) and information from data.

The term data science was phrased by William Cleveland in 2001 to describe an academic discipline bringing statistics, business, and computer science closer together.

Teams working on AI projects will undoubtedly be working with data, whether little or big in volume. In the case of big data, real-time data usually demands real-time analytics. In most business cases, a data scientist or data engineer will be required to perform many technical roles related to the data including finding, interpreting, and managing data; ensuring consistency of data; building mathematical models using the data; and presenting and communicating data insights/findings to stakeholders. Although you cannot do big data without data science, you can perform data science without big data—all that is required is data.

Because data exploration requires statistical analysis, many academics see no distinction between data science and statistics. A data science team (even if it is a team of one) is fundamental to deploying a successful project.

The data science team typically performs two key roles. First, beginning with a problem or question, it is seeking to solve it with data; and second, it is using the data to extract insight and intelligence through analytics (Figure 1-3).

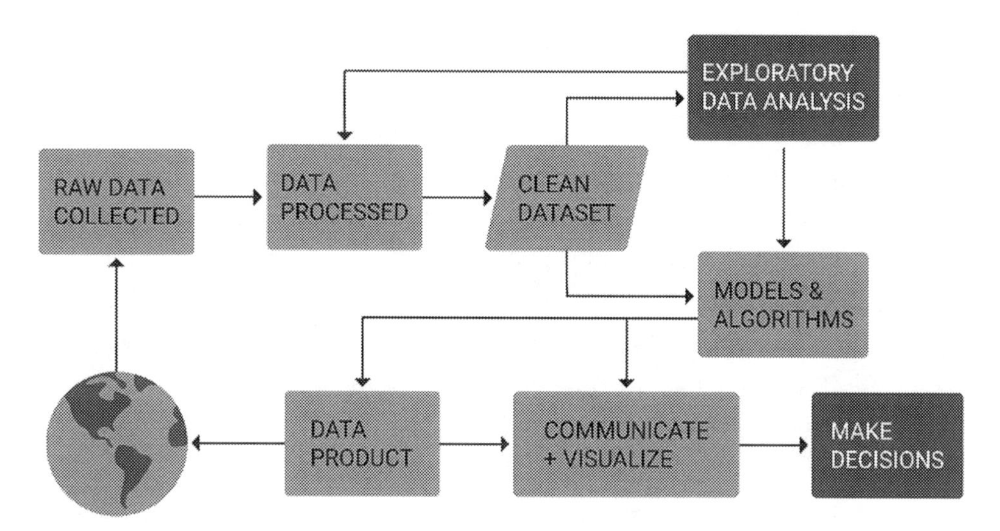

Figure 1-3. *Data science process*

Learning from Real-Time Big Data

AI had previously been stifled by the technology and data at its disposal. Before the explosion of smartphones and cheaper computing, datasets remained limited in respect to their size, validity, and representative rather than real-time nature.

Today, we have real-time, immediately accessible data and tools that enable rapid analysis. Datafication of modern-day life is fueling machine learning's maturity and is facilitating the transition to an evidence-based, data-driven approach. Datafication refers to the modern-day trend of digitalizing (or datafying) every aspect of life. Technology is now agile enough to access these huge datasets to rapidly evolve machine learning applications.

Both patients and healthcare professionals generate a tremendous amount of data. Phones collect metrics such as blood pressure, geographical location, steps walked, nutritional diaries, and other unstructured data such as conversations, reactions, and images.

It's not just digital or clinical data either. Data extraction techniques can be applied to paper documentation, or images scanned, to process the documents for synthesis and recall.

Healthcare professionals collect health biomarkers alongside other structured metrics. Regardless of the source of data, to learn, data must be turned into information. For example, an input of a blood glucose reading from a person with diabetes into an app has far more relevance when blood glucose level targets are known by the system so that it can be understood whether the input met the recommended target range or not.

In the twenty-first century, almost every action leaves some form of transactional footprint. The Apple iPhone 6 is over 32,000 times faster than the Apollo-era NASA computers that took man to the moon. In fact, a simple Wi-Fi router or even a USB-C charger has more than enough computing power to send astronauts to the moon [13]. Not only are devices smaller than ever but they are also more powerful than ever. With organizations capitalizing on sources of big data, there has been a shift toward embedding learnings from data into every aspect of the user experience—from buying a product or service to the user experience within an app, and interfaced with in a number of different ways.

The value of data is understood when it is taken in its raw form and converted into knowledge that changes practice. This value is driven by project and context. For example, the value may be as the result of faster identification of shortfalls in adherence, compliance, and evidence-based care. It may be better sharing of data and insights within a hospital or organization or more customized relationships with patients to drive adherence and compliance and boost self-care. Equally, it may be to avoid more costly treatments or costly mistakes.

Applications of AI in Healthcare

It is highly unlikely artificially intelligent agents will ever completely replace doctors and nurses, but machine learning and AI are transforming the healthcare industry and improving outcomes. And it's here to stay.

Machine learning is improving diagnostics, predicting outcomes, and beginning to scratch the surface of personalized care (Figure 1-4).

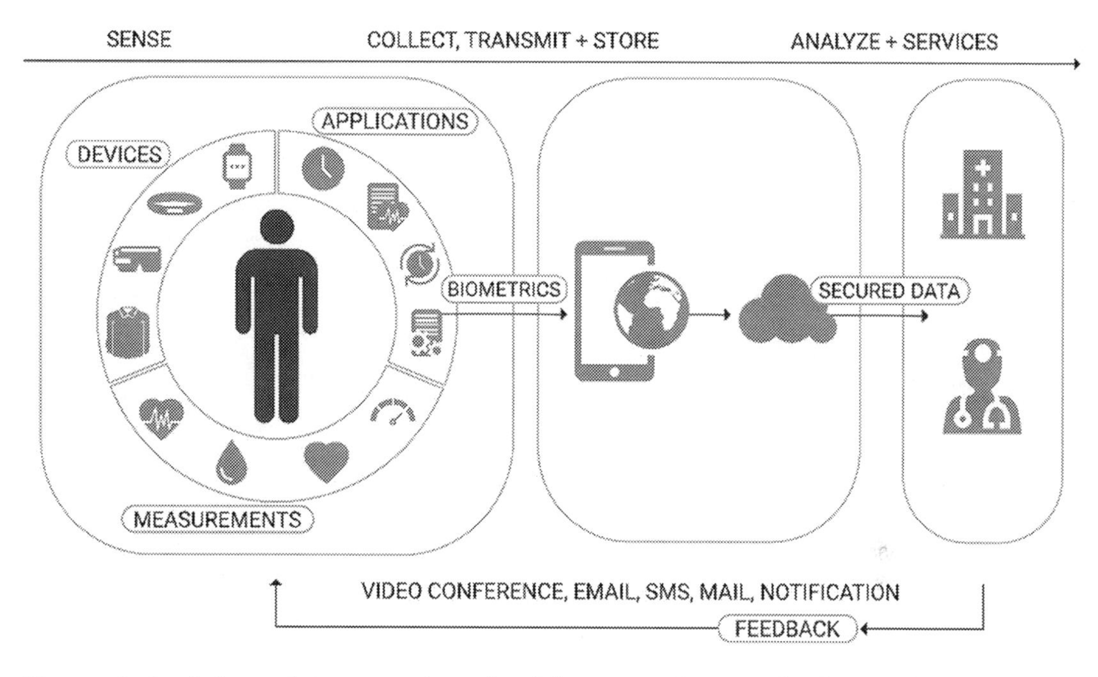

Figure 1-4. *A data-driven, patient-healthcare professional relationship*

Imagine a situation where you walk in to see your doctor (or via a teleconferencing app) with pain around your heart. After listening to your symptoms, perhaps shared as a video or through a health IoT device, the doctor inputs them into their computer, which pulls up the latest evidence base they should consult to efficiently diagnose and treat your problem. You have an MRI scan, and an intelligent computer system assists the radiologist in detecting any concerns that would be too small for your radiologist's human eye to see. They may even suggest using a particular smartphone app alongside a device you're wearing or have access to, to measure heart metrics and the risk of a number of potential health conditions.

Your watch may have been continuously collecting your blood pressure and pulse, while a continuous blood glucose monitor has a real-time profile of your blood glucose readings. Your ring may be measuring your temperature. Finally, your medical records and family's medical history are assessed by a computer system that suggests treatment pathways precisely identified to you. Data privacy and governance aside, the implications of what we can learn from combining various pools of data are exciting.

Prediction

Technologies already exist that monitor data to predict disease outbreaks. This is often done using real-time data sources such as social media as well as historical information from the Web and other sources. Malaria outbreaks have been predicted with artificial neural networks (ANNs), analyzing data including rainfall, temperature, number of cases, and various other data points [14].

Diagnosis

Many digital technologies offer an alternative to nonemergency health systems. Considering the future, combining the genome with machine learning algorithms provides the opportunity to learn about the risk of disease, improve pharmacogenetics, and provide better treatment pathways for patients [15].

Personalized Treatment and Behavior Modification

Digital therapeutics are an emerging area of digital health. Digital healthcare can be personalized to user experience to engage people to make sustainable behavior change [16].

A digital therapy from Diabetes Digital Media, Gro Health, is an award-winning, evidence-based behavior change platform that tackles non-communicable diseases by addressing the four key pillars of health (modifiable risk). Education and support in the areas of nutrition, activity, sleep, and mental health are powered by AI to tailor experience to user goal and focus, disease profile, ethnicity, age, gender, and location with remote monitoring support used in population health management. The app provides personalized education and integrated health tracking, learning from the user's and wider community's progress.

At the end of 8 weeks, Gro Health demonstrates improvements in mental health including 23% reduction in perceived stress at 8 weeks, 32% reduction in generalized anxiety, and 31% reduction in depression [17].

Drug Discovery

Machine learning in preliminary drug discovery has the potential for various uses, from initial screening of drug compounds to predicted success rate based on biological factors. This includes R&D discovery technologies like next-generation sequencing.

Drugs do not necessarily need to be pharmacological in their appearance. The use of digital therapeutics and aggregation of real-world patient data are providing solutions to conditions once considered to be chronic and progressive. The Low Carb Program (LCP) app, for example, used by over 300,000 people with type 2 diabetes places the condition into remission for 26% of the patients who complete the program at 1 year [16, 18].

Follow-Up Care

Hospital readmittance is a huge concern in healthcare. Doctors, as well as governments, are struggling to keep patients healthy, particularly when returning home following hospital treatment. Digital health coaches aid care, similar to a virtual customer service representative on an ecommerce site. Assistants can prompt questions about the patient's medications and remind them to take medicine, query them about their condition symptoms, and convey relevant information to the doctor [16]. In some locations such as India, there is a market for "butlers," or tech-savvy health coaches, who set up digital care for elderly or reduced-mobility patients. By setting up Zoom calls or other services, butlers enable patients to engage with their healthcare team, removing any inertia in the process.

Realizing the Potential of AI in Healthcare

For AI and machine learning to be fully embraced and integrated within healthcare systems, several key challenges must be addressed.

Understanding Gap

There is a huge disparity between stakeholder understanding and applications of AI and machine learning. Communication of ideas, methodologies, and evaluations are pivotal to the innovation required to progress AI and machine learning in healthcare. Data, including the sharing and integration of data, is fundamental to shift healthcare toward realizing precision medicine.

Developing data science teams, focused on learning from data, is key to a successful healthcare strategy. The approach required is one of investment in data. Improving value for both the patient and the provider requires data and hence data science professionals.

Fragmented Data

There are many hurdles to be overcome. Data is currently fragmented and difficult to combine. Patients collect data on their phones, Fitbits, and watches, while physicians collect regular biomarker and demographic data. At no point in the patient experience is this data combined. Nor do infrastructures exist to parse and analyze this larger set of data in a meaningful and robust manner. In addition, electronic health records (EHRs), which at present are still messy and fragmented across databases, require digitizing in a mechanism that is available to patients and providers at their convenience.

COVID-19 certainly expedited the linking of data sources; however, progress toward a real-life, useful data fabric is still in infancy. Data fabric refers to a unified environment comprising architecture, technologies, and services to help an organization manage and improve decision-making. For instance, a data fabric for health within a particular country may have all of the health data relating to a patient accessible by all through a common route.

Although this sounds utopian, the bulk of health apps and services use secure, cloud-based hosting providers such as Amazon Web Services, Microsoft Azure Cloud, and Google Cloud. Linkage by vendors to create a data fabric, for instance, may be closer than we think.

Appropriate Security

At the same time, organizations face challenges of security and meeting government regulation specifically with regard to the management of patient data and ensuring its accessibility at all times. What's more, many healthcare institutions are using legacy versions of software that can be more vulnerable to attack. The NHS (National Health Service) digital infrastructure was paralyzed by the wrath of the ransomware WannaCry. The ransomware, which originated in America, scrambled data on computers and demanded payments of $300–600 to restore access [19].

Hospitals and GP surgeries in England and Scotland were among over 16 health service organizations hit by the "ransomware" attack. The impact of the attack wasn't just the cost of the technological failure; it had a bearing on patients' lives. Doctors' surgeries and hospitals in some parts of England had to cancel appointments and refuse surgeries. In some areas, people were advised to seek medical care in emergencies only. The NHS was just one of many global organizations crippled through the use of hacking tools; and the ransomware claimed to have infected computers in 150 countries.

During COVID-19, the NHS faced ridicule for rejecting Apple and Google's plans for COVID-19 tracing, opting instead to build their own which had considerable security flaws [20].

Security and safety are the primary considerations of health technologies, whether digital or otherwise.

Data Governance

With data security comes the concept of data governance. Medical data is personal and not easy to access. It is widely assumed that the general public would be reluctant to share their data because of privacy concerns. However, a Wellcome Foundation survey on the British public's attitude to commercial access to health data found that 17% of people would never consent to their anonymized data being shared with third parties [21].

Adhering to multiple sets of regulation means disaster recovery and security is key, and network infrastructure plays a critical role in ensuring these requirements can be met.

Healthcare organizations require modernization of network infrastructure to ensure they are appropriately prepared to provide the best patient care possible.

During COVID-19, some healthcare practices became able to contact patients. This may sound simple, but it transformed care across the world [22].

Healthcare professionals feel an obligation to act if healthcare data is seen. This has led to a reluctance to seeing identifiable patient data, particularly during nonworking hours. So, for patients to email their healthcare team, the fact responses were received transformed care delivery. COVID-19 made it acceptable for everyone to perform tasks in their own time—moving expectancy from time based to task based.

Bias

A significant problem with learning is bias. As AI becomes increasingly interwoven into our daily lives—integrated with our experiences at home, at work, and on the road—it is imperative that we question how and why machines do what they do. Within machine learning, learning to learn creates its own inductive bias based on previous experience. Essentially, systems can become biased based on the data environments they are exposed to.

It's not until algorithms are used in live environments that people discover built-in biases, which are often amplified through real-world interactions.

This has expedited the growing need for more transparent algorithms to meet the stringent regulations on drug development and expectation. Transparency is not the only criteria; it is imperative to ensure decision-making is unbiased to fully trust its abilities. People are given confidence through the ability to see through the black box and understand the causal reasoning behind machine conclusions.

Bias can lead to health inequalities. Frighteningly, some bias is only made apparent on chance, reflection, and/or inspiration.

Only in 2020, a medical student from St George's, University of London, created a handbook of medical diagnoses based on black and Asian skin tones, as there were none available that used nonwhite skin types [23].

Software

Traditional AI systems had been developed in Prolog, Lisp, and ML. Most machine learning systems today are written in Python due to many of the mathematical underpinnings of machine learning that are available as libraries.

Algorithms that "learn" can be developed in most languages, including Perl, C++, Java, and C. This is discussed in more depth in further chapters.

Conclusion

The potential applications of machine learning in healthcare are vast and exciting. Some of them are becoming apparent in traditional healthcare services, but the future is only getting started.

Intelligent systems that can help us reverse disease, detect our risk of cancers, and suggest courses of medication based on real-time biomarkers exist—separately. The potential of AI is limitless—particularly as services are unified to create a pervasive data fabric. This will transform any industry, but health has never been more important. Age, obesity, and ethnicity were demonstrated to be key risk factors during COVID-19 [24, 25, 26].

There has never been a better incentive to live well. Providing personalized digital treatments to engage older populations, obese people, and people who are typically perceived "hard to reach" can truly transform healthcare.

With this also comes tremendous responsibility and questions of wider morality. We don't yet fully understand what can be learned from health data. As a result, the ethics of learning is a fundamental topic for consideration.

On the basis that an intelligent system can detect the risk of disease, should the system tell the patient it is tracking the impending outcome? If an algorithm can detect your risk of pancreatic cancer—an often-terminal illness—based on your blood glucose and weight measurements, is it ethical to disclose this to the patient in question? Should such sensitive patient data be shared with healthcare teams—and what are the unintended consequences of such actions?

The explosion in digitally connected devices and applications of machine learning are enabling the realization of these once-futuristic concepts, and conversation around these topics is key to progress.

And if a patient opts out of sharing data, is it then ethical to generalize based on known data to predict the same illness risk? And what if I can't get life insurance without this pancreatic cancer check? There are considerable privacy concerns associated with the use of a person's data and what should be private or not and equally as to what data is useful.

Invariably, the more data available, the more precise a decision can be made—but exactly how much is too much is another question. The driving factor is determining the value of data.

The ethics of AI are currently without significant guidelines, regulations, or parameters on how to govern the enormous treasure chest of data and opportunity.

Many assume that AI has an objectivity that puts it above questions of morality, but that's not the case. AI algorithms are only as fair and unbiased as the learnings, which come from the environmental data. Just as social relationships, political and religious affiliations, and sexual orientation can be revealed from data, learning on health data is revealing new ethical dilemmas for consideration.

Data governance and disclosure of such data still require policy, at the national and international levels. In the future, driverless cars will be able to use tremendous amounts of data in real time to predict the likelihood of survival if involved in a collision. Would it be ethical for the systems to choose who lives or dies or for a doctor to decide whom to treat based on the reading from two patients' Apple Watches? And where should one engage with health services—in the local clinic or hospital or in a local takeaway, supermarket, or convenience store?

This is just the beginning.

As technologies develop, new and improved treatments and diagnoses will save more lives and cure more diseases. With the future of evidence-based medicine grounded in data and analytics, it begs the question as to whether there will ever be enough data. And even if this is to be the case, one wonders just what the consequences are of collecting it.

CHAPTER 2

Data

The plural of anecdote is data.

—Raymond Wolfinger

Data is everywhere. Global disruption and international initiatives are driving datafication. Datafication refers to the modern-day trend of digitalizing (or datafying) every aspect of life [27]. This data creation is enabling the transformation of data into new and potentially valuable forms. Entire municipalities are being incentivized to become smarter. In the not too distant future, our towns and cities will collect thousands of variables in real time to optimize, maintain, and enhance the quality of life for entire populations. They may even be connected to your Amazon or Google accounts and aware of all significant events in our health and day-to-day living as human–computer interaction becomes even more embedded in smart care.

One would reasonably expect that as well as managing traffic, traffic lights may also collect other data such as air quality, visibility, and speed of traffic. One wonders whether a speeding fine may be contested based on the temperature from someone's smartwatch or the average heart rate from a smart ring.

Data is everywhere. The possibilities are endless. Big data from connected devices, embedded sensors, and the IoT has driven the global need for the analysis, interpretation, and visualization of data. COVID-19 itself proved to be a great example of data sharing between countries and clinicians—particularly around the risk factors, comorbidities, and complications of the novel coronavirus.

What Is Data?

Data itself can take many forms—character, text, words, numbers, pictures, sound, or video. Each piece of data falls into two main types: structured and unstructured.

© Arjun Panesar 2021
A. Panesar, *Machine Learning and AI for Healthcare*, https://doi.org/10.1007/978-1-4842-6537-6_2

At its heart, data is a set of values of qualitative or quantitative variables. To become information, data requires interpretation. Information is organized or classified data, which has some meaningful value (or values) for the receiver. Information is the processed data on which decisions and actions should be based.

Healthcare is undergoing a data revolution, thanks to two huge shifts: the need to tame growing cost pressures, which is generating new incentives and reimbursement structures, and digital health, which is democratizing healthcare by empowering people through their digital devices and innovative technologies.

Clinical trends are changing. The advent of social media and immediate access to information through the Internet and health communities mean that patients are clued up about their health. Patients are generating data constantly and want to own and transfer their data between services. Concurrently, evidence-based medicine is striving for personalized care, led by an evidence-based approach to decision-making.

Patients and healthcare professionals alike are vocal in their appetite for all available clinical data to make more effective treatment decisions based on current evidence. A Diabetes.co.uk study in 2019 demonstrated that one in three patients would like to share their data with their healthcare professional, but only one in five patients can do so [9].

Aggregating individual datasets into more significant populations also provides more robust evidence because subtleties in subpopulations may be infrequent in that they are not readily apparent in small samples. For instance, aggregating patients with cardiovascular disease and high blood pressure may present health conditions such as high cholesterol with more confidence than separate datasets.

The incorporation of data in modern healthcare provides an opportunity for significant improvements within a plethora of areas. As medicine evolves to become evidence based and personalized, data can be used to improve

- Patient and population wellness
- Patient education and engagement
- Prediction of disease and care risks
- Medication adherence
- Remote monitoring
- Disease management
- Disease reversal/remission

- Individualization and personalization of treatment
- and care
- Financial, transactional, and environmental forecasting, planning, and accuracy (of particular relevance to any operations and finance teams within healthcare organizations)
- User experience, which can be used to expedite all of the preceding items

Data facilitates information. Information powers insights, and insights lead to the making of better decisions.

Types of Data

Structured data typically refers to something stored in a database—in a structure that follows a model or schema. Almost every organization will be familiar with this form of data and may already be using it effectively. Most organizations will store at least some form of data in Excel spreadsheets, for example. Within a clinical setting, EHRs may also take a similar, structured form.

Readings from embedded sensors, smartphones, smartwatches, and IoT devices are typically forms of structured data—whether that be the provision of blood glucose readings, steps walked, calories burned, heart rate, or blood pressure.

Structured data is similar to a machine language. Highly organized in its format, structured data facilitates simple, straightforward search and information retrieval operations. Structured data would typically be stored in a relational database for this purpose.

Unstructured data refers to everything else. Unstructured data does not have a predefined model or schema. Data that is unstructured has no identifiable structure within it, and this presents problems for querying and information retrieval. Emails, text messages, Facebook posts, Twitter tweets, and other social media posts are good examples of unstructured data. Unstructured data can unleash a treasure trove of information and insight.

The Gartner report has indicated that data volume is set to grow 800% over the next 5 years, and 80% of this data will be in the form of unstructured data [29].

The problem that unstructured data presents is one of volume. The lack of structure makes compilation and interpretation a resource-intensive task in regard to time and energy. Any organization would benefit from a mechanism of unstructured data analysis to reduce the resource costs unstructured data adds. Unless the data is well understood, a subject matter expert may be required.

Structured data is traditionally easier to analyze than unstructured data due to unstructured data's raw and unorganized form. However, analytics of unstructured, informal datasets is improving, with the use of data science and machine learning methods such as natural language processing (NLP) assisting in the understanding and classification of sentiment.

It may not always be possible to transform unstructured data into a structured model. For instance, the transmission of information through an email or notification holds data such as the time sent, subject, and sender as uniform fields. However, the contents of the message would not be easily dissected and categorized.

Semi-structured data fits the space between structured and unstructured data. Semi-structured data does not necessarily conform with the formal structures of data schemas associated with relational databases or data tables. However, semi-structured data may contain tags or markers to separate semantic elements and enforce grouping of records and fields within the data.

Languages such as JSON (JavaScript Object Notation) and XML (Extensible Markup Language) are all forms of semi-structured data. Both have been instrumental in facilitating data exchange.

Data is the fuel required to learn in an intelligent system, the most critical component for successful AI.

It is categorized into five sources:

- Web and social media data: Clicks, history, health forums
- Machine-to-machine data: Sensors, wearables
- Big transaction data: Health claim data, billing data
- Biometric data: Fingerprints, genetics, biomarkers driven from wearables
- Human-generated data: Email, paper documents, electronic medical records

When thinking about typical Excel spreadsheets, and in the domain of machine learning, the following definitions are also useful:

- Instance: A single row of data or observation.
- Feature: A single column of data. It is a component of the observation.
- Data type: This refers to the kind of data represented by the feature (e.g., Boolean, string, number).
- Dataset: A collection of instances used to train and test machine learning models.
- Training dataset: Dataset used to train the machine learning model.
- Testing dataset: Dataset used to determine
- accuracy/performance of the machine learning model.

Big Data

The term big data is given to the collection of voluminous, traditional, and digital data that are sources for discovery and analysis.

Big data is a popular term that defines datasets that are too big to be stored and processed in a conventional relational database system. In this way, the term big data is vague—while size is undoubtedly a part of big data, scale alone doesn't tell the whole story of what makes big data truly big.

Through the analytics of big data, we can uncover hidden patterns; unknown correlations, trends, and preferences; and other information that can help stakeholders make better and more informed decisions. Machine learning provides a toolbox of techniques that can be applied to datasets for this very purpose.

Big data was first described by Laney in 2001 as having the following characteristics, also known as the three Vs (Figure 2-1) [30]:

- Volume: The quantity of generated and stored data. Big data is typically high in volume. The sheer size of data brings with it its intricacies in the form of storage, indexing, and retrieval.
- Variety: Big data is varied in the types and nature of data, requiring efficient storage and analysis as well as systems for processing such data.
- Velocity: Big data is received at a speed that brings its demands and challenges.

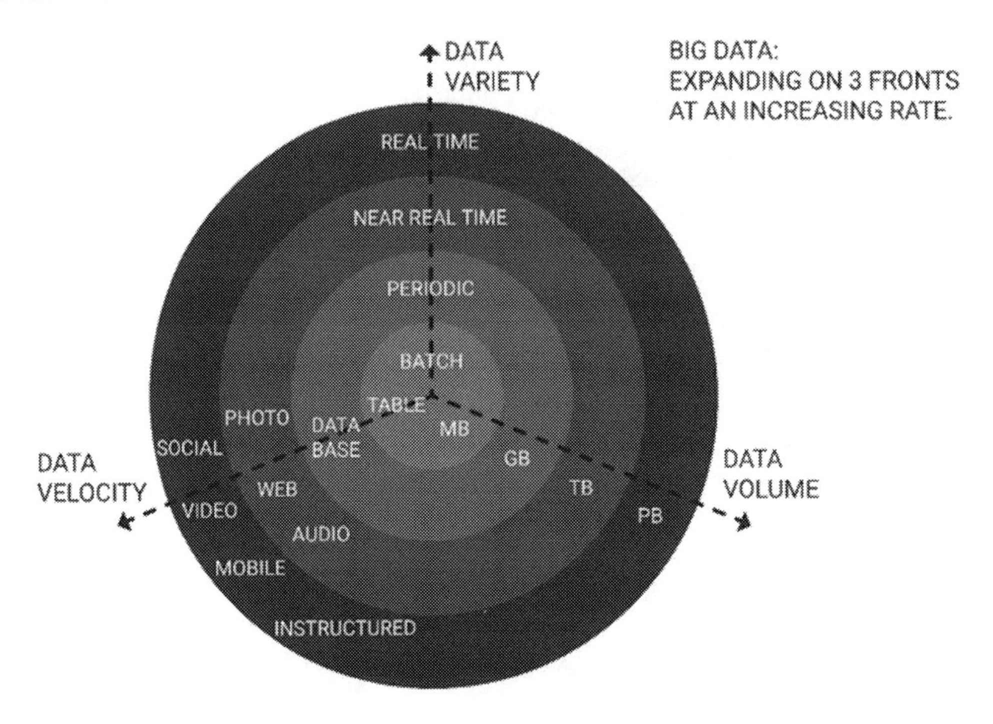

Figure 2-1. *The three Vs of data*

Data science has been discussing the three Vs of big data for some time. However, there are two other aspects of data to consider that are perhaps more important than the three Vs of big data: the concepts of data veracity and data value. Mark van Rijmenam proposed four more Vs to further understand the incredibly complex nature of big data [31].

Further still, there are a total of ten Vs (Figure 2-2) [32]. In reality, differing subsets of Vs are important for organizations to keep in mind when developing a data strategy.

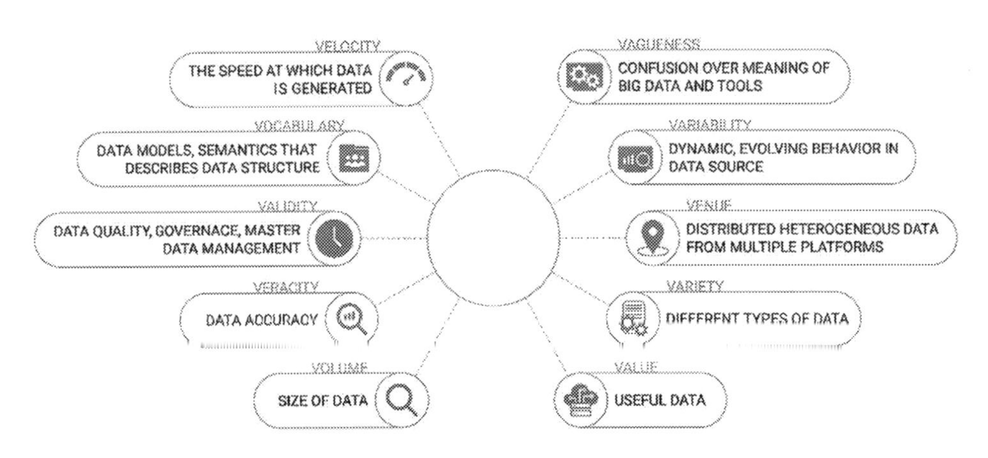

Figure 2-2. *The ten Vs of data*

A subset of the many Vs now proposed have been included here.

A Gartner survey of 199 members demonstrated that investment in big data was increasing; but the investigation showed signs of slowing growth, with fewer companies having a future intent to invest in big data initiatives. Only 15% of businesses reported deploying their big data project, indicating that the adoption curve is still maturing and that talent and expertise is in high demand [33].

Patients and physicians are driving the adoption of big data. Patients drive data through clinical data, wearable devices, mobile health apps, and telehealth/remote healthcare services. Physicians meanwhile leave an exhaust of clinical data (such as written notes, imaging, and insurance data), patient records, and machine-generated data.

Volume

Laney defined a key characteristic, the first V, of big data in 2001—the proposition that there is a lot of it.

Key inertia in big data adoption has historically been the associated demand for storage and its respective costs. The movement of data centers from being locally housed to existing in the cloud has quashed this concern. Better still, it comes with benefits—primarily safety and scalability. This has significantly reduced storage costs as well as providing flexibility in data storage, collaboration, and disaster recovery. The cloud, or cloud computing, refers to distributed computing delivered through the Internet.

Moore's Law has not just affected storage capacity; it has also benefited cost. Moore's Law is a statement made by Gordon Moore, the Intel cofounder, in 1965. Moore states that the number of transistors that are able to fit onto an integrated circuit doubles approximately every 18 months. In 1981, the price of a gigabyte of storage was US $300,000. In 2004, this was $1.00; and it was $0.10 in 2010. Today, 1 GB can be rented in the cloud for $0.023 per month, with the first year of storage free [34].

Healthcare, whether that be delivered through public health services, governments, pharmaceutical organizations, or corporates, is no stranger to big data. Data is continuously collected both digitally and non-digitally and spans a variety of patient, behavioral, epidemiological, environmental, and medical information.

Data is consumed, managed, and generated for use in patient care, transactional records, research, compliance, processes, and regulatory requirements.

As of 2020, a patient's EHR is not readily available digitally in a secure and transparent means. Many organizations are striving to place this information on the blockchain, which revolutionizes access to health data through creating decentralized network systems.

Patients and healthcare organizations are adopting smartphones, wearables such as smartwatches and fitness trackers, home sensors, intelligent personal assistants, and social networks within their environments. Data is plentiful, accessible, and determined to be reliable. However, it is currently not used by most healthcare providers, highlighting the following:

- Patient data generation happens outside of the healthcare system.
- Patient-generated data needs to be integrated and explored within a more extensive patient record or health timeline.
- Patient data extends beyond medical markers to more holistic markers including behavioral and environmental health.
- There is an enormous interest for healthcare organizations still to be realized from opportunity and cost benefits.

A recent initiative in the United Kingdom saw the prevalence of type 1 diabetes in children geographically mapped through the use of social media hashtags #facesofdiabetes and #letstalk. The sentiment of users was also profiled, with appropriate support provided to users through the use of machine learning tools to classify post types [35].

People-powered data and its use in near real time is the holy grail of personalized care. Datafication is well under way in the finance, energy, and retail sectors. With the prevalence of the health wearables and mobile healthcare, every individual and organization has the potential to tap into this growing data volume.

Coping with Data Volume

For big data to work for your organization, it is vital that storage is efficient and cost-effective. Choosing the most appropriate storage solution is dependent on several factors:

- Approach: Data storage costs increase alongside storage requirements. Hence, determining whether specific data is to be collected or whether a catchall approach to data is required depends on the project and regulatory requirements. Low-cost cloud storage providers make catchall approaches feasible. Beginners to data science and machine learning may wish to determine a set of key data metrics for evaluation for the sake of simplicity.
- Types of data: Consider whether structured or unstructured data is to be stored—and the source—for example, audio or video content, text, images, and so forth. Video, audio, or image storage will require more resources compared to text.
- Deployment: Determine how the solution will be deployed, which could include locally, internally, or cloud based.
- Access: Determine how the solution will be accessed—through an app, web interface, intranet, and so forth.
- Operations: This includes data structure, architecture, archiving, recovery, event logging, and legal requirements.
- Future use: This involves planning for further sources of data, extensions to the system, and potential future use cases.

The distributed nature of datasets has meant that inertia in the adoption of big data has previously referred to the fragmented nature of such systems.

Variety

The second V of data is variety. This refers not only to variations in the types of data but also sources and use cases. Twenty years ago, we used to store data in the form of spreadsheets and databases. Today, data may be in the form of photographs, sensor data, tweets, encrypted files, and so on. This variety of unstructured data creates problems for storage, mining, and analyzing data. This is also an area that machine learning can significantly assist humans in.

Data is available in the form of structured data (in the form of clinical records and results), unstructured data (e.g., in the form of communication and interaction data), and also semi-structured data (e.g., an X-ray with written annotations).

The ever-increasing digitalization of services and the broad adoption of wearables and cyber-physical devices are continuously enabling new and exciting sources of data for discovery and analysis. This enables the creation of new risk models powered by sensor data, clinical records, communication and engagement, demographics, and billing records—which can make accurate predictions and save precious time and resources.

The Internet of Things

The Internet of Things refers to the rapidly increasing variety of smart, interconnected devices and sensors and the volumes of data they generate (Figure 2-3).

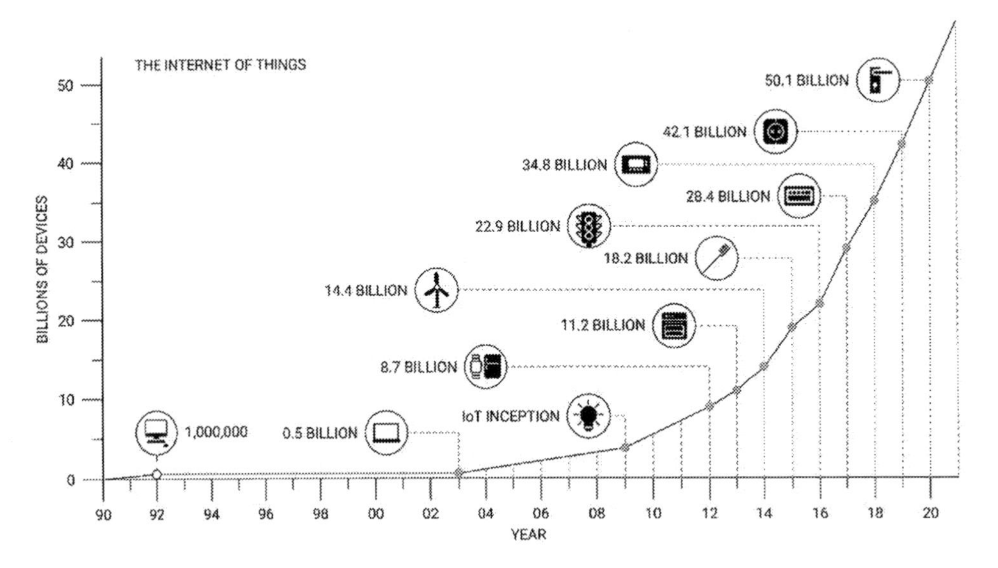

***Figure 2-3.** Evolution of the Internet of Things*

Data is transferred between devices, services, and, ultimately, people. You may also hear similar terms such as the Industrial Internet of Things and the Healthcare Internet of Things, which refer to their respective vertical, or industry, within the Internet of Things.

Today, a variety of devices monitor every sort of patient behavior—from glucose monitors to fetal monitors to electrocardiograms to blood pressure monitors. Many of these measurements require a follow-up visit with a healthcare professional.

Smarter monitoring devices communicating with other devices, services, and systems could greatly refine this process, possibly lessening the need for direct clinical intervention and perhaps replacing it with a phone call from a nurse. Current innovations in smart devices include medication dispensers that can detect whether the medication has been taken—with data transmission occurring via Bluetooth. If it is determined that the user has not taken their medication, they are contacted via telephone to discuss and encourage adherence to their medication regime. There are apps used to detect skin cancer and provide urine analysis services. There are a wealth of opportunities enabled by the Internet of Things to not only improve patient care but to reduce healthcare costs at the same time.

A variety of big data includes traditional data sources as well as newer sources of both structured and unstructured data. As the variety of data grows, so too does the potential of what algorithms and machine learning tools deployed in healthcare can achieve.

As with a variety of sources, variety in big data can come in the form of the following:

- Data types: Text, numbers, audio, video, images, and so forth.
- Function: Use cases and user requirements.
- Value of data: Is the data fit for a purpose? This would be a focus on quality in the context of the data's use. More data does not necessarily mean better data.

Just as there are a variety of data sources, values, types, and use cases, there are also a variety of applications for data. These include access points (i.e., Web, mobile, SaaS [software as a service] also sometimes known as MDaaS [medical device as a service], API [application programming interface]) and users (typically either humans or machines).

Data variety is increasing exponentially. As new healthcare IoT devices develop, it will become imperative to identify useful data features from noise.

Legacy Data

Legacy data, whether computer based or non-computer based, can also be used in your projects. An increasing number of services enable digitalization for natural language processing or classification purposes. By using legacy data, stakeholders can maximize the insights from underutilized data. Legacy, or traditional, data may also be referred to as little or small data, discussed later in this chapter.

Typically, legacy data is fragmented and incomplete, particularly for the criteria desired today. For example, many records from more than 10 years ago may not have an associated email address or correct phone number.

Velocity

The third V of big data is velocity, referring to the speed at which data is created, stored, and prepared for analysis and visualization. The velocity of data imposes unique demands on underlying computer hardware infrastructures.

A benefit of cloud computing has been the ability to quickly store and process the volume and variety of big data that would typically overwhelm a traditional server. Cloud computing is the preferred route for big data projects due to flexibility regarding storage and cost. Cloud providers can store petabytes of data and scale up thousands of servers in real time to requirements. Perhaps more valuable is that computational power is also inexpensive and distributable.

After the Haiti earthquake in 2010, Twitter data was a quicker way of detecting and tracking the deadly cholera outbreak compared to traditional methods. A subsequent research study determined that social media platforms outperformed official methods of monitoring disease prevalence in both speed and accuracy of detecting the progress of cholera [36].

In the big data era, data is created in real time or near real time. The ubiquity of cyber-physical devices, embedded sensors, and other devices means that data transfer can occur at the moment it is created, bar a MAC address and Internet access.

The speed at which data is created is unimaginable. It is widely understood that the data generated in the previous 2 years is greater than the data created from the start of time up until then [37]. The challenge organizations have is to cope with the enormous speed at which data is created and consumed in real time.

As the penetration of wearable devices and sensors continues, so too will their application within clinical healthcare (Figure 2-4). The key to maximum clinical value is in the integration of various and heterogeneous data sources.

Data Type	Example	Characteristics
Patient behavior and sentiment	Social media Smartphones Web forums	Most data is unstructured or semi-structured. Data is large and in real time. Opportunities for learning, engagement, and understanding. Typically open source.
Patient health data	Sensors Blood glucose meters Smartphones Fitness trackers Images	Owned by patient. Data is typically device reported. Storage requirements typically low. Image storage requirements vary depending on data required for storage. Sensor data is high velocity. Structured data enables ease of classification and pattern detection. Some items may require digitalization.
Pharmaceutical and R&D data	Clinical trial data Population data	Encompasses population and disease data owned by pharmaceutical companies, research, academia, and the government.
Health data on the Web	Patient portals (Diabetes.co.uk) Digital interventions (e.g., Low Carb Program, Gro Health, Hypo Program)	Useful metadata. Typically device reported rather than clinical.
Clinical data	EHR Patient registries Physician data Images (scans and others)	Structured in nature. Owned by providers.
Claims, cost, administrative data	Claims data Expenses data	Structured in nature. Encompasses any datasets relevant for reimbursement.
Transactional data	Health information exchanges	May require preprocessing to become useful data.

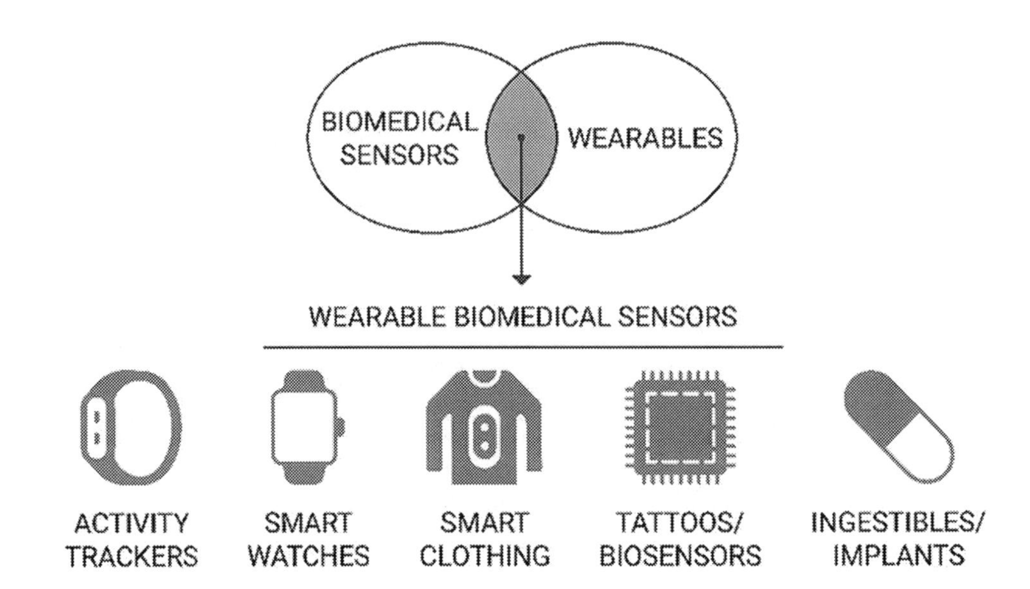

Figure 2-4. *Wearable biomedical sensors*

The traditional three Vs give an insight into the scale of data and the rapid speeds at which these vast datasets grow and multiply. However, data variety only begins to scratch the surface of the depth and challenges of big data.

The power of big data was best demonstrated by monolith Google in 2009 when it was able to track the spread of influenza with only a one-day delay across the United States, faster than the Centers for Disease Control and Prevention (CDC), through the analysis of associated search terms [38].

However, in 2013, Google Flu Trends got it wrong—which spurred questions on the concepts of value and validity of data [39].

In a clinical setting, data value and veracity would inherit an elevated position, as clinical decisions are made on data with lives at risk and go far beyond transactional implications.

Value

Value refers to the usefulness of data. Key to building trust in your data is ensuring its accuracy and usability. In healthcare, this would be evaluated through qualitative and quantitative analysis. Value can be circumscribed through a variety of factors, primarily clinical impact on patient outcomes or behavior modification, patient and stakeholder engagement, impact on processes and workflows, and monetization (cost saving, cost benefits, etc.).

McKinsey stated that the potential annual value of big data to US healthcare is $300 billion. They also mention that big data has a potential annual value of €250 billion to Europe's public sector administration [40].

Data's value comes from the analyses conducted on it. It is this analysis of data that enables it to be turned into information, with the aim of eventually turning it into knowledge. The value lies in how organizations use available data and become information centric and data driven for purposes of decision-making. Value is not found; it is made. The key is to make the data meaningful to your users and your organization by managing it well.

Google did a perfect job in tracking the searches for flu symptoms in 2013. However, it had no way of knowing that its results weren't valid regarding predicting flu prevalence. Google Flu Trends, at least in this instance, wasn't providing valuable results for this use case. To add insult to injury, Google was never to know that real-world cases of flu failed to match the search distribution and frequency—so it could never do anything about it. Perhaps if this project was to be extended, analyzing social media posts might have been more valuable in determining flu prevalence. Social media and wider unstructured data sources provide a wealth of often unexplored information.

Identifying a population at risk of type 2 diabetes, predicting atrial fibrillation (AFIB), assisting patients with recovery from health events, and recommending the latest evidence-based medicine therapy to a patient are tasks with a clear and valuable return on investments.

Veracity

Worthless data, no matter how plentiful, is always going to be worthless. Veracity refers to data truthfulness and whether data is of optimal quality and suitability in the context of its use relating to the biases, noise, and abnormality in data.

Various factors can affect the veracity of data, including the following:

- Data entry: Was data entered correctly? Have there been any errors or events? Is there an audit trail of data entry?
- Data management: What is the integrity of the data moving through the system?
- Integration quality: Is data appropriately referenced, commented on, and unique?

- Staleness: Is data appropriate for use? Is it timely and relevant?
- Usage: Is the data available in a way that is actionable? Will the data be useful toward achieving business goals? Is it ethical to use this data?

The six Cs of trusted data can help assess whether the data you are using for your projects has veracity. It is essential to ensure your datasets are clean, complete, current, consistent, and compliant.

- Clean data: This is the result of good data quality procedures such as deduplication, standardization, verification, matching, and processing. Clean data enables robust decisions to be made on quality,
- non-tainted data. Clean data is the biggest challenge for user trust. Clean data is the product of thorough data preparation.
- Complete data: This is the result of consolidated data infrastructures, techniques, and processes that enable robust decision-making.
- Current data: Fresh data is normally considered more trustworthy over stale data. The question then becomes, at what point is data not current? As real-time, current data grows, deciphering the signal from noise can become a more challenging task.
- Consistent data: Data must be consistent (thus enabling machine readability). This is required for cross-compatibility of systems and applies to metadata also.
- Compliant data: Compliance regulations can come from various sources such as stakeholders, clients, legal legislation, or as the result of a new policy. Although many data infrastructures and standards exist, data science as a category is still coming up with first-time problems for which governance and legislation are required. Compliance can mean different things to internal and external stakeholders. Internally, there will be standards to ensure data is compliant with quality, security, and privacy procedures. All stakeholders need to trust that data has been

accessed and distributed by internal and external regulations. Typically, organizations may have a governance board for data and information.

The sixth C is **collaborative data**, which refers to collaboration over data to ensure that data management and business management goals are aligned. The sixth C refers more to the approach to data than the data itself. While all of the Cs are required for trusted data, the purpose is to have clean data that represents the real world with ideally no to little bias.

While most advocate complete veracity in data for clinical care, the gold standard of clinical care is harder to achieve in real-world scenarios with such a varied input of data. One would question that if data isn't being used effectively already, why there would be a need to add even more. Enforcing data quality may be considered too difficult to achieve cost-effectively. In this case, training an intelligent system to learn to estimate for parameters not seen may be of use. Veracity can refer to ensuring sizable training samples for rich model building and validation, which empowers whole-population analytics.

Veracity can be applied to the analyses performed on the data, ensuring they are correct. Not only does data need to be trustworthy but so too do the algorithms and systems interpreting it.

Data entry can be a risky point that goes unnoticed without good data science. This is typically a human-based problem. Even in small data settings, humans make mistakes. To be clinically relevant, data requires maximum veracity. As the veracity of data improves, machine learning on this data enables more veracious conclusions.

Validity

Similar to big data veracity, there is also an issue of data validity, referring to whether data is correct and accurate for the intended use. In clinical applications, validity may be considered to be the prioritized V—to ensure that only useful and relevant data is used.

Truthfulness or veracity of data is absolute, whereas validity is contextual. Valid data represents the real world without bias.

Variability

Big data is variable. Variability defines data where the meaning is regularly changing. Variability is very relevant in performing sentiment analysis. For example, in a series of tweets, a single word can have a completely different meaning.

Variability is often confused with variety. To illustrate, a florist may sell five types of roses. That is variety. Now, if you go to the florist for two weeks in a row and buy the same white rose every day, each day it will have a subtly different form and fragrance. That is variability.

To perform proper sentiment analysis, algorithms need to be able to understand the context of texts and be able to decipher the exact meaning of a word in that particular context. This is still very difficult, even with progress in natural language processing abilities.

Visualization

Visualization is the final of the eight Vs, referring to the appropriate analyses and visualizations required on big data to make it readable, understandable, and actionable.

Visualization may not sound complex; however, overcoming the challenge of visualizing complex datasets is crucial for stakeholder understanding and development. Quite often, it is data visualization that becomes the principal component in transferring knowledge learned from datasets to stakeholders.

Massive Data

The term massive data is applied to datasets that are enormous collections of records. EHRs would be considered to be massive data. Massive datasets may be simple two-dimensional relational tables. The resources required for massive data surpass simple spreadsheet analysis and are computationally intensive due to the enormous matrices involved in computation.

Small Data

Little data (or small data) is a direct contrast to big data. Big data is distributed, varied, and comes in real time, whereas small data is data that is accessible, informative, and actionable as a result of the format and volume.

Examples of small data include patient medical records, prescription data, biometric measurements, a scan, or even Internet search histories. In comparison to organizations and services such as Google and Amazon, the amount of data is far smaller.

Analytics has gone through a rapid evolution spurned by the adoption and demands of big data. Going back to traditional datasets and applying more modern techniques such as machine learning should not be overlooked and is a great place for a data scientist to start. It's not the size of the data; it's what you do with it that counts.

Metadata

Metadata is data about data—which is descriptive data about each asset, or individual piece, of data. Metadata provides granular information about a single file supporting the facility to discover patterns and trends from metadata as well as the data it is supporting.

Metadata gives information about a file's origin, date, time, and format; and it may include notes or comments. It is key to investing resources, predominantly time, in strategic information management to make sure assets are correctly named, tagged, stored, and archived in a taxonomy consistent with other assets in the collection. This facilitates quicker dataset linkage and maintains consistent methodology for asset management to ensure files are easy to find, retrieve, and distribute.

Big data hype has overshadowed metadata. However, with big data comes big metadata, enabling organizations to generate knowledge and leverage value. For example, Google and Facebook use taxonomy languages such as Open Graph [41] that help create a more structured Web, enabling more robust and descriptive information to be provided to users. This, in turn, provides human-friendly results to users, optimizing click-through and conversion.

Metadata can be very useful in any data-led project. For instance, a machine learning algorithm could use the metadata belonging to a piece of music, rather than (or alongside) the actual music itself, to be able to suggest further relevant music. Features of the music such as genre, artist, song title, and year of release could be obtained from metadata for relevant results.

Healthcare Data: Little and Big Use Cases

Healthcare stakeholders understand they are surrounded by masses of data from patients, professionals, and transactions. It is key to know how to drive value and meet KPIs (key performance indicators). The following are a selection of exciting healthcare data use cases.

Predicting Waiting Times

In Paris, France, four of the hospitals that comprise the Assistance Publique-Hôpitaux de Paris (AP-HP) teamed up with Intel and used data from internal and external sources, including 10 years of hospital admissions records, to determine day- and hour-based predictions of the number of patients expected to enter their facility [42].

Time series analysis techniques were used to predict admission rates at different times. This data was made available to all surgeries and clinics, and demonstrates an immediate way data could be used to improve efficiency and empower stakeholders.

Most, if not all, clinics across the world have access to similar data, which demonstrates just how healthcare is only beginning to scratch the surface of data's application.

Reducing Readmissions

The same approach to waiting times can be used to help manage hospital costs. Using data analytics, at-risk patient groups can be identified based on medical history, demographics, and behavioral data. This can be used to provide the necessary care to reduce readmission rates.

At the UT Southwestern hospital in the United States, EHR analytics led to a drop in the readmission rate of cardiac patients from 26.2% to 21.2% through successful identification of at-risk patients [23].

Predictive Analytics

The preceding examples could very well use static data (i.e., non real time small data) and be fairly accurate in predicting waiting times and readmission intervals. The same concept of analyzing data can be used at scale for prediction of disease and the democratization of care.

In the United States, Optum Labs has collected the EHRs for over 30 million patients, creating a database for predictive analytics tools to improve the delivery of care. The intention is to enable doctors to make data-driven, informed decisions with proximity and therefore improve patients' treatment [44].

The robustness that 30 million health records provide allows models to be trained and validated to find people who fit predictive risk trends for certain diseases such as hypertension, cardiovascular disease, stroke, type 2 diabetes, heart disease, and metabolic syndrome.

By analyzing patient data including age, social and economic demographics, fitness, and other health biomarkers, providers can improve care at both an individual and population level through not only predicting risk but through the delivery of treatments for optimal patient outcomes.

Electronic Health Records

EHRs haven't quite come to fruition as yet. The idea is theoretically simple: that every patient has a digital health record consisting of their details, demographics, medical history, allergies, clinical results, and so forth. Records can be shared, with patient consent, via secure computer systems and are available for healthcare providers from both public and private sectors. Each record comprises one modifiable file, which means that doctors can implement changes over time with no danger of data replication or inconsistencies.

EHRs make perfect sense; however, complete implementation across a nation is proving a task. In the United States, up to 94% of eligible hospitals use EHRs according to HITECH research. Europe is further behind. A European Commission directive has set the task of creating a centralized European health record system by 2020 [46].

In the United States, Kaiser Permanente has implemented a system that shares data across all their facilities and made it easier to use EHRs. A McKinsey report highlighted how the data sharing system achieved an estimated $1 billion in savings as the result of reduced office visits and lab tests. The data sharing system improved outcomes in cardiovascular disease [47].

The EHR is evolving into the blockchain, which seeks to decentralize and distribute access to data.

Value-Based Care/Engagement

No longer are patients considered passive recipients of care. Healthcare is now obliged to engage patients in their health, healthcare decision-making, care, and treatment. Engagement can be maintained through digital means. Note that patient engagement is not to be confused with patient experience, which is the pathway (journey) a patient may take.

Financial drivers have already seen healthcare practices becoming increasingly incentivized to best engage with each patient, to ensure services received are satisfactory and of quality.

One of the key drivers of data-driven solutions is the demand for patient engagement and the transition toward value-based care. Better patient engagement enhances trust between patients, treatment providers, and bill payers. Moreover, it leads to better health outcomes and cost savings (or some other benefit) to the provider.

Pioneering health insurance initiatives seek to engage patients with the goal of better health outcomes through entwining premiums to good health. Health innovators at Diabetes Digital Media in Warwick, England, are working with global insurance providers and health organizations to provide scalable digital health technologies for users to optimize their health and well-being [18].

Blue Shield of California is improving patient outcomes by developing an integrated system that connects doctors, hospitals, and health coverage to the patient's broader health data to deliver evidence-based, personalized care [48]. The aim is to help improve performance in disease prevention and care coordination.

Healthcare IoT: Real-Time Notifications, Alerts, Automation

Millions of people use devices that datafy their lives toward the quantified self. Devices connected to the Internet currently include weighing scales; activity monitors (such as Fitbit, Apple Watch, Microsoft Band) that measure heart rate, movement, and sleep; and blood glucose meters, all of which send metrics in real time and track user behavior in up-to-the-second fashion. There is even a fetal heartbeat wearable [50].

The data recorded could be used to detect the risk of disease, alert doctors, or request emergency services depending on the biometrics received. Many integrated devices are going beyond pulse and movement to measure sweat, oxidation levels, blood glucose, nicotine consumption, and more.

With sophisticated devices come sophisticated solutions to novel problems. For instance, now that heart rate monitoring is cheaper and more pervasive, conditions such as atrial fibrillation can be detected far easier and far earlier than ever before. Fluttering of one's heart rate above 300 bpm (rather than 60–80 bpm) could be a symptom of atrial fibrillation. Patients diagnosed with the condition are 33% more likely to develop dementia; and more than 70% of patients will die from a stroke [51, 52]. The treatment can often be simple: anticoagulants and blood thinners, which are up to 80% effective.

Healthcare providers are moving slowly toward sophisticated toolkits to utilize the massive data stream created by patients and to react every time the results appear disturbing. This adoption is being driven by established digital organizations and start-ups in conjunction with health bill payers.

In another use case, an innovative program from the University of California, Irvine, gave patients with heart disease the opportunity to return home with a wireless weighing scale and weigh themselves at regular intervals. Predictive analytics algorithms determined unsafe weight gain thresholds and alerted physicians to see the patient proactively before an emergency readmittance was necessary [31].

What is interesting to recognize is that these connected health devices do not necessarily motivate the audiences or populations most at risk of adverse health events. According to several randomized trials, Fitbit wearers do exercise more, but not enough to guarantee weight loss and improved fitness.

In fact, some studies have determined they can be demotivating [55]. Then there is the question of accuracy. A Cleveland Clinic study found that heart rate monitors from four brands on the market were reporting inaccurate readings 10–20% of the time, which demonstrates that there is some way to go in the precision of the technology [56].

While these devices show potential, there is still the problem of an average abandonment rate of more than 30% after a period of engaged usage. The software application is now the channel for engagement, used as an intelligent layer on top of devices and IoT, where they can also introduce behavior change psychology for sustainable behavior change and usage.

Offering users of devices such as these tangible incentives like discounts on health or life insurance could become more mainstream and drive prevention of many chronic lifestyle-related diseases similar to incentivized car insurance with embedded black box sensors.

Real-time alerting can also be used to notify patients of adverse effects of medications they are prescribed. Currently, patients wouldn't receive any notification unless they were registered with the treatment provider. Healthcare providers can use public feeds to notify their patients of potential adverse effects. Emails or text messages would suffice as a form of engagement and could provide as much instruction as required, reducing time with the clinician.

EHRs can also trigger warnings, alerts, and reminders for when a patient should get a new lab test or track prescriptions to see if a patient has been following treatment instructions.

Movement Toward Evidence-Based Medicine

Evidence-based medicine is a term denoting treatment given based on proven scientific methods in pursuit of the best possible outcomes. Clinical trials work on a small scale, testing new treatments in small groups with internal validity (i.e., no other conditions or concerns other than those specified) and looking at how well treatments work and establishing if there are any side effects.

With growing datafication, there is also increasing "real-world evidence" or data, which can be analyzed at an individual level to create a patient data model and aggregated across populations to derive larger insights around disease prevalence, treatment, engagement, and outcomes. This approach improves quality of care, transparency, outcomes, and value and, at its core, democratizes healthcare delivery.

Using data for similar groups of patients, we can understand the treatment plans patients with similar profiles have taken, allowing the best treatment plan to be recommended based on how others in the population responded to any given treatment. The recommended treatment pathway would be best for the patient and could be explained to both the patient and healthcare professional as to why it was the best pathway to follow. This has roots in predictive analytics and goes beyond identification of populations and treatments.

By mining real-world patient data, including clinical patient records and real-time, personal, patient health data—and with demographic and health data on disease prevalence, treatment pathways, and outcomes—we can learn about optimal treatment plans for individuals, facilitating precision medicine, which is the pinnacle of evidence based medicine.

By connecting real-world patient and clinical data to the genome, it can be personalized down to the genetic makeup of patients and populations for the following:

- Prescribing of medications
- Adverse effects and reactions
- Prevention strategies
- Likelihood risk of future diseases

Public Health

Analysis of disease patterns and outbreaks allows public health to be substantially improved through an analytics-driven approach.

Big data can assist in determining needs, required services, or treatments; and it can predict and prevent future crises to benefit the population. By mapping patient location, it would be possible to predict outbreaks, such as influenza, that could spread within an area, making it easier to formulate plans for dealing with patients, vaccinations, and care delivery.

In West Africa, mobile phone location data proved invaluable in tracking the spread of the population, and as a result, helped to predict the Ebola virus's expanse [57].

After the Haiti earthquake in 2010, a team from Karolinska Institute in Sweden and Columbia University in the United States analyzed calling data from 2 million mobile phones on the Digicel Haiti network [58]. Phone records were used to understand population movements and for the United Nations to allocate resources more efficiently. The data was also used to identify areas at risk of the subsequent cholera outbreak.

Evolution of Data and Its Analytics

Proximity is driving the future. As people, patients, or agents in a digital world, we expect a relationship regarding nearness in time between data that is given and feedback that is received. As the scale of data increases, so too does the demand that the time between data creation and insight and action be reduced. There is substantial economic value in reducing the time from detected event to automatic response. This could help spot fraudulent transactions before a transaction completes or even serve information matched to an interest or need.

Just as data has evolved, so too has the analytics of data (Figure 2-5).

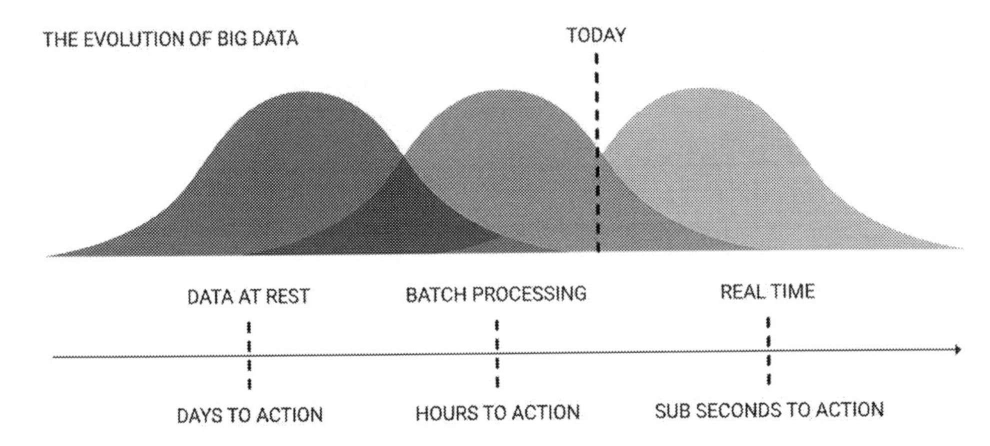

Figure 2-5. *The evolution of big data and its analytics*

As a concept, the analysis of data, or analytics, has gone through several iterations, driven by the demand for access to data to drive decision-making.

Analytics 1.0 took the form of traditional analytics on small, slower-velocity datasets, often in the form of spreadsheets and static documents. This is suitable for little data. Typical analytics were descriptive and report based.

Data would usually be structured in nature. This drove business intelligence in the 1990s with predefined queries and detailed or historic views—typically driven by web dashboards leveraging structured data like customer data, behavioral data, sales data, patient records, and so forth. Data about production processes, sales, interactions, and transactions were collected, aggregated, and analyzed within traditional relational databases. In the Analytics 1.0 era, data scientists spent much of their time preparing data for analysis rather than performing analytics themselves.

Analytics 2.0 marked the integration of big data with traditional analytics, with interfaces to support the querying of big datasets in real time. Big data analytics began in the 2000s, which facilitated competitive insight through complex queries combined with predictive views leveraging both structured and unstructured data such as social media, behavioral data, and the growing variety of user data. Big data that couldn't fit or be analyzed fast enough on a centralized platform would be processed with frameworks such as Hadoop, an open source software framework for fast batch data processing across parallel servers both in the cloud or on-premises.

Unstructured data would be supported through more than just SQL (Structured Query Language) databases, supporting key and value, document, graph, columnar, and geospatial data. Other big data technologies introduced during this period include "in-memory" analytics for fast analysis whereby data is managed and processed in memory rather than on disk.

Today, vast amounts of data are being created at the edge of the network, and the traditional ways of doing analytics are no longer viable. Organizations that engage with consumers—whether they make things, move things, consume things, or work with customers—have increasing amounts of data on those devices and activities. Each device, shipment, and person leave a trail referred to as a data exhaust.

Analytics 3.0 marked the stage of maturity in which organizations realized measurable business impact from the combination of traditional analytics and big data. The approach supports agile insight discovery and real-time analysis at the point of decision. Analytics 3.0 is mostly a combination of conventional business intelligence, big data, and the IoT distributed throughout the network. Organizations can analyze those datasets for the benefit of particular audiences to achieve their objectives—whether that's improving health outcome or monetization. They also can embed advanced analytics and optimization in near real time into every product or service decision made on the front lines of their operations.

Turning Data into Information: Using Big Data

Data's value is demonstrated in the ability to convert it into information that can drive actionable insight to drive behaviors and workflows. Data analysis/analytics is a subset of data science that seeks to draw insights from sources of raw data.

Today's advances in analyzing big data allow researchers to decode human DNA in minutes, predict where terrorists plan to attack, and determine which gene is most likely to be responsible for specific diseases and, of course, which ads you are most likely to respond to on Facebook.

Analysis of big data can help organizations understand the information contained in their data as well as identifying data most important to business objectives, outcomes, and decisions. Stakeholders typically want the knowledge that comes from analyzing data, which is determined based on the type of results produced and the reasoning approach.

The majority of big data is unstructured. IBM estimates that 80% of the world's big data is unstructured, which includes photos, audio, video, documents, and interactions [59].

Data needs to be processed to be of value. For this to take place, resources are required—in the form of time, talent, and money. Herein is one of the inertias in adopting big data approaches: is processing the data into something with usable insight worth the capital and risk of doing so?

There are four streams within the data science analytics taxonomy (Figure 2-6). The classifications are based on the types of results that are produced.

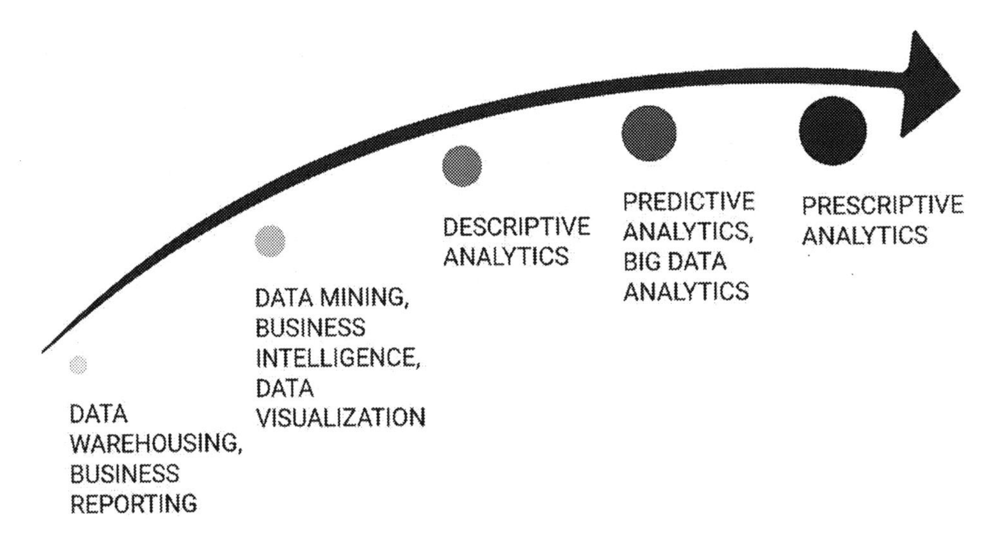

Figure 2-6. *Streams of analytics*

Descriptive Analytics

Descriptive analytics brings insight to the past, focusing on the question of what happened, through analyzing data from history. Descriptive analytics uses techniques such as data aggregation and data mining to provide historical understanding. There is much to be learned from descriptive analytics.

Common examples of descriptive analytics are reports that provide the answers to questions such as how many patients were admitted to a hospital last July, how many were readmitted within 30 days, and how many caught an infection or suffered from a mistake in care. Descriptive analytics serves as the first form of big data analysis.

This basic level of reporting is still lacking for many organizations. Proprietary data standards maintain unhelpful pits of information; a lack of available human expertise or proper organizational buy-in can leave data collecting and unused.

Data can become more valuable through dataset linking. For instance, linking physician appointments to hospital admissions, to education attendance, to medication adherence, and to access to services can provide substantial insight and understanding for health delivery optimization and cost savings.

Limitations of descriptive analytics are that it gives limited ability to guide decisions because it is based on a snapshot of the past. Although this is useful, it is not always indicative of the future.

Diagnostic Analytics

Diagnostic analytics is a form of analytics that examines data to answer the question of why something happened. Diagnostic analytics comprises techniques such as decision trees, data discovery, data mining, and correlations.

The next two types of analytics, predictive and prescriptive analytics, require the use of learning algorithms and are covered in more depth in following chapters.

Predictive Analytics

Predictive analytics allows us to understand the future and predict the likelihood of a future outcome. Predictive analytics uses the data that you have at your disposal and attempts to fill in missing data with best-guess estimates.

Predictive analytics is characterized by techniques such as regression analysis, multivariate statistics, data mining, pattern matching, predictive modeling, and machine learning. Predictive analytics uses historical and current data to forecast the likelihood of future events or actionable outcomes.

Predictive analytics skills are in demand as healthcare providers seek evidence-based ways to reduce costs, take advantage of value-based reimbursements, and avoid the penalties associated with the failure to control chronic diseases and adverse events that are within their power to prevent. Predictive analytics is in the midst of disruption. The last 5 years of innovation has made significant progress with measurable impacts to the lives of patients. Wearable technology and mobile apps now make it possible to spot conditions such as asthma, atrial fibrillation, and COPD [60].

Predictive analytics remains elusive, as it requires access to real-time data that allows near-real-time clinical decision-making. To support this, medical sensors and connected devices must be fully integrated to provide up-to-the-moment information on patient health.

Besides, clinicians need to be experienced in using such data. As well as using the individual patient's data, more accurate diagnoses and treatments must draw from as much patient data as is available, including population-level data.

As data sources and technologies develop, advanced decision support from cognitive computing engines, natural language processing, and predictive analytics can help providers pinpoint diagnoses that may otherwise elude them. Population health management tools can highlight those most at risk of hospital readmission, at risk of developing costly chronic diseases, or responding adversely to medication.

Use Case: Realizing Personalized Care

Metabolic health has been demonstrated to influence the risk of a variety of health conditions, including type 2 diabetes, hypertension, some dementias, and some cancers. While acting as a digital intervention for type 2 diabetes, the Low Carb Program app used a variety of features in the form of health biomarkers and demographics including blood glucose, weight, gender, and ethnicity to score a patient's metabolic health. As the app expanded to integrate with wearable health devices, increased weight and blood glucose data enabled the algorithm to extend to determine the likely risk of pancreatic cancer. This is communicated to patients with their healthcare team where relevant and is referenced through cohort data comparison [16].

Users receive supported notifications that their data does not seem to fit within the demographic that is expected of them and prompts the user to speak to their healthcare professional. As demonstrated in this example, as further biosensors and algorithms are developed, the ethical implications of predictive analytics are significant.

Use Case: Patient Monitoring in Real Time

A typical hospital ward would see ward nurses manually visit patients to monitor and confirm vital health signs. However, there is no guarantee that the patient's health status won't deviate, for better or worse, between the time of scheduled visits. As a result, caregivers often react to problems as the result of adverse events, whereas arriving earlier in the process can make a significant difference to patient well-being. Wireless sensors can capture and transmit patient vitals far more frequently than human beings can visit the bedside.

Over time, the same data could be applied to predictive analytics techniques to determine the likelihood of an emergency before it can be detected with a bedside visit.

Prescriptive Analytics

Prescriptive analytics strives to make decisions for optimal outcomes, that is, to use all available data and analytics to inform and evolve a decision of what action to take—that is, smarter decisions. Prescriptive analytics attempts to quantify the effect of future decisions to advise on possible outcomes before decisions are made.

At its best, prescriptive analytics predicts not only what will happen but also why it will happen to provide recommendations regarding actions that will take advantage of the predictions.

Prescriptive analytics uses a combination of AI techniques and tools such as machine learning, data mining, and computational modeling procedures. These techniques are applied against input from many different datasets including historical and transactional data, real-time data feeds, and big datasets.

Prescriptive analytics performs an in-context prediction, taking into account the available evidence. For example, typical hospital readmission analytics displays a forecast of patients likely to return in the next 14–30 days. A more useful predictor may integrate the same dashboard with projected costs, real-time bed counts, available education material, and follow-up care. With additional information, clinicians can determine patients with a high risk of readmission and take actions to minimize this risk, resulting in improved patient outcomes and reduced consumption of hospital resources. Population management approaches could consider patients who are obese, adding filters for factors such as high cholesterol or heart disease risk to determine where to focus treatment and which treatments to target. Pharmaceutical companies can use prescriptive analytics to expedite drug development through identifying audiences that are most suitable for clinical trials. This includes patients presumed to be compliant and those not expected to have complications.

Prescriptive analytics enables the optimization of population health management through identification of appropriate intervention models for individuals from risk-stratified populations—combining clinical, patient, and wider available health data.

Analytics is undergoing a further transformation, from prescriptive analytics through to contextual analytics. Contextual analytics considers wider data—such as environmental, location, and situational data—when prescribing the best course of action.

Use Case: From Digital to Pharmacology

Different treatment pathways work differently for different people. Collected on a large scale, physicians can prescribe therapies based on the latest evidence-based and predictive and prescriptive analyses of patient populations.

Incorporating digital education as well as pharmacological therapies, delivery of treatments through such resources democratizes healthcare access by ensuring the most precise treatment is always prescribed, with trackable adherence and progress insight alongside ensuring resources are focused where required.

Reasoning

A system can determine conclusions with information available in a knowledge base using three main methods: deduction, induction, and abduction. Although a background in logic is not required, it is useful to understand the differences between the different reasoning approaches. Any data scientist should be familiar with these reasoning techniques.

Deduction

Deductive reasoning allows you to make statements that are necessitated by facts that you know. For example, you are given two facts:

1. It rains every Saturday.
2. Today is Saturday.

Deductive reasoning allows you to determine the true statement that it is going to rain today if today is Saturday.

In deductive reasoning, one infers a proposition q, which is logically sensical from a premise p. Most reporting systems and business intelligence software are deductive.

Induction

Inductive reasoning enables you to make statements based on the evidence you have accumulated until now. However, the key point is that evidence is not the same as fact. Substantial evidence for something only suggests, however strongly, that some "thing" is a fact. Consequently, statements that are determined are only very likely to be true in an inductive approach, rather than absolute.

With inductive reasoning, one attempts to infer a proposition q, which follows from p, but which is not necessarily a logical consequence of p.

For example, if it has rained in Warwick, England, during December for a recorded history of over 50 years, you would have strong evidence (data) for the inductive statement that it will rain in the coming December in Warwick. However, this does not mean it is going to happen; it is not a fact.

Statistical learning is about inductive reasoning: looking at some data, scientifically guessing at a general hypothesis, and making statements or predictions on test data based on this premise.

Abduction

Abduction is an adaptation of inductive reasoning. Abductive reasoning attempts to use a hypothesis p to explain a proposition q. In abduction, reasoning flows in the opposite direction to deductive reasoning. The best hypothesis, that is, the one that most effectively explains the data, is inferred to be the most probably correct one.

A classic example is the following:

- (q) Observation: When I woke up, the grass outside my window was wet.
- (p) Knowledge base including information: Rain can make the grass wet.
- Abductive inference: It probably rained at night.

There are often many up to infinite possible outcomes, so abductive reasoning is useful in determining how to prioritize explanations to investigate. Big datasets provide opportunities for induction and inference through machine learning.

How Much Data Do I Need for My Project?

Although big data is defined by the fact it is stored in distributed systems, when it comes to creating machine learning algorithms, it's about quality and not quantity. For instance, if you have over 1,000 data points for a patient, how are you to understand the most important attributes from what is noise? Further still, is it likely that you would get better outcomes from 5,000, 10,000, or 100,000 data points? This problem is best addressed through data science and is entirely dependent on the domain and the quality of your data samples.

The future of evidence-based medicine is the realization of data-driven, personalized healthcare. Data-driven healthcare occurs when there is a synergetic partnership between healthcare providers, patients, and data obtained from all clinical documentation. This results in analytics that drives actionable insights in real time, allowing for an enhanced patient experience and streamlined clinical and administrative workflows.

Challenges of Big Data

There are several challenges in the development of big data initiatives.

Data Growth

By the very nature of big data, storage is a challenge. There is broad agreement that the size of the digital universe will double every 2 years, which equates to a fiftyfold growth from 2010 to 2020.

Most big data is unstructured, meaning that it doesn't reside in a traditional database schema. Documents, photos, audio, videos, and other unstructured data can be difficult to search, analyze, and retrieve. Hence, management of unstructured data is a growing challenge. Big data projects can evolve as quickly as the data used within them.

Infrastructure

Big data consumes technical resources in the form of infrastructure, storage, bandwidth, databases, and so forth. The challenge is not technical so much as one of finding reliable vendors of services and support and of getting the correct economic model of remuneration.

A solution delivered from the cloud will have greater scalability, cost-effectiveness, and efficiency compared to an on-premises solution.

Expertise

Finding good-quality expertise (or a deep stack expertise) in the disciplines of data analysis and data science is a challenge faced by many organizations. Compared to the amount of data generated, there are very few data scientists in proportion.

Data Sources

A significant challenge with big data is the volume and velocity at which data is generated and delivered. Managing the plethora of incoming data sources is a task in itself.

Quality of Data

Data quality is certainly not a new concern, but the ability to store each piece of data generated in real time compounds the problem. Common causes of dirty data must be addressed and include user input errors, duplicate data, and incorrect data linking. As well as maintaining data, big data algorithms can also be used to help clean data.

Security

There are significant risks in the security and privacy of patient data. With sensitive health data, there is rightly an increased sensitivity to security and privacy concerns. The tools used for analysis, storage, management, and utilization of data are from a variety of sources, which expose the data to risk. Cybersecurity is a concern. Significant data breaches have occurred in the healthcare industry. In the United Kingdom, the Information Commissioner's Office (ICO) fined the Brighton and Sussex University Hospitals NHS Trust after it was found that the sensitive data of several thousand people were resident on hard drives placed up for auction on eBay [61, 62].

To date, the most substantial healthcare breach recorded has been of American health insurance company Anthem [63]. The breach exposed the personal records, including home addresses and personal information, of over 70 million current and former members. Data loss, including data theft, is a genuine problem that is continuously tested to its limits.

Security challenges include authentication for every team and team member accessing datasets, restricting access based on user needs, recording data access histories, and meeting compliance regulations and ensuring data is encrypted to protect from snoopers and any malintent.

Security of medical devices poses a unique threat because of their technological diversity. Medical devices, from health applications on a smartphone to insulin pumps, are becoming increasingly networked, leaving unique openings for hackers [64]. This kind of activity is regarded as dangerous, although for those more technical, it is understood that this can be utilized for positive benefit. Insulin pumps and continuous glucose monitors have been "hacked" to function as an artificial pancreas by enthusiasts way before pharmaceutical companies or digital companies had. However, the key concern is that, if exploited, similar device vulnerabilities could lead not only to data breaches but fatalities in people relying on medical devices. Today, the balance between disruption and progress is one of real-world experience. Yesterday's disruptors are today's innovators.

Then there is the topic of third-party services and vendors. Most health devices and health apps offer API access. It's useful to remember that external vendor services and APIs are only as good as the developers who have developed them. Evidenced trust and reputation is all that one truly relies on.

Resistance

Lack of understanding of value and outcomes from machine learning projects can leave internal stakeholder resistance. You can only manage a project efficiently if you can ascribe clear value to results and prioritize clear explanations of outputs to stakeholders.

Over and above inertia, once key patterns have been identified, organizations must be prepared to make necessary actions and implement changes to derive value from them. This is often a primary inertia, so precise and jargon-free documentation and governance is useful for understanding the project and governance purposes.

Policies and Governance

The vast amount of big data captured through wearable devices and sensors raises questions as to how data is collected, how data is treated, how data is analyzed, and how data should affect the policy-making process. Besides, data privacy and security are legal areas in a state of constant development and evolution.

Fragmentation

Most organizational data is highly fragmented. Hospital teams have their patient data, as do physicians and tertiary care teams. This creates challenges at several levels: syntactic (i.e., defining common formats across teams, sites, and organizations), semantic (agreeing on common definitions), and political (determining and confirming ownership and responsibilities). This is before the patient's own data, collected through phones, health apps, wearables, and so forth.

Lack of Data Strategy

To benefit from data science investments, organizations must have a coherent data strategy. It's key to establish the need for the data and identify the right type and sources of data for use. What is the desired goal with the data and what are the objectives? It is important to establish the data needs, to save wasting resources and efforts collecting data that are of no use.

Within a hospital emergency department, for instance, collecting the time of patient admission, triage, and consultation is required to understand how long it takes for patients to be seen, but it is of negligible use to determine quality of care or outcomes.

Visualization

To truly benefit from data science investments, stakeholders must understand the outputs of data analysis and exploration. Poor visualization of data is connected to lack of data strategy: if specific use cases and objectives are not determined, visualization of results from data can prove challenging due to a lack of clear direction.

Timeliness of Analysis

Linked to proximity, the value of data may decrease over time. Applications such as fraud detection in healthcare, just like with banking, require data to be delivered as close to real time as possible to be effective.

Ethics

Data ethics is emerging as a topic as datasets become big enough to raise practical rather than theoretical concerns about ethics. Patients are leaving a constant data trail through sensors, wearables, transactions, social media, and transportation. For instance, the genetic testing company 23andMe sells reports on genetic risk disposition for over ten diseases including Alzheimer's disease and Parkinson's disease. This spurns challenging conversations about what is and what is not appropriate, discussed more in Chapter 8.

Data and Information Governance

Data is a precious asset to any business or organization; but in healthcare, data governance is paramount to ensure consistency, safety, and progression of any organization toward becoming data driven and analytically driven.

The need for data governance is driven primarily by the requirement for accountability in a risk-averse industry. As the value of data is realized, data governance too is evolving to define the approach, management, and utilization of data in an organization. Processes and standards such as ISO (International Organization for Standardization) 10027 and GDPR (General Data Protection Regulation) have been used to bring the entire data industry in line with best practice.

Data governance and information governance are often used interchangeably but do differ.

Governance of data refers to the practice of managing data assets throughout their lifecycle to ensure that they meet standards of quality and integrity. The goal of data governance is to provide user trust in the data through optimal data validity and accountability. It sets the framework for creating high-quality data and using data in a secure, ethical, and consensual manner. Data governance in healthcare seeks to improve efficiency, create accountability, and establish a pool of data that providers and patients can use to make proper health decisions.

Often the best approach to data governance is to govern to the least extent necessary to achieve the most common good. In several cases, governance is being set for circumstances that don't require governing yet.

Organizations can waste tremendous amounts of time through management and decision-making, including constraints on data, which can hinder project development and direction. Healthcare organizations should be motivated by the acquisition of more data to learn—to gather information to manage risks better and improve outcomes.

Data governance is required to uphold a variety of criteria to return quality information.

Data Stewardship

The steward is accountable for the data as defined and its appropriate use. The objective of this aspect of data governance is to assure quality regarding data accuracy; data accessibility; and data consistency, completeness, and updating. This is typically the organization running the project. Further still, most organizations would have an appointed officer(s), responsible for accessing the data, with elevated privileges.

Data Quality

Ensuring data quality is arguably the most important function of data governance besides security. Poor data quality has a detrimental impact on accuracy, particularly for learning projects. Governance should seek to enforce quality, accuracy, and timeliness of data.

Data Security

Security of data is essential. First and foremost, data governance sets out how the data is protected, typically how it is encrypted, who has access to it, processes for handling data—including how the data may be used—and what is done in case of a breach. This is motivated by several drivers: regulatory requirements that demand action (such as GDPR taking place in Europe), the growing risk of cybersecurity, and the fact that perceptions of digital weakness can harm reputation and trust. This is demonstrated in how the number of websites that operate with a secure https:// protocol has risen by 4,000% since 2012 [65]. Patients expect their data to be secure.

Having said that, the Garmin ransomware attack of 2020 showed that people will be patient with services that they rely on, with messages of support and jubilation as services were returned online [66]. The situation was similar with Babylon Health which shared the recordings from other GP consultations in public view and access to a minority of patient users [67].

People are aware data security is a constant concern for all involved, and the Garmin cyberattack showed people will be patient with services and brands they prefer or that give them use or value. It is noteworthy that even when founders and executives from technology companies are brought before governors, time has shown there is very little understanding of the technologies involved or intricacies of use from those responsible for questioning.

Data Availability

Fragmentation of data is one of the biggest problems healthcare organizations face. Ensuring stakeholders', providers', and patients' data access is critical.

People "own" their data; and as a result, access needs to be transparent, quick, and efficient for all. Data access governance may set permissions and authentication levels for user roles to systems.

Data Content

The data and information governance for a project may state the types of data collected, including health data; metadata; and location, profile, and behavior data. Data governance may also detail what the data is used for.

Master Data Management (MDM)

As data transfer becomes a priority, the MDM acts as a master reference to ensure consistent use of data across organizations. The goal of MDM is to create a common point of reference that can be shared within an organization.

Use Cases

In almost all cases, data governance policies will include use cases for data access and "what would happen if" scenarios. It is useful to have all team members required in action to be involved in dummy runs to ensure that data access or access requests are not met with periods of silence.

Use cases would include the set of processes that would take place in light of a request for data from a patient, request for deletion, ensuring appropriate roles were able to access the data, ensuring appropriate transactional logs were kept, and so forth.

Data governance provides several benefits over and above those that are internal, including the following:

- Protecting the interests of data
- stakeholders—particularly the data "giver"
- Standardizing procedures and processes for streamlined repetition and minimization of error
- Reducing costs and improving effectiveness
- Greater transparency and accountability between data transfer parties

Information governance has a slightly different purpose. Information governance is the management and control of information. This information is formed through the use of data assets (Figure 2-7).

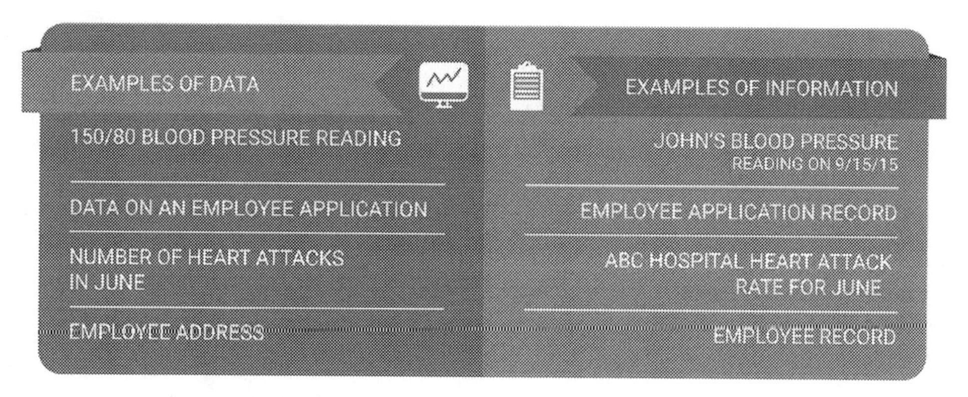

Figure 2-7. *Data vs. information—what's the difference?*

A single data point, such as a blood glucose reading, without any additional context or metadata, offers very little. However, a series of blood glucose readings over the past 3 months can show whether the patient has good or bad blood glucose control, for example, which can be a risk factor for comorbidities. It pays attention to personal and sensitive information linked to patients and employees.

In most organizations, data and information governance is consolidated; but it is useful to understand the distinction.

A resolute data governance program would involve a governing body or council, a defined set of procedures, and the way in which said procedures are upheld and verified.

As learning systems develop, information governance is becoming more critical. Digital health apps and integrated bio-human health devices will soon be able to predict and determine risk and time to disease as datasets grow infinitely large.

Regardless of the question of accuracy, legislation on what can and can't happen with this information is required.

Conducting a Big Data Project

Getting started with any big data project involves three steps. Big data projects do not have to be part of a bigger machine learning project:

1. Understand how the project will impact your organization.

   This involves identifying the business case; objectives (what is it I want to achieve?); current and required infrastructures; data sources; and quality, tools, and KPIs regarding measuring success. Identifying quantitative objectives is key to measuring success and performance. Determining a measurable ROI to determine the success of the project is useful to demonstrate value to stakeholders. Identifying goals with members from technical, medical, and operational teams helps to maintain clarity of focus and direction. Identify potential use cases and create use case scenarios and personas if time allows and if desired.

2. Find the skills and technology you need.

   Identification of the use cases for the project can help define the technology, infrastructure, and capabilities required for the project. Skills in Python and R (open source programming languages) are most likely going to be required, with analytics skills also desired for data science. This can be outsourced, contracted in, or developed into full-time roles.

3. Implement the project.

   Identify the data you wish to include and make a job of identifying data that will also be excluded from use cases. Identify the types of analyses required on the data to gather the desired output from the project. Explore the requirements necessary for collecting and preparing data into usable formats for algorithms; the governance required for holding and using the data; and how it will be presented to stakeholders and, most importantly, the end users. Architectural decisions such as database structures, platform providers, and type of models may be made at this point. It's at this point you can consider the types of analytics applied to the data.

You may wish to consider phasing of the project and the gaps between current and future capabilities that require addressing.

Develop the project in a test environment and present results to users in a meaningful way.

Big Data Tools

There are various open source and paid-for tools that can be used for big data collection and machine learning projects. The fundamentals begin with Hadoop and NoSQL (non-relational) databases.

Hadoop is synonymous with big data and is a Java-based, open source software framework, or software ecosystem, for distributed storage of large datasets on computer clusters. Hadoop can process big and complex datasets for insights and answers. Hadoop Distributed File System (HDFS) is the storage system used by Hadoop that separates big data and distributes it across nodes within a cluster. This can also replicate data in a cluster, providing high availability. For those not wanting to delve into Java themselves, Microsoft Azure, Cloudera, Google Cloud Platform, and Amazon Web Services all provide big data hosting services powered by Hadoop.

MongoDB is a modern approach to databases that is useful for managing data that changes frequently or data that is unstructured or semi-structured in nature. The architecture has better write performance, takes less storage space with compression, and reduces operational overhead. MongoDB is part of a growing movement known as NoSQL databases, which are used to store unstructured data with no particular schema. Rows can all have their own set of unique column values. NoSQL has been driven by the requirements to provide better performance in storing big data.

R is an open source software and programming language used for data mining and statistical analysis and graphical techniques.

Python is another popular open source programming language used in machine learning and data mining for data preprocessing, computing, and modeling.

Conclusion

Big data and healthcare mean significant progress toward realizing a plethora of opportunities. This includes remote patient monitoring, precision medicine, preventing readmissions, and advancing disease understanding.

From mundane tasks of data entry and record keeping to advanced and functional applications like observing and understanding people's blood glucose patterns, data is becoming part of our day-to-day fabric.

The potential for positive impact that big data–driven solutions can bring seemingly outweighs any negative sentiment. The growing trend toward centralization of medical data, toward the custodianship of the patient, is causing concern; yet as long as privacy and security are maintained, it is likely to facilitate the development of new treatments and contribute toward the evolution of medical understanding. The healthcare industry remains well within its infancy of leveraging big data for clinical and business use. The explosion of wearables and health IoT has allowed the measurement of heart rate or steps to be taken during a day, which may seem like a relatively ordinary dataset; but the potential impact on preventative medicine and the general health of a population is immense.

Big data provides the opportunity for precision medicine through the delivery of data-driven, evidence-based medicine. The benefits are plentiful, first and foremost through the delivery of more precise treatments for patients.

With this comes cost savings through fewer mistakes, fewer hospital admissions, decreased mortality, and improved patient satisfaction. This also saves money by optimizing the use of the physician and wider healthcare team's time, as more time is spent with patients in need.

CHAPTER 3

What Is Machine Learning?

Any fool can know. The point is to understand.

—Albert Einstein

The first Checkers program from machine learning pioneer Arthur Samuel debuted in 1956, demonstrating artificial "intelligence" capabilities [68].

Since then, not only has the application of AI grown; there is now a velocity, volume, and variety of data that has never been seen before. Samuel's software ran on the IBM 701, a computer the size of a double bed. Data was typically discrete.

Almost 70 years later, data is ubiquitous, and computers are now more powerful such that 100 IBM 701s can fit into the palm of our hand. This has facilitated the domains of machine learning and deep learning within AI (see Figure 3-1).

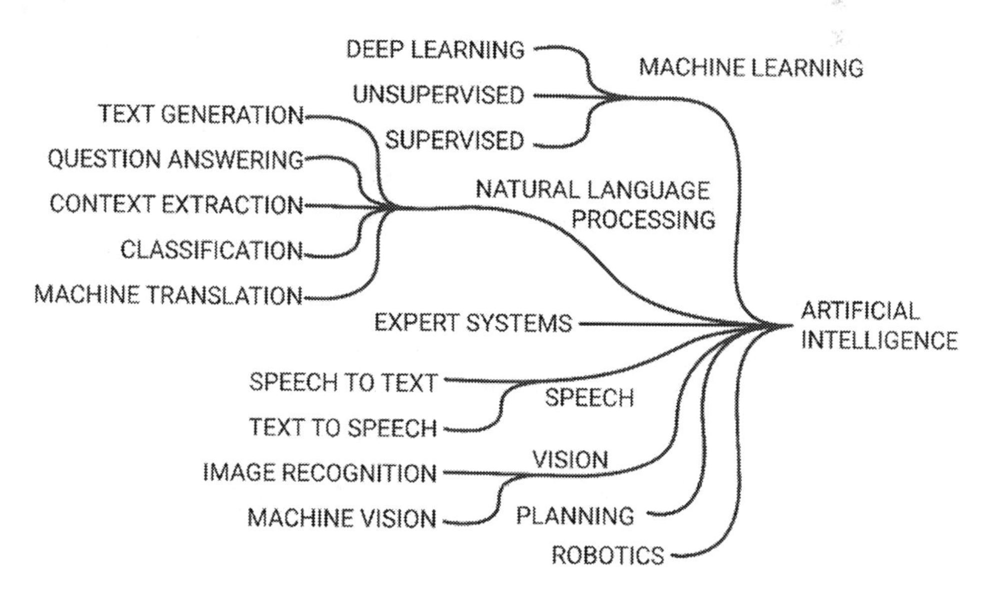

Figure 3-1. *Components of AI*

© Arjun Panesar 2021
A. Panesar, *Machine Learning and AI for Healthcare*, https://doi.org/10.1007/978-1-4842-6537-6_3

We see and engage with "artificial" intelligence, whether developed by machine learning or otherwise, in our everyday lives: in ordering an Uber, ordering food on a digital menu, looking for an Airbnb, searching on eBay, asking Alexa a question, scrolling through Facebook, and engaging in digital therapeutics.

A case study on the use of AI-powered digital therapeutics is detailed further in Chapter 10.

Pondering on the AI humans are able to develop is limited only by the human imagination. New and innovative devices are increasingly connected to the cloud and increasing the wealth of opportunities in healthcare.

With a wealth of data at its helm as an industry, AI is shifting its focus onto the power of machine learning and deep learning, where machines can learn for themselves—and this is accelerating current generation breakthroughs in the AI space.

If Samuel's Checkers program demonstrated simple AI, machine learning is best demonstrated through AlphaGo Zero, which learned to play checkers simply by playing games against itself, surpassed human level of play, and defeated the previous version of AlphaGo by 100 games to 0 [69].

Machine learning systems are not uncommon in healthcare. Researchers at Houston Methodist Research Institute in Texas have developed an agent that is enabling review and translation of mammograms 30 times faster than a human doctor and with 99% accuracy [70].

According to the American Cancer Society, 50% of the 12.1 million mammograms performed annually yield false results, meaning that healthy women get told they have cancer [70]. The software, which can be understood as an autonomous entity, learned from millions of mammogram records, reducing the risk of false positives.

There are many books and resources on the field of AI and machine learning. Although a comprehensive and mathematically grounded discipline, this book aims to cover the topic for the discerned reader, without delving too deep into the semantics of linear algebra, probability, and statistics.

This chapter will cover many of the topics within machine learning, beginning with an introduction to the basics of AI and terminologies and then moving on to the various forms of learning and data mining.

Algorithms are covered in the next chapter.

Basics

Before delving into the depths of AI and machine learning, it is useful to understand several key definitions. A brief guide to useful terminology is included in the following, with a more in-depth glossary at the end of this book.

Agent

An agent is anything that can be seen as perceiving its environment through sensors and acting on them through effectors. This takes Russell and Norvig's approach to agents. As a result, this includes robots, humans, and software programs.

There are many different types of agents, which can be classified into categories based on their characteristics. *Artificial Intelligence: A Modern Approach* (AIMA) is a comprehensive textbook on artificial intelligence written by AI experts Stuart J. Russell and Peter Norvig [71].

How should the agent behave? Rational agents are agents that do the right thing. This can be represented by a mathematical measure, or evaluation of the agent's performance, which the agent can seek to maximize.

An agent can be understood as a mapping between percept sequences and actions. For example, a medical diagnosis system may treat a patient in a hospital as its environment: perceiving symptoms, findings, and patient's answers. It may then act on these through questions, tests, and treatments with the goal of having a healthy patient.

The definition of an agent applies to a software program or algorithm.

Autonomy

In AI, autonomy refers to the extent to which the agent's behaviors are determined by its own experience.

If there is no autonomy, all decisions are made from a static dataset or knowledge base. In a completely autonomous situation, an agent would make decisions at random. Agents have the explicit ability to become more autonomous over time.

Software, just like humans, can learn from experiences over time. Systems can learn from experiences in their environment (at its most basic, "learning" from some new data) to become more autonomous in it.

Interface

An interface agent is one that provides an interface to a complex system.

Performance

Performance is the measure used to evaluate the behavior of the agent in the environment. It answers the question, "Does the agent do what it's supposed to do in the environment?"

Goals

Goals refer to what the agent is trying to achieve. In the field of machine learning, the goal is not to be able to get the agent to "learn." Instead, this refers to the overall goal or objective.

Goals can be broken down into smaller tasks. For example, to accomplish the goal of a Caesarean section, the tasks could be considered cutting open the abdomen, bringing the baby out, checking the baby is healthy, and stitching the abdomen back up again.

Utility

Utility refers to the agent's own, internal performance assessment—that is, the agent's own measure of performance at any given state. This may differ from the performance of the agent. It is noteworthy that the utility function is not necessarily the same as the performance measure, although it can be.

An agent may have no explicit utility function, whereas there is always some performance measure. A utility function is used to map a state to a real number, which describes the associated degree of happiness or progress toward a goal or objective.

This allows rational decisions in cases where there are several paths to the same goal, allowing an agent to distinguish the paths better than others.

Knowledge

Knowledge is acquired by an agent through its sensors or knowledge about the environment.

Knowledge can be used to decide how to act; if it is stored, it can be used to store previous knowledge states (or, in other words, a history); and it can be used to determine how actions may affect the environment.

Machine learning is data driven. Thus, the dawn of ubiquitous data and pervasive computing is fantastic for developing systems with real-world application.

Environment

The environment is the state of the world for an agent. An environment has several characteristics, following Russell and Norvig's lead.

Accessibility

This refers to how much of the environment the agent has access to. In cases of missing information, agents may need to make informed guesses to act rationally.

Determinism

Environments can be deterministic or nondeterministic. Deterministic environments are environments where the exact state of the world can be understood. In this instance, utility functions can understand their performance. In nondeterministic environments, the exact state of the world cannot be determined, which makes utility functions rely on making best-guess decisions.

Episodes

Episodic environments are those where the agent's current action choice does not rely on its past actions. Non-episodic environments require agents to plan for the effects of their current actions.

Type of Environment

Static environments are where the environment doesn't change while an agent is making a decision. Dynamic environments can change during processing, so agents may need to feed back into the environment or anticipate changes between the time of input and output.

Flow of Data to Environment

One critical aspect of agent design is how the data is received. In a chess game, for example, there are only a finite number of moves that the agent can select. This would be a discrete environment, as there are only so many moves that can be performed. On the other hand, on the operating table, for example, things can change at the last millisecond. Continuous data environments are environments in which data seems infinite or without end.

An intelligent agent is fundamentally one that can display characteristics such as learning, independence, adaptability, inference, and to a degree, ingenuity.

For this to occur, an agent must be able to determine sequences of actions autonomously through searching, planning, and adapting to changes in its environment and from feedback through its past experiences. Within machine learning, there are further terminologies.

Training Data

Training data is the data that will be used by the learning algorithm to learn possible hypotheses. An example is a sample of x including its output from the target function.

Target Function

This is the mapping function f from x to f(x).

Hypothesis

This is an approximation of f. Within data science, a machine learning model approximates a target function for mapping input x to outputs (y).

Learner

The learner is the learning algorithm or process that creates the classifier.

Validation

Validation includes methods used within machine learning development that provide a method of evaluation of model performance.

Dataset

A dataset is a collection of examples.

Feature

A feature is a data attribute and its value. As an example, skin color is brown is a feature where *skin color* is the attribute and *brown* is the value.

Feature Selection

This is the process of choosing the features required to explain the outputs of a statistical model while excluding irrelevant features.

What Is Machine Learning?

Machine learning is a subset of AI where computer models are trained to learn from their actions and environment over time with the intention of performing better.

In 1959, Checkers' creator Arthur Samuel defined machine learning as a "field of study that gives computers the ability to learn without being explicitly programmed" [72].

Machine learning was born from pattern recognition and the theory that computers can learn without being programmed to perform specific tasks—that is, systems that learn without being explicitly programmed. As a result, learning is driven by data—with intelligence acquired through the ability to make effective decisions based on the nature of the learning signal or feedback. The utility of these decisions is evaluated against the goal.

Machine learning focuses on the development of algorithms that adapt to the presentation of new data and discovery. Machine learning exemplifies principles of data mining but is also able to infer correlations and learn from them to apply to new algorithms. The goal is to mimic the ability to learn in a human, through experience, and achieving the assigned task without, or with minimal, external (human) assistance.

Just as with learning in humans, machine learning is composed of many approaches. At its most basic, first, things are memorized. Second, we learn from extracting information (through reading, listening, learning new things). And third, we learn from example.

For example, if we were being taught square numbers and if students were shown a set of numbers Y = {1, 4, 9, 16, ...} and taught that Y = n*n, they would typically understand the notion of square numbers without being explicitly taught every square number.

How Is Machine Learning Different from Traditional Software Engineering?

Traditional software engineering and machine learning share a common objective of solving a problem or set of problems. The approach to problem-solving is what distinguishes the two paradigms.

Traditional software engineering or programming refers to the task of computerizing or automating a task such that a function or program, given an input, provides an output. In other words, writing a function f, such that given input x, output y = f(x).

This is done with logic, typically if–else statements, while loops, and Boolean operators.

Machine learning differs from conventional programming in that rather than providing instructions about the function f, the computer is provided input x and output y and expected to determine or predict function f (see Figure 3-2).

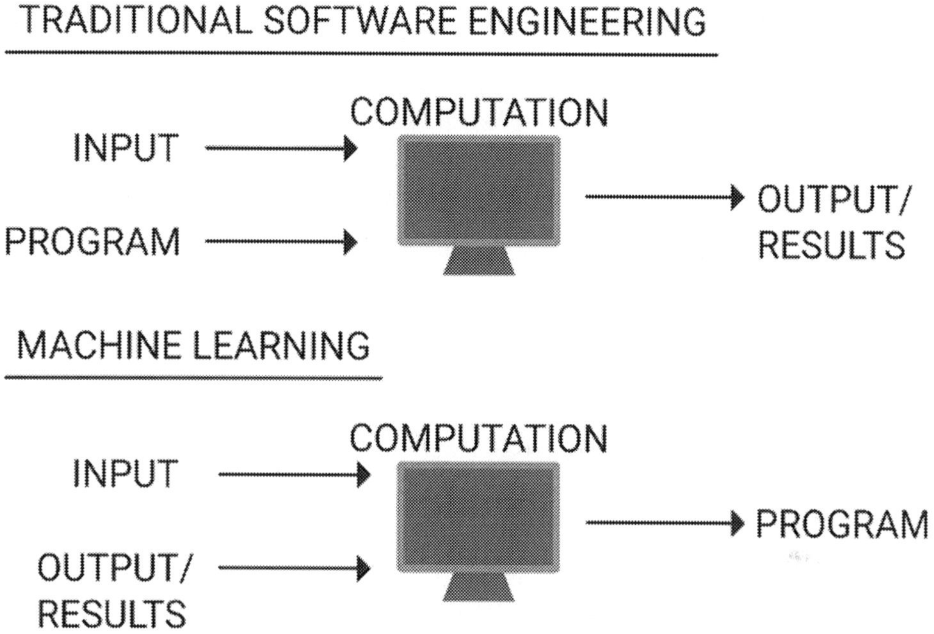

Figure 3-2. *How machine learning works*

Whereas traditional programs have been written by humans to solve a problem, machine learning programs learn through reasoning to solve a problem from examples, rules, and information. Many classifiers in machine learning assist with prediction.

Machine learning programs can also learn to generalize and help with issues of uncertainty due to the use of statistics and probability-driven techniques. Models can learn from previous computations or experience to produce reliable, repeatable decisions and results.

The availability of data, information on the Internet, and leaps in the amount of digital data being produced, stored, and made accessible for analysis are enabling engineers to realize machines that can learn what we prefer and personalize to our requirements.

Machine learning can be seen everywhere if you look hard enough—in web search rankings, drug design, personalization of preferences and social network feeds, music recommendations, logistics, web page design, self-driving cars, fridges, washing machines, TVs, and virtual assistants.

Machine Learning Basics

Tasks for machine learning algorithms are broadly categorized as the following (see Figure 3-3):

- Supervised learning (also known as inductive learning)
- Unsupervised learning
- Semi-supervised learning
- Reinforcement learning

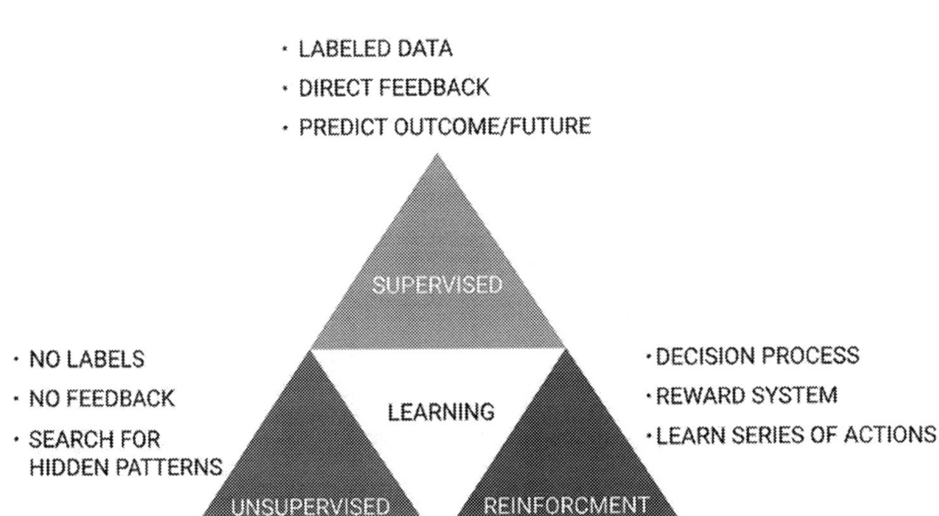

Figure 3-3. *Methods of learning*

Supervised Learning

In supervised learning, algorithms are presented with example inputs and the subsequently desired outputs in the form of training data, with the goal to learn general rules that map inputs to outputs (see Figure 3-4). Input data is called training data and has a known output (result) associated with it.

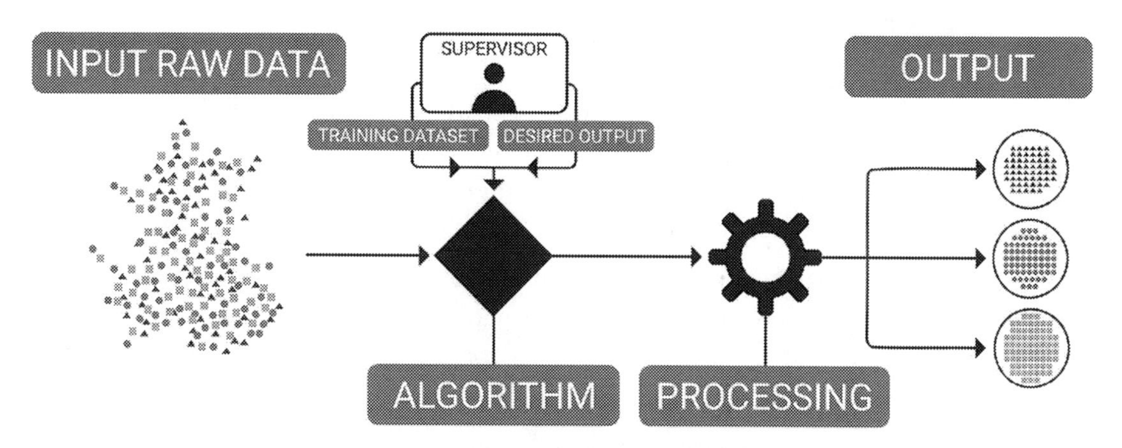

Figure 3-4. *Supervised learning: how it works*

The training data guides the development of the algorithm. A model is created through a training process in which the model makes predictions and is corrected when predictions are incorrect.

Training continues until the model achieves the desired level of accuracy on the training data. Supervised algorithms include logistic regression. Supervised machine learning algorithms can apply what has been learned in the past to new data using labeled data examples to predict future events.

The learning algorithm can also compare the output with the correct output to find any errors and modify the model accordingly.

Supervised learning problems can be employed in the following forms.

Classification

This is to predict the outcome based on a training dataset where the output variable is in the form of distinct categories. Models are built through inputting training data in the form of prelabeled data.

Classification techniques define decision boundaries and include

- Support vector machines
- Naïve Bayes
- Gaussian Bayes
- k-Nearest neighbors (kNNs)
- Logistic regression

An example of classification is the label (or, in real-world application, diagnosis) of someone as sick or unhealthy based on a set of symptoms.

Regression

Regression is very similar to classification. The only difference between classification and regression is that regression is the outcome of a given sample where the output variable is in the form of real values (i.e., the temperature as a value, rather than a classification of hot or cold).

Examples include height, body temperature, and weight. Linear regression, polynomial regression, support vector machine (SVM), ensembles, decision trees, and neural networks are examples of regression models.

Forecasting

This is the method of making predictions based on past and present data. This is also known as time series forecasting. An ensemble is a type of supervised learning that combines multiple, different, machine learning models to predict an outcome on a new sample. In an ensemble, models become features.

Unsupervised Learning

When people reference systems that can learn for themselves, they are actually referencing unsupervised learning (see Figure 3-5). In unsupervised learning, the learning algorithm doesn't receive labels for data, leaving the algorithm to find structure from the input.

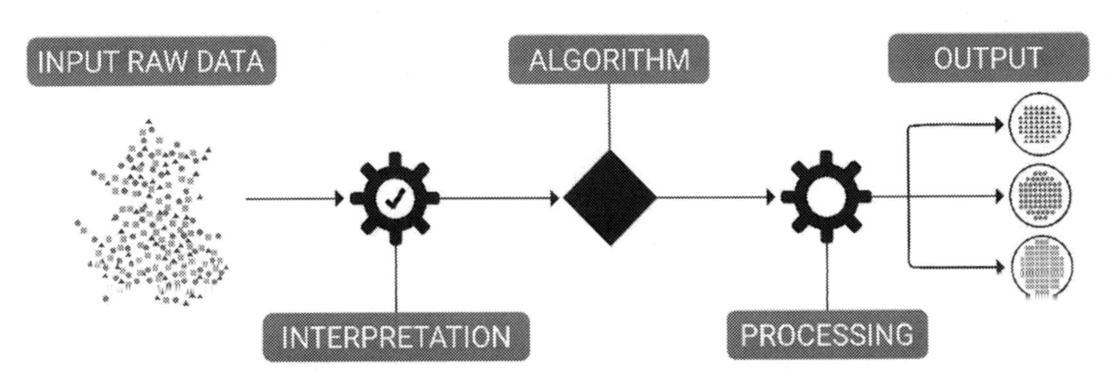

***Figure 3-5.** Unsupervised learning: how it works*

As data is unlabeled, there is no evaluation of the accuracy of the structure that is output by the algorithm.

Data may be missing both classifications and labels. Thus, the model is developed through interpretation: through finding hidden structures and inferences in the input data. This may be through the extraction of rules, reducing data redundancy, or as the result of organizing data.

This includes clustering, dimensionality reduction, and association rule learning. The algorithm may never find the right output but instead models the underlying structure of the data.

There are three types of unsupervised learning problems as defined in the following.

Association

Association is the discovery of the probability of the co-occurrence of items in a collection. This is used extensively in marketing as well as healthcare to inform decisions.

For instance, this would be the percentage likelihood of developing any form of cancer if diagnosed with obesity. Association is similar to classification; however, any attribute can be predicted in association, whereas classification is binary.

Clustering

Clustering refers to grouping items such that items within the same cluster are more like each other than to items from another cluster.

Dimensionality Reduction

Dimensionality reduction can be achieved through feature selection and feature extraction. Dimensionality reduction may mathematically re-represent or transform data.

Feature Extraction

Feature extraction performs data transformation from a high-dimensional to a low-dimensional space. This can involve reducing the variables of a dataset while maintaining data integrity and ensuring that the most important information is represented.

Feature extraction methods can be used for dimensionality reduction. Typically, a new set of features would be created from the original feature set. An example would be combining all of a patient's clinical test results into a health risk scoring from the results that are demonstrated to affect mortality.

Reducing the number of dimensions allows easier visualization, particularly in two or three dimensions, as well as reducing time and space required for storage. Algorithms include

- Apriori algorithm
- FP growth (also known as frequent pattern growth)
- Hidden Markov Model
- Principal component analysis (PCA)
- Singular value decomposition (SVD)
- k-Means
- Neural networks
- Deep learning

Deep learning utilizes deep neural network architectures, which are types of machine learning algorithms.

Semi-supervised

Semi-supervised learning is a hybrid where there is a mixture of labeled and unlabeled data in the input. Although there may be a desired outcome, the model also learns structures to organize data and make predictions. Example problems are classification and regression.

Reinforcement Learning

This is where systems interact with a dynamic environment in which an agent must perform a specific goal. Reinforcement learning acts as an intersection between machine learning, behavioral psychology, ethics, and information theory.

The algorithm is provided feedback concerning rewards and punishments as it navigates the problem. Reinforcement learning allows the agent to decide the next best action based on its current state and by learning behaviors that will maximize the reward. Optimal actions (or an optimal policy) are typically learned through trial and error and feedback. This allows the algorithm to determine ideal behaviors within context.

Reinforcement learning is typically used in robotics, for instance, a robotic hoover that learns to avoid collisions by receiving negative feedback through bumping into tables, chairs, and so forth—although today, this also involves computer vision techniques.

Reinforcement learning differs from standard supervised learning in that correct examples are never presented nor are suboptimal decisions explicitly corrected. The focus is on performance in the real world.

Similar to how humans learn, reinforcement learning assumes rational behavior regarding learning from one's experience and as such is assisting in understanding how humans learn from experience to make good decisions (see Figure 3-6).

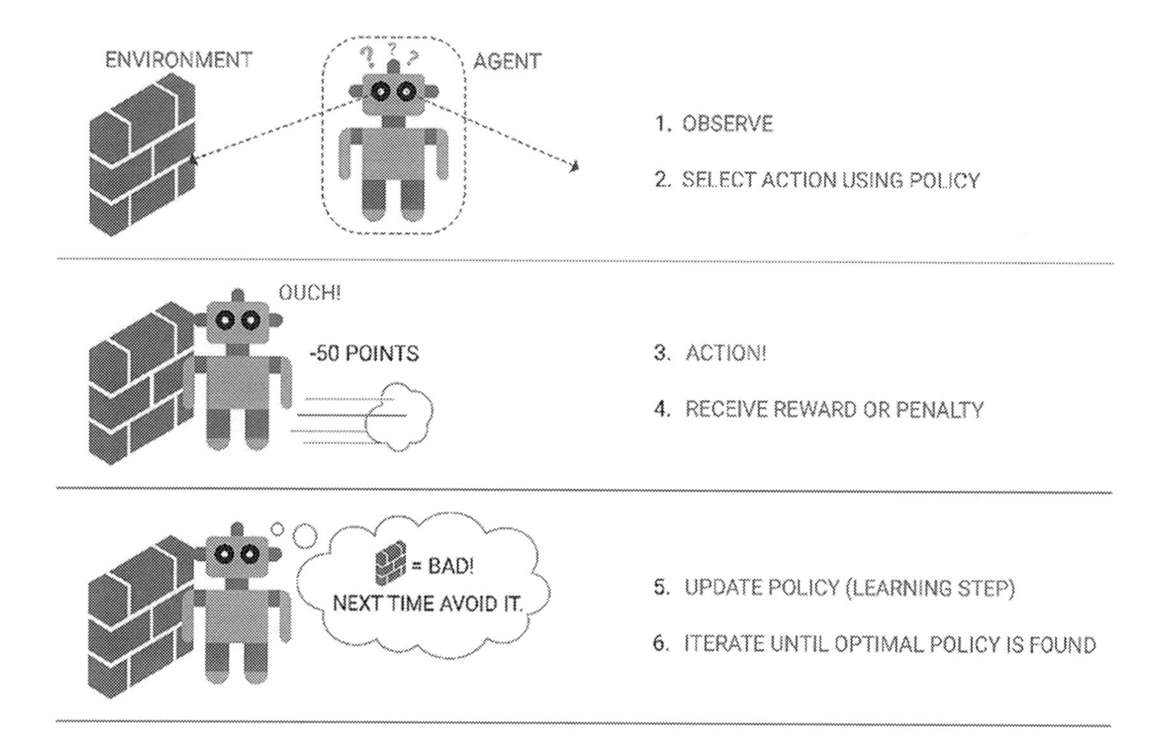

Figure 3-6. *Reinforcement learning: how it works*

DeepMind, for example, exposed their AlphaGo agent to master the ancient Chinese game of Go through showing "AlphaGo a large number of strong amateur games to help it develop its own understanding of what reasonable human play looks like. Then we had it play against different versions of itself thousands of times, each time learning from its mistakes and incrementally improving until it became immensely strong" [73].

Reinforcement learning is used in autonomous vehicles. In the real world, the agent requires consideration of aspects of the environment including current speed, road obstacles, surrounding traffic, road information, and operator controls.

An agent learns to act in a way (or develop a policy) that is appropriate for the specific state of the environment. This leads to new ethical dilemmas posed by AI and machine learning for agents with reward-driven programming.

Data Mining

Data mining, also known as knowledge discovery in databases, is defined as the process of exploring data to extract novel information (including patterns and relationships) using sophisticated algorithms. This can be applied to every aspect of life and industry and enables the following:

- Pattern predictions based on trends and behaviors
- Prediction based on probable outcomes
- Analysis of large datasets (particularly unstructured)
- Clustering through identification of facts previously unknown

Data mining uses the power of machine learning, statistics, and database techniques and is at the heart of analytics efforts (see Figure 3-7). Data mining typically uses large amounts of data and can also give simpler, descriptive analytics. For instance, in healthcare, data mining applications can be used to identify conditions or medical procedures that are often seen together.

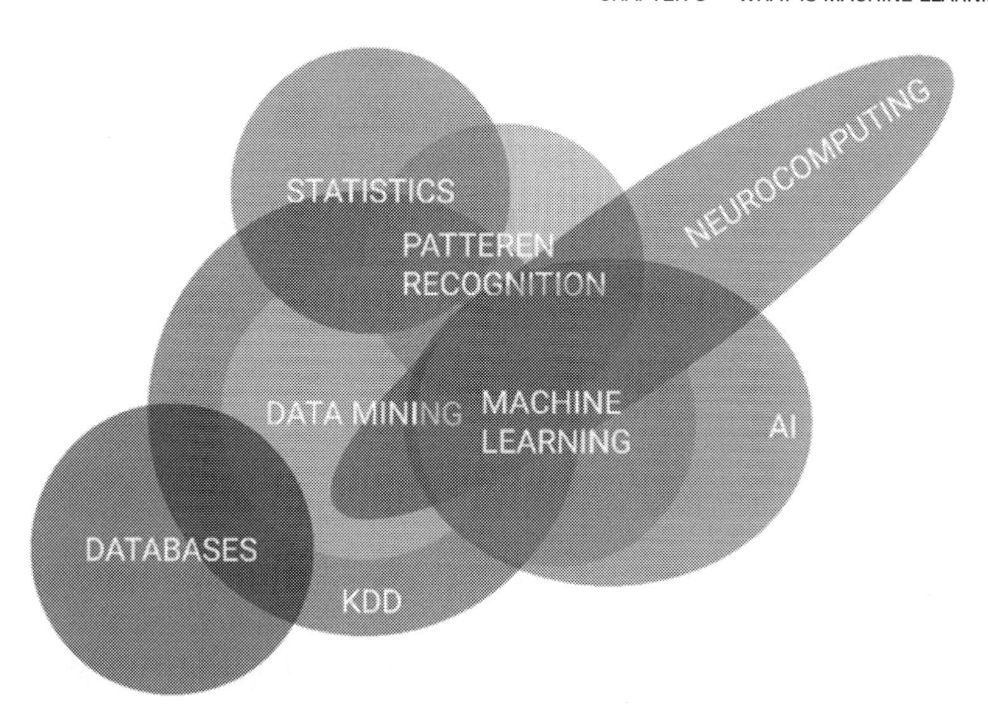

Figure 3-7. *Data mining and its wider synergies*

Data mining is most often seen in day-to-day sentiment analysis of user, customer, or patient feedback.

Data mining also allows forecasting, through identifying relationships within the data. This can help insurance companies detect risky customer behavior patterns.

Machine learning can be used for data mining. Data mining will inform decisions such as the kind of learning models best suited for solving the problem at hand. However, data mining also encompasses techniques besides those encapsulated within machine learning. In data mining, the task is of unknown knowledge discovery, whereas machine learning performance is evaluated with respect to reproducing known knowledge.

For instance, if you have a dataset of a patient's blood pressures, you can perform anomaly detection, which is considered a data mining task to identify previously unknown outliers. The task may make use of machine learning techniques, perhaps a k-means algorithm for cluster analysis, to identify the anomalous data and assist the algorithm's learning.

Data mining is not the same as natural language processing (NLP). NLP may be a technique used in data mining to help an agent understand or "read" text. Data mining algorithms include

- C4.5
- k-Means
- Support vector machines
- Apriori
- EM (expectation-maximization)
- PageRank
- AdaBoost
- kNN
- Naive Bayes
- CART (Classification and Regression Trees)

Parametric and Nonparametric Algorithms

Algorithms may be parametric or nonparametric in form. Algorithms that can be simplified to a known, finite form are labeled parametric, whereas nonparametric algorithms learn functional form from the training data.

In other words, with nonparametric algorithms, the complexity of the model grows with the amount of training data. With parametric models, there is a fixed structure or set of parameters making parametric models faster than nonparametric models.

Nonparametric models can provide better accuracy if they are supplied enough training resources in terms of training data and time (see Figure 3-8).

	PROS	CONS
PARAMETRIC ALGORITHMS	SIMPLER SIMPLER TO UNDERSTAND AND INTERPRET FASTER QUICK TO FIT TO DATA LESS DATA REQUIRES LESS DATA TO PERFORM WELL	LIMITED COMPLEXITY SUITED TO SIMPLER PROBLEMS WHERE STRUCTURE IN THE DATA CAN BE INFERRED
NONPARAMETRIC ALGORITHMS	FLEXIBILITY CAN FIT VARIETY OF FUNCTIONAL FORMS, WHICH DO NOT NEED TO BE ASSUMED PERFORMANCE PERFORMANCE WILL LIKELY BE HIGHER THAN PARAMETRIC ALGORITHMS AS SOON AS DATA STRUCTURES GET COMPLEX	SLOWER COMPUTATIONS ARE SIGNIFICANTLY LONGER MORE DATA REQUIRES SIZEABLE DATA TO LEARN OVERFITTING AFFECTS MODEL PERFORMANCE

Figure 3-8. *Parametric vs. nonparametric algorithms*

When to Use Parametric and Nonparametric Algorithms

Parametric algorithms simplify a function to a known form and learn function coefficients from trained data. As such, strong assumptions are made about the data. Parametric algorithms are useful in that they are quick to learn from data and do not require much of it and can be used in scenarios that require easy to explain results.

Nonparametric algorithms use a varying number of parameters. k-Nearest neighbors, neural networks, and decision trees are nonparametric and suited to scenarios where there is more data and requirement for higher performance.

How Machine Learning Algorithms Work

Supervised learning is the computerized task of inferring a function from labeled training data.

Y = f(x),

which can be understood as

Output = function(Input).

Inputs and outputs can be referred to as variables and are typically vector in format: f refers to the function, which you are trying to infer from the Input.

Take the example of a patient who is to be prescribed a particular treatment. In real life, the process may include the patient looking at research describing qualities about the medication or treatment they are recommended.

For example, if they were to see that the research mostly consists of words like "good," "great," "safe," and so forth, then we would conclude from the sentiment that the treatment is a good treatment, and we can feel confident in accepting it. Whereas if words like "bad," "not good quality," and "unsafe" were to appear regularly, then we would conclude that it is probably better to look for another treatment. The research helps us perform an action based on the pattern of words that exist in the treatment research.

Machine learning attempts to understand this human decision-making process through algorithms. It focuses on the development of algorithms (or programs) that can access data and use it to learn for themselves. Machine learning is very good for classification and prediction tasks.

Machine learning is most commonly used for predictive analytics—learning the mapping of Y = f(X) to make predictions of Y for new X.

Practical examples of machine learning are the following:

- Fraud detection x is the property of the transaction f(x) is whether a transaction is fraudulent.
- Disease diagnosis x are the properties of the patient f(x) is whether a patient has disease.
- Speech recognition x are the properties of a speech input f(x) is the response given to the instruction contained within the speech input.

The process of machine learning begins with input data provided as examples, direct experience, or instructions—to identify patterns within the data and make better decisions in the future based on the data that was provided.

The aim is to enable the program to learn automatically without human intervention or assistance and adjust rational actions accordingly.

Conclusion

Readers should have an understanding of machine learning and data mining, approaches to performing learning tasks and key considerations when using these techniques. Most importantly, it has covered how machine learning algorithms work.

The next chapter will use this foundation knowledge to focus on algorithm choice and design, model considerations, and best practice.

CHAPTER 4

Machine Learning Algorithms

Nature doesn't feel compelled to stick to a mathematically precise algorithm; in fact, nature probably can't stick to an algorithm.

—Margaret Wertheim

You do not need a background in algebra and statistics to get started in machine learning. The previous chapter introduced key foundation principles. Be under no illusions; mathematics is a huge part of machine learning. Mathematics is key to understanding how the algorithm works and why coding a machine learning project from scratch is a great way to improve your mathematical and statistical skills.

Not understanding the underlying principles behind an algorithm can lead to a limited understanding of methods or adopting limited interpretations of algorithms. If nothing else, it is useful to understand the mathematical principles that algorithms are based on and thus understand best which machine learning techniques are most appropriate.

There is an overwhelming number of machine learning algorithms in the public domain. Many are variations (typically faster or less computationally expensive) on several prominent themes. Learners are the nucleus of any machine learning algorithm, which attempts to minimize a cost function—also referred to as an error function or loss function—on training and test data.

Many of your machine learning projects will use defined methods from popular libraries such as numpy, pandas, Matplotlib, SciPy, scikit-learn, scrapy, NLTK (Natural Language Toolkit), and so forth.

The most common machine learning tasks are classification, regression, and clustering.

© Arjun Panesar 2021
A. Panesar, *Machine Learning and AI for Healthcare*, https://doi.org/10.1007/978-1-4842-6537-6_4

Several algorithms can work for functions of discrete prediction, such as k-nearest neighbors, support vector machines, decision trees, Bayesian networks, and linear discriminant analysis.

This chapter provides a comprehensive analysis of machine learning algorithms, including programming libraries of interest and examples of real-world applications of such techniques.

Defining Your Machine Learning Project

Tom Mitchell provides a concise definition of machine learning: "A computer program is said to learn from experience E with respect to some class of tasks T and performance measure P, if its performance at tasks in T, as measured by P, improves with experience E" [75].

This definition can be used to aid machine learning projects in assisting us to think clearly about what data is collected and utilized (E), what the task at hand is (T), and how we will evaluate the results (P).

Task (T)

The task refers to what we want the machine learning model to do. The function the model learns is the task. This does not refer to the task of actually learning but rather the task at hand. For example, a robotic hoover would have the task of hoovering a surface. Often a task is broken down into smaller tasks.

Performance (P)

Performance is a quantitative measure of a machine learning model's ability. Performance is measured using a cost function, which typically calculates the accuracy of the model on the task.

Performance can be measured by an error rate, which is the proportion of examples for which the model gives the wrong outputs. The learning method aims to minimize the error rate, ideally without falling into local minima or maxima.

Experience (E)

Experience refers to the amount of labeled data that is available as well as the amount of supervision required by the machine learning model.

Mitchell's reference proves useful in defining machine learning projects. This definition helps understand the data required for collection (E), which develops as the knowledge base; the problem at hand that needs a decision (T); and how to evaluate the output (P).

For example, a prediction algorithm may be assigned task T to predict peak emergency hospital admissions based on experience data E of past hospital admissions and their respective dates and times. The performance measure P could be the accuracy of prediction; and as the model receives more experience data, it would ideally become better in its forecast. It is noteworthy that one criterion does not necessarily determine performance.

In addition to accuracy of the prediction, a peak hospital admissions method could provide extra information by predicting peaks that would be the most costly or resource intensive. Table 4-1 gives examples of tasks and the experience which would be created, along with performance metrics.

Table 4-1. *Understanding a learning problem by its components*

Problem	Task	Performance	Experience
Learning how to perform surgical stitching	Stitching a patient's head	Accuracy (perceived pain and/or time could be a feature)	Stitching of patients' heads and feedback on performance measures
Image recognition	Recognizing a person's weight from an image	Accuracy of weight prediction	Training dataset of images of people and their respective weights
Robotic arms used for patient medication delivery	Moving the correct medication into the correct patient packet	Percentage of correctly placed medications	Training examples, real-life experience
Predicting the risk of disease	Diagnosing likelihood of type 2 diabetes	Percentage of correctly diagnosed patients	Training dataset of patient health records, real-life experience in prediction and feedback

Machine learning is used to answer questions such as "Is it likely I have type 2 diabetes?", "What object is this in this image?", "Can I avoid the traffic?", and "Will this recommendation be right for me?"

Machine learning is employed favorably in each of these problem domains, yet the complexity of real-world problems means that it is currently infeasible for a specialized algorithm to work faultlessly every time, in every setting, for everyone—whether in healthcare or any other industry.

A common aphorism in math comes from British statistician George Box: "All models are wrong, but some are useful." The purpose of machine learning is not to make perfect guesses, as there is no such thing as an ideal model. Instead, machine learning builds on foundations in statistics to provide useful, accurate predictions that can be generalized to the real world.

When training a machine learning algorithm, the training dataset must be a statistically significant random sample. If this is not the case, there is a risk of discovering patterns that don't exist, or the noise. The memorization of training data is known as overfitting. The model remembers the training data, performing well in training but not with previously unknown data. Equally, if the training data is too small, the model can make inaccurate predictions.

The goal of machine learning, whether supervised or unsupervised, is generalization—the ability to perform well based on past experiences.

Common Libraries for Machine Learning

Python is increasingly popular within the data science and machine learning industry. All of the libraries are open source on GitHub, which also provides metrics for popularity and robustness:

- GitHub is a service based on the Git version control system. The Git approach to storing data is stream-like. Git takes a snapshot of your files and stores references to these snapshots. Git has three states: committed, modified, and staged, which refer to data storage in your local file system, local file changes, and flagged files to be committed, respectively. The benefits of using metadata are evident when using GitHub.

- Numerical Python, or numpy, is an essential Python extension library. It enables fast, N-dimensional array objects for vector arithmetic, mathematical functions, linear algebra, and random number generation. Basic array arithmetic performance is noticeably slower without this library.
- SciPy is a library for scientific processing that builds on numpy. It enables linear algebra routines as well as signal and image processing, ordinary differential equation methods, and special functions.
- Matplotlib is a Python library of similar data plotting and visualization techniques. Matlab is a standard tool used for data processing. However, the simplicity and usefulness of Matplotlib helps Python to become a viable alternative to Matlab.
- Pandas is a tool for data aggregation, data manipulation, and data visualization. Within pandas, one-dimensional arrays are referred to as series, with multidimensional arrays referred to as dataframes.
- Scikit-learn contains image processing and machine learning techniques. This library is built on SciPy and enables algorithms for clustering, classification, and regression. This includes many of the algorithms discussed in this chapter such as naïve Bayes, decision trees, random forests, k-means, and support vector machines.
- NLTK, or Natural Language Toolkit, is a collection of libraries that are used in natural language processing. The NLTK enables the foundations for expert systems such as tokenization, stemming, tagging, parsing, and classification—which are vital for sentiment analysis and summarization.
- Genism is a library for use on unstructured text.
- Scrapy is an open source data mining and statistics designed initially for scraping websites.

- TensorFlow is a Google Alphabet-backed open source library of data computations optimized for machine learning. It enables multilayered neural networks and quick training. TensorFlow is used in many of Google's intelligent platforms.
- Keras is a library for building neural networks that builds on TensorFlow.

Supervised Learning Algorithms

In many supervised learning settings, the goal is to finely tune a predictor function, f(x), or the hypothesis. The learning comprises of using mathematical algorithms to represent the input data x within a particular domain. For example, x could take the value of the time of day, and f(x) could be a prediction of waiting time at a particular hospital.

In reality, x typically represents multiple data points. Hence, inputs are expressed as a vector.

For instance, in the preceding example, f(x) takes in input x as the time of the week. In improving the predictor, we could use not only day (x_1) but time (x_2) of day, weather (x_3), place within the queue (x_4), and so on—on the premise the data was available. Deciding which inputs to use is an integral part of the machine learning design process.

Each record i in the dataset is a vector of features x(i). In the case of supervised learning, there will also be a target label for each instance, y(i). The model is trained with inputs in the form (x(i), y(i)). The training dataset can be represented as {(x(i), y(i)); i = 1, ... , N}, where x represents the input and y the output.

The model has the simple form:

> f(x) = ax + b, where a and b are constants and b refers to the noise in the data.

Machine learning aims to find optimal values of a and b such that the predictions f(x) are as accurate as possible. The method of learning is inductive. The model trawls the training dataset to learn patterns in the data. The model induces a pattern, or hypothesis, from a set of examples. The output of a machine learning prediction model differs from the actual value of the function due to bias, noise, and variance, as discussed in Chapter 3.

Optimization of f(x) takes place through training examples. Each training example {x_1 ... x_n} has an input value x_training and a corresponding output, Y.

Each example is processed, and the difference between the known and correct value y (where $y \in Y$) and the predicted value f(h_training) is computed. After processing a suitable number of samples from the training dataset, the error rate of f(x) is calculated, and values a and b are used as factors to improve the correctness of the prediction.

This also addresses inherent randomness or noise in the data, also referred to as irreducible error.

Selecting the Right Features

There are two different main approaches when it comes to feature selection. First is an independent assessment based on the general characteristics of the data. Methods belonging to this approach are called filter methods, as the feature set is filtered out before model construction.

The second approach is to use a machine learning algorithm to evaluate different subsets of features and finally select the one with the best performance on classification accuracy. The latter algorithm will be used in the end to build a predictive model. Methods in this category are called wrapper methods because the arising algorithm wraps the entire feature selection process.

The law of large numbers theorem applies to machine learning prediction. The theorem describes how the result of a large number of identical experiments will eventually converge or even out.

This process is iterated until the system converges on the best values for a and b. As such, the machine learns through experience and is ready for deployment. Typically, we are concerned with minimizing the error of a model or making the most accurate predictions possible.

Within supervised machine learning, there are two main approaches: regression-based systems and classification-based systems.

Classification

Classification is the process of determining a category for an item (or items, x) from a set of discrete predefined categories (or labels, y) through approximating a mapping function, f. Classification algorithms use features (or attributes) to determine how to classify an item into at least a binary class or two or more classes.

Training data is supplied as an input vector. The input vector consists of labels for item features. For instance, an item may read the following:

```
{"HbA1c, 7.8%", "Diabetic"} in the form {x_training, output}
```

It is from the data that classification algorithms learn to classify. Classification problems require examples to be labeled to determine patterns; hence, a model's classification accuracy is best validated through the correctly classified examples of all predictions made.

Algorithms that perform classification include decision trees, naïve Bayesian, logistic regression, kNNs (k-nearest neighbors), support vector machine, and so forth.

Common use cases of classification algorithms are the following:

- Suggesting patient diagnosis based on health
- profile—that is, does patient have retinopathy?
- Recommending the most suitable treatment pathway for a patient
- Defining the period in the day in a clinic that there are peak admissions
- Determining patient risk of hospital readmission
- Classifying imagery to identify disease—for example, determining whether a patient is overweight or not from a photograph

Regression

Regression involves predicting an output that is a continuous variable. As a regression-based system predicts a value, performance is measured by assessing the number of prediction errors.

The coefficient of determination, mean absolute error, relative absolute error, and root mean square error are statistical methods used for evaluating regression models.

Regression algorithms include linear regression, regression trees, support vector machines, k-nearest neighbor, and perceptrons.

It is often possible to convert a problem between classification and regression. For example, patient HbA1c can also be useful if stored as a continuous value. A continuous value of HbA1c = 6.9% can be classified into a discrete category, diabetic, on rules defined on HbA1c as Diabetic (Set 1), HbA1c ≥ 6.5%, and Non-diabetic (Set 2), HbA1c < 6.5%.

Time series refers to input data that is supplied in the order of time. The sequential ordering of the data adds a dimension of information. Time series regression problems are commonplace.

Common use cases of regression algorithms are the following:

- Estimating patient HbA1c from previous blood glucose readings (time series forecasting)
- Predicting the days before hospital readmission (which may be received as a time series regression problem)
- Approximating patient mortality risk based on
- clinical data
- Forecasting the risk of developing health complications
- Forecasting when the office printer will be queued with requests from other employees

Decision Trees

Decision trees are flowcharts that represent the decision-making process as rules for performing categorization. Decision trees start from a root and contain internal nodes that represent features and branches that represent outcomes. As such, decision trees are a representation of a classification problem.

Decision trees can be exploited to make them easier to understand. Each decision tree is a disjunction of implications (i.e., if-then statements), and the implications are Horn clauses that are useful for logic programming. A Horn clause is a disjunction of literals.

On the basis that there are no errors in the data in the form of inconsistencies, we can always construct a decision tree for training datasets with 100% accuracy. However, this may not roll out in the real world and may indicate overfitting, as we will discuss.

For instance, take the simplified dataset in Table 4-2, which can be represented as the decision tree as shown in Figure 4-1.

Table 4-2. *Risk of type 2 diabetes*

Record	Unhealthy	BMI	Ethnicity	Outcome
Person 1	Yes	>25	Indian	*High risk*
Person 2	Yes	<25	Indian	*Low risk*
Person 3	Yes	>25	Caucasian	*High risk*
Person 4	No	<25	Caucasian	*Low risk*
Person 5	No	>25	Chinese	*Medium risk*

Classifying an example involves subjecting the data to an organized sequence of tests to determine the label. Trees are built and tested from top to bottom as so:

1. Start at the root of the model.
2. Wait until all examples are in the same class.
3. Test features to determine best split based on a cost function.
4. Follow the branch value to the outcome.
5. Repeat number 2.
6. Leaf node output.

The central question in decision tree learning is which nodes should be placed in which positions, including the root node and decision nodes.

There are three main decision tree algorithms. The difference in each algorithm is the measure or cost function for which nodes, or features, are selected. The root is the top node. The tree is split into branches, evaluated through a cost function; and a branch that doesn't split is a terminal node, decision, or leaf.

Decision trees are useful in the way that acquired knowledge can be expressed in an easy to read and understandable format (see Figure 4-1). It mimics human decision-making whereby priority—determined by feature importance, relationships, and decisions—is clear. They are simple in the way that outcomes can be expressed as a set of rules.

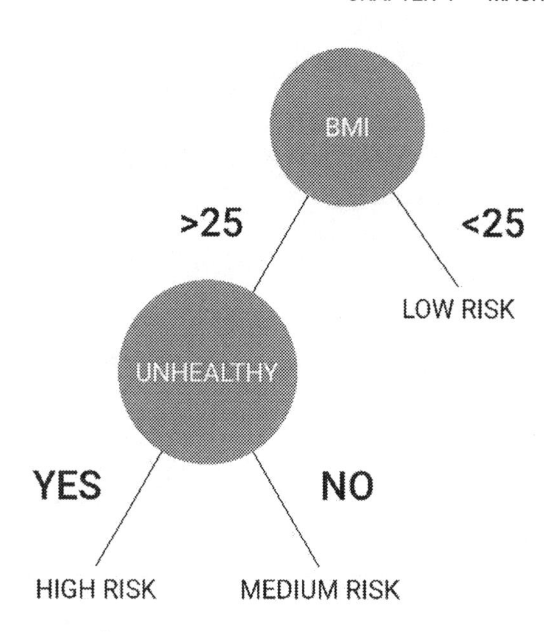

Figure 4-1. *Decision tree of n = 2 nodes*

Decision trees provide benefits in how they can represent big datasets and prioritize the most discriminatory features. If a decision tree depth is not set, it will eventually learn the data presented and overfit. It is recommended to set a small depth for decision tree modeling. Alternatively, the decision tree can be pruned, typically starting from the least important feature or the incorporation of dimensionality reduction techniques.

Overfitting is a common machine learning obstacle and not limited to decision trees. All algorithms are at risk of overfitting, and a variety of techniques exist to overcome this problem. Random forest or jungle decision trees can be extremely useful in this.

Pruning reduces the size of a decision tree by removing features that provide the least information. As a result, the final classification rules are less complicated and improve predictive accuracy.

The accuracy of a model is calculated as the percentage of examples in the test dataset that is classified correctly:

- True positive/TP: Where the actual class is yes and the value of the predicted class is also yes.
- False positive/FP: Actual class is no, and predicted class is yes.
- True negative/TN: The value of the actual class is no, and the value of the predicted class is no.

- False negative/FN: When the actual class value is yes, but predicted class is no.
- Accuracy: (Correctly predicted observation)/(Total observation) = (TP + TN)/(TP + TN + FP + FN).
- Precision: (Correctly predicted Positive)/(Total predicted Positive) = TP/TP + FP.
- Recall: (Correctly predicted Positive)/(Total correct Positive observation) = TP/TP + FN.

Classification is a common method used in machine learning; and ID3 (Iterative Dichotomizer 3), C4.5 (Classification 4.5), and CART (Classification and Regression Trees) are common decision tree methods where the resulting tree can be used to classify future samples.

Iterative Dichotomizer 3 (ID3)

Python, scikit-learn; method, decision-tree-id3

Ross Quinlan has contributed significantly to the area of decision tree learning, developing both ID3 and C4.5 [76]. The ID3 algorithm uses the cost function of information gain, based on work from "the father of information theory" Claude Shannon. Information gain is calculated through a measure known as entropy.

Entropy determines how impure, or disordered, a dataset is. Given an arbitrary categorization C, into categories {c1, ... cn}, and a set of examples S, for which the proportion of ci is pi, the entropy of sample S is the following:

$$H(S) = \sum_{X \in X} -p(x)\log_2 p(x)$$

Information gain is calculated using entropy. This value is given for attribute A, with respect to examples S. The values attribute A can take {t1, ... tn}, representing the totality of set T.

$$IG(A,S) = H(S) - \sum_{t \in T} p(t)H(t)$$

ID3 selects the node with the highest information gain to produce subsets of data that are then applied the ID3 algorithm on recursively. The information gain of an attribute can be understood as the expected loss in entropy as the result of learning the value of attribute A. The algorithm terminates once it exhausts the attributes or the decision tree classifies the examples completely. ID3 is a classification algorithm that is unable to cater for missing values.

For example, if the following row were to be received, the model would categorize the outcomes as high risk. There is a chance that ethnicity is a determining factor in the outcome. However, there is not enough data for the model to learn the potential outcome.

Record	Unhealthy	BMI	Ethnicity	Outcome
Person 5	No	<25	Chinese	?

C4.5

Python, scikit-learn; method, C45algorithm

C4.5 is an enhancement built on ID3 that takes number and size of branches into account when choosing an attribute. C4.5 can handle continuous and discrete labels and uses gain ratio as its splitting criteria. It improves on the execution time and accuracy as the size of the dataset increases.

C4.5 introduces pruning—a bottom-up technique also known as subtree raising—and subtree replacement to ensure the model doesn't overfit the data.

CART

Python, scikit-learn; method, cart

CART, which stands for Classification and Regression Trees, is a decision tree algorithm which uses the Gini index as the cost function. Represented as a binary tree, as with C4.5, CART can be used for regression and classification problems. While entropy is used for exploratory analysis, Gini is to minimize misclassification. Root nodes represent input variable (x), and leaf nodes contain the output variable (y) used to make a prediction.

Ensembles

An ensemble is a form of machine learning algorithm and is a popular technique employed in the field of machine learning. Ensemble methods use a collection of machine learning models or learners. As a result, a group of weak learning models is combined to create a more accurate learning model.

There are two types of ensemble techniques—bagging and boosting.

Bagging

Bagging is an ensemble technique that involves creating multiple independent models with datasets using the bootstrap sampling technique (see Figure 4-2). The technique aids in reducing error by reducing variance. In general, bagging improves predictions as anomalous behaviors are averaged out. In comparison, a collection of iterative predictive models is known as boosting.

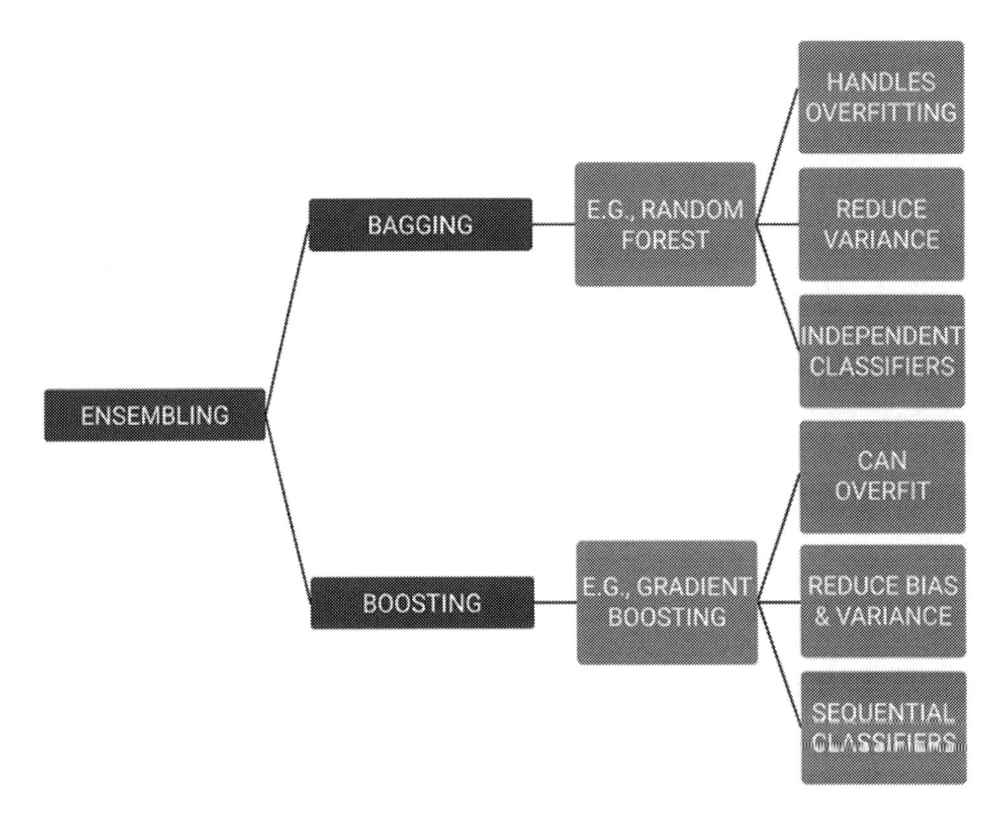

Figure 4-2. *Bagging vs. boosting*

In bagging, n′ < n samples are taken from dataset D with replacement. A random selection of features for constructing the best split is chosen each time.

An average, majority vote, or statistical method, is used to determine the average prediction from a model. Random forest decision trees (see Figure 4-3) extend the concept of bagging, as they use a random subset of features too, resulting in less correlation among predictions from learners.

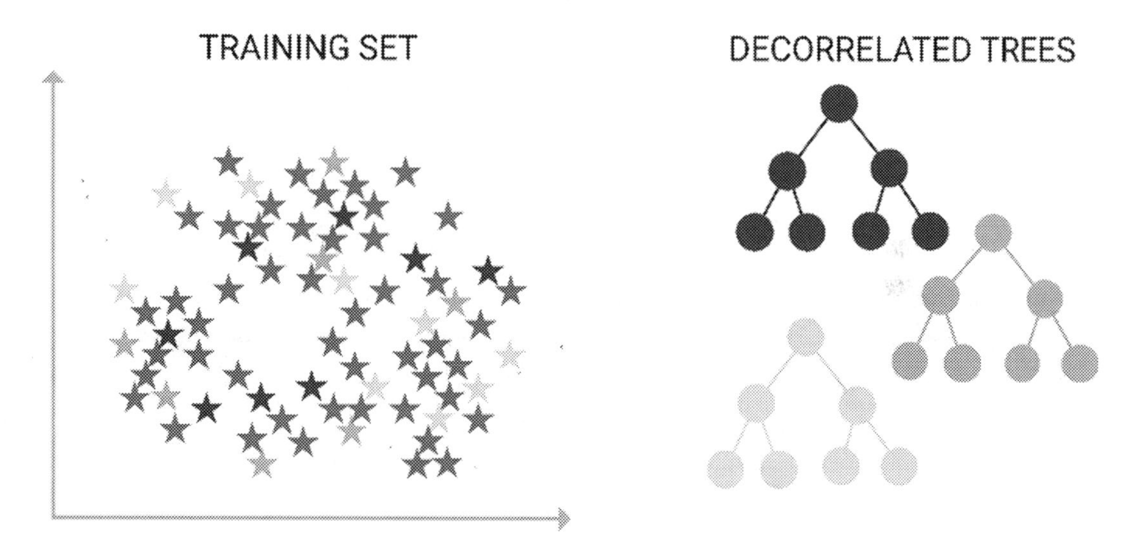

Figure 4-3. *Random forest decision trees*

Random Forest Decision Trees

Python, scikit-learn; method, RandomForestClassifier

Random forest decision trees are an example of bagging. Random forests are created through training multiple decision trees at the same time and refer to a collection of decision tree learners (hence the "forest"). The purpose of random forests is to prevent overfitting. The more decision trees in the random forest, the more accurate results are.

Each random forest takes a sample of the dataset and a random subset of features for decision.

Random forest decision trees can be used for classification and regression problems. With regression problems, an average of the outcomes is taken; whereas with classification problems, it is the majority category. A random forest decision tree is an ensemble of different models. This is an ensemble method due to the procedure reducing variance in the algorithm through combining predictions from multiple models.

The random forest model seems to work particularly well when data is limited to classification or decision problems by bootstrapping existing data. A benefit of random forest classifiers is that they can also handle missing values.

Boosting

Boosting is an ensemble method that produces a collection of predictive models iteratively. The concept involves each new model learning from the errors of previous models.

Boosted methods are typically implemented when there is a plethora of data for prediction.

Through iteratively creating new models to compensate for error, the process "boosts" or converts weak learners into strong learners—reducing their bias and variance. Each iteration of the technique learns relationships in the data and is then subsequently analyzed for errors, with incorrect classifications given a weighting. The aim is to minimize the error from the previous model. The underlying machine learning method used for boosting algorithms can be anything.

Boosting works through the following process:

1. Data is represented.
2. Decision stump is made on the most significant cut of features.
3. Weightings are given to misclassified observations.
4. The process is repeated and all stumps combined to achieve the final classifier.

Gradient Boosting

Python, scikit-learn; method, GradientBoostingClassifier, GradientBoostingRegressor

Gradient boosting is known by many names including multiple additive regression trees, stochastic gradient boosting, and gradient boosting machines. Gradient boosting is a technique derived from decision trees when Jerome Friedman applied the gradient boosting concept to decision trees in 2001 [77]. AdaBoost (Freund and Shapire, 2001) is a gradient boosting variant. In gradient boosting, the model is trained sequentially. Each iterative model seeks to minimize the loss function [78].

Similar to random forest decision trees, gradient boosting is made from weak predictors. Each new tree is subsequently trained with the data previously incorrectly classified by the previous tree. This iterative procedure expedites the model by focusing on more challenging data as easier predictions are made early on in learning.

XGBoost, or extreme gradient boosting, is a variation of gradient boosting. XGBoost adds regularization and takes advantage of computing power with distributed, multithreaded processing for greater speed and efficiency.

Adaptive Boosting

Python, scikit-learn, numpy; method, AdaBoostClassifier, AdaBoostRegressor

Adaptive boost is a popular method that iteratively learns from mistakes. Decision tree stumps are weak learners, splitting data by one rule only. The decision tree stump is improved by focusing on samples that were incorrectly classified. These are identified through weightings that are associated with the data. The model can learn from its mistakes and provides a final solution with lower bias than a single decision tree.

Linear Regression

Python, scikit-learn; method, linear_model.LinearRegression

Linear regression is a continuous statistical technique to understand the relationship between input (x: predictors, explanatory or independent variables) and output (y: response or dependent variable). As the name suggests, linear regression (see Figure 4-4) requires that y be calculated from a linear combination of input variables.

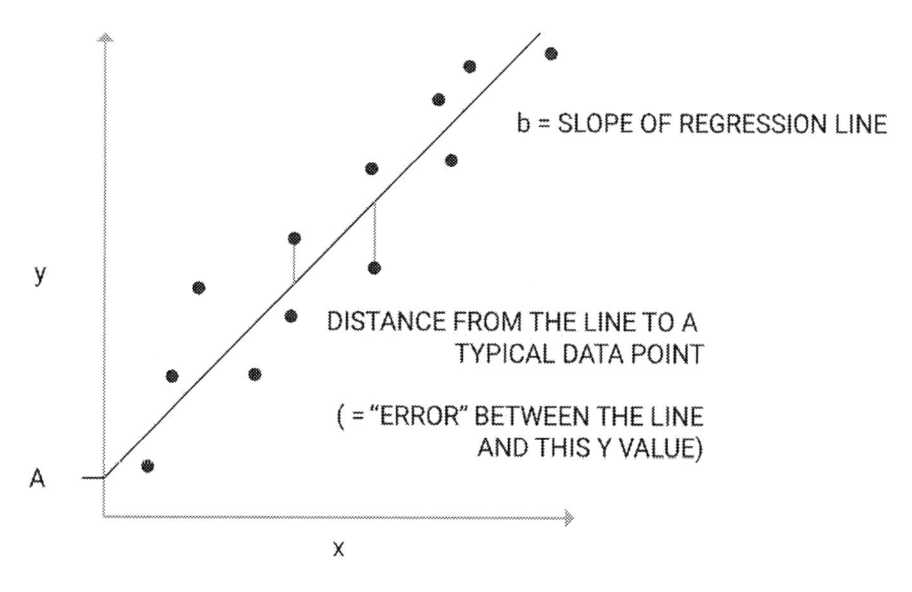

Figure 4-4. *Linear regression*

The aim with linear regression is to place a line of best fit through all of the data points, which enables easy understanding of predictions by minimizing margin or residual variation.

$$\min \sum \left(y_i - \hat{y}_i\right)^2,$$

where y_i is the actual observed response value, $\hat{y}_i$ is the predicted value of the response variable, and their subtraction is the residual variation or difference between observed and predicted values of y.

- Criterion variable: Outcome variable we are predicting
- Predictor: What variable predictions are based on
- Intercept: Where the line of best fit intercepts the y-axis

When there is one input variable x, this technique is known as simple linear regression. Multiple linear regression evaluates two or more predictor variables.

$$Y = \beta_0 + \beta_1 X$$

$$\beta_1 = \frac{\sum_{i=1}^{m}\left(x_i - \bar{x}\right)\left(y_i - \bar{y}\right)}{\sum_{i=1}^{m}\left(x_i - \bar{x}\right)^2},$$

$$\beta_0 = \bar{y} - \beta_1 \bar{x}$$

where y is the output variable, x is the predictor, B0 and B1 are coefficients that are estimated to move the line of best fit, $\bar{x}$ refers to the mean of x, B0 is the bias or intercept (where it intercepts the y-axis), and B1 is the slope or rate at which y increases/decreases per unit of x.

Once the model has learned the parameters, they are used to predict values of y given new, unseen input X.

Popular techniques used to estimate the coefficients are ordinary least squares and gradient descent. Given a line of best fit placed through a sample dataset with a linear relationship, ordinary least squares attempts to minimize the sum of the squared distances from each data point to the regression line. The intention is to minimize the sum of squared residuals. Coefficient values are iteratively optimized to reduce the error of the model.

Ordinary least squares starts with random values for B0 and B1. The sum of squared errors is calculated for each (input, output) pair. A learning rate is chosen, which acts as a scale factor; and the coefficients are optimized toward minimizing the error until no further gains can be made.

A learning rate is a hyperparameter that has its value determined by experimentation. Different values are tried, and the value that gives the best results is used.

Criticisms of linear regression focus on its simplicity, resulting in it being unable to capture complex relationships within the data. Noise in the data can also be learned by the model, resulting in overfitting.

Relationships between variables are not always linear, and therefore, a linear regression model can perform poorly on data with nonlinear relationships. However, variables can sometimes be transformed to fit a linear regression model. Linear regression also assumes variables are homoscedastic, which means all variables in the vector have the same variance.

The aforementioned is a perfect demonstration as to why when training a model, one should use a variety of machine learning algorithms.

Logistic Regression

Python, scikit-learn; method, linear_model.LogisiticRegression

Linear regression applies to continuous variables, whereas logistic regression predictions apply to discrete values or classification problems after the application of a transformation function.

Developed by statistician David Cox in 1958, logistic regression (or logit regression) is often used for binary classification problems. For example, logistic regression can be used to predict whether a patient will experience an adverse event given a particular medication, whether a patient is likely to be readmitted to a hospital or not, or whether a patient has a particular disease diagnosis.

Unlike linear regression, the output of the model comes in the form of a probability between 0 and 1. The predicted output is generated by log-transforming the x input and using the logistic function h(x) = 1/(1 + e^ - x). A specified threshold is used to transform this probability into a binary classification.

Logistic regression uses the sigmoid function, an S-shaped curve (see Figure 4-5) that can take any continuous value and map it into a probability value between 0 and 1 for a default class.

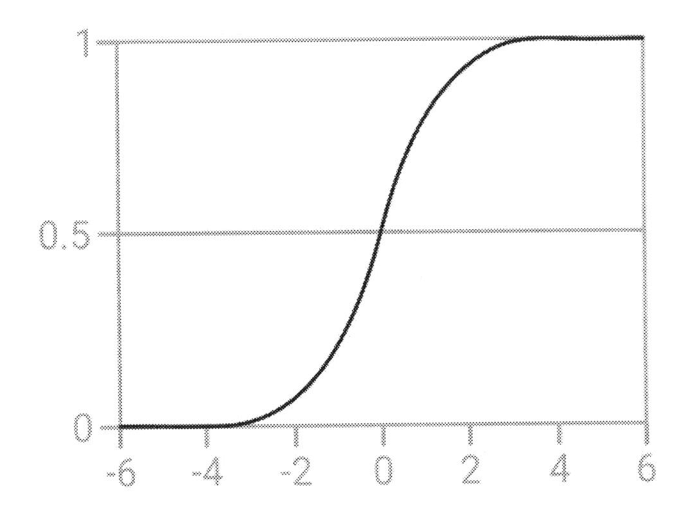

***Figure 4-5.** Logistic function*

Sigmoid function: 1/(1 + e^–value), where e is Euler's number and value is the value requiring transformation.

Logistic regression uses an equation P(x) = e ^ (b0 + b1*x)/(1 + e^(b0 + b1*x)), which can be transformed into ln(p(x)/1–p(x)) = b0 + b1*x). As with linear regression, b0 and b1 are values learned from the training data.

As an example, take a model that determines whether a growth on the skin surface is benign or not. The input variable, x, could be the size, depth, or texture of the growth; and the default variable y = 1 indicates a benign growth. As depicted, the logistic function transforms the x-value of the various instances into a probability value between 0 and 1. If the probability crosses the threshold of 0.5, the tumor is classified as benign.

Logistic regression seeks to learn values of b0 and b1 through training that minimize the error between predicted and actual outcomes. Maximum likelihood estimation is an iterative procedure used for this purpose.

Maximum likelihood estimation starts with a random value as the best weight for each predictor variable and then adjusts these coefficients iteratively until there is no improvement in the ability to predict the outcome value. The final coefficients have a value that minimizes the error in the probabilities predicted.

As with linear regression, logistic regression assumes a linear relationship between input variables and output. Feature reduction may be required to transform data into linear models. Logistic regression models are also prone to overfitting, which may be overcome through the removal of highly correlated inputs.

Finally, it is possible that coefficients fail to converge, which can happen if data is sparse or highly correlated.

SVM

Python, scikit-learn; method, svm.SVC, svm.LinearSVC

Support vector machine (SVM) is a nonprobabilistic binary linear classifier used with both classification and regression problems. SVM is often used along with NLP methods to analyze text for topic modeling and sentiment analysis. It is also used in image recognition problems and handwriting digit recognition.

The algorithm finds a hyperplane, or line of best fit, between two classes determined by the support vectors. Support vectors are the data points nearest the hyperplane that would alter the position of the hyperplane if removed. The greater the margin value, or distance from the data points to the hyperplane, the more confidence there is that the data is appropriately classified. The line of best fit is learned from an optimization procedure that seeks to maximize the margin.

SVM uses something known as kernelling to map data to higher-dimensional feature spaces. Data is mapped using the kernel trick in iteratively more dimensions until a hyperplane can be formed to classify it.

Take the problem in Figure 4-6. On the left, we see that it would be impossible to find a line of best fit in two dimensions. SVM uses the kernel trick to map data into three dimensions and can define a hyperplane to classify the data.

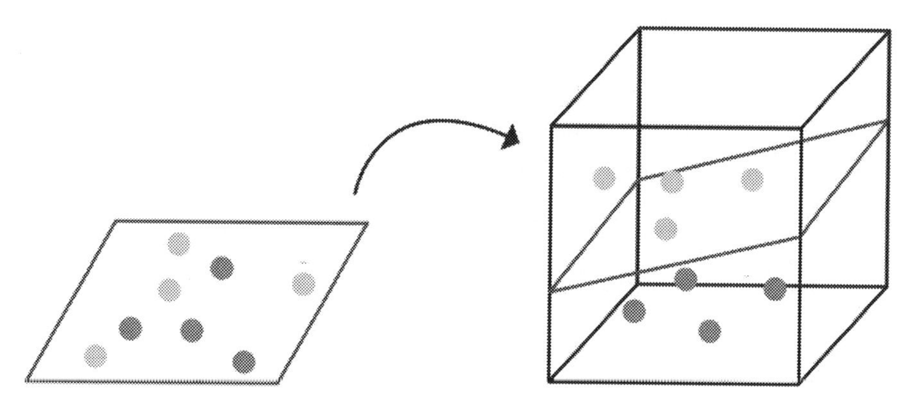

Figure 4-6. *SVM visualization*

SVMs take numerical inputs and work well on small datasets. However, as dimensionality increases, the ability to understand and explain the model reduces. As dataset size increases, so too can training time. SVMs are also less capable on noisy data.

The SVM can be expressed as a sum over the support vectors (see Figure 4-7).

$$f(x) = \sum_i \alpha_i y_i (\mathbf{x}_i{}^\top \mathbf{x}) + b$$

SUPPORT VECTORS

Figure 4-7. *SVM expressed as a sum over the support vectors*

Naive Bayes

Python, scikit-learn; method, GaussianNB, MultinomialNB, BernoulliNB

Naïve Bayes uses Bayes's theorem to calculate the probability an event will occur given another event has already occurred. The algorithm is acknowledged as naïve because it assumes all variables are independent of each other, which is atypical of real-world examples. The Bayesian classifier is often used when input dimensionality is high.

For example, given a specified threshold, this method can be used to probabilistically classify whether a vector is in one class or another:

P(B|A) = (P(A|B) * P(B))/P(A)

- **P(B|A)** = Posterior probability (Figure 4-8): Probability of hypothesis B being true, given the data A, where P(B|A)= P(a1|b) * P(a2|b) * ... * P(an|b) * P(A)
- **P(A|B)** = Likelihood: The probability of data A given that the hypothesis B was true
- **P(A)** = Class prior probability
- **P(B)** = Predictor prior probability

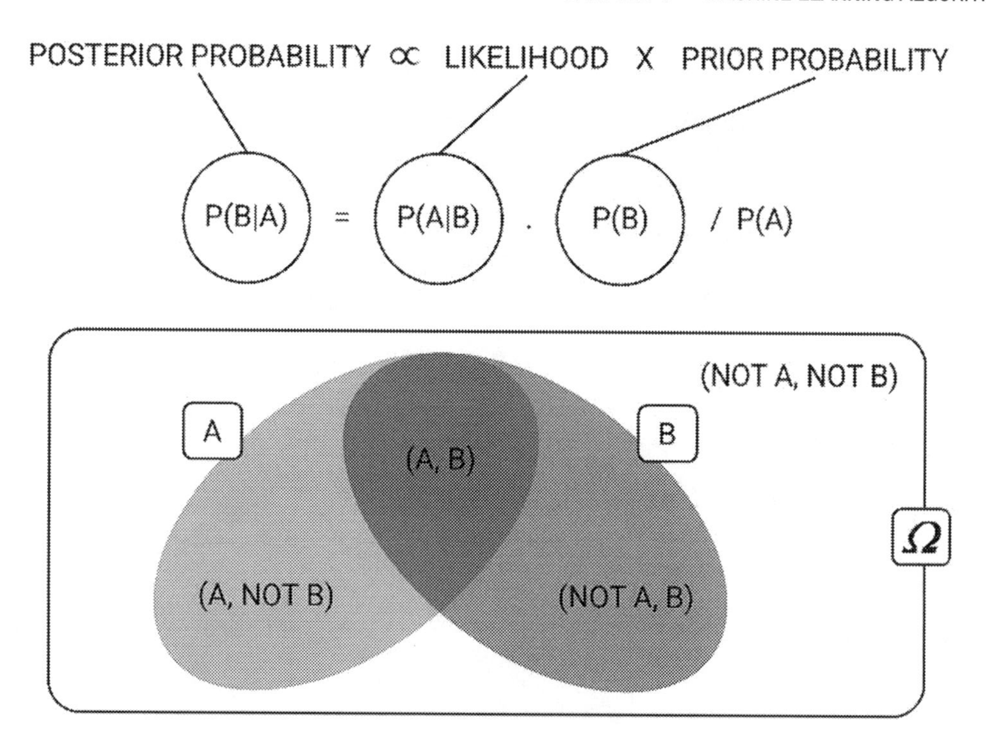

Figure 4-8. *Posterior probability*

The prior distribution P(B) and likelihood probability P(B|A) can be estimated from the training data. To calculate which class an example belongs to, the probability of being in each class is calculated. The class assigned to an example is the class that produces the highest probability for the example.

Using Table 4-3 as a simple example, we can learn the probability of a patient's disease risk from whether they are unhealthy or not. To determine the outcome for unhealthy = yes, we calculate P(high risk|unhealthy) and P(low risk|unhealthy) and choose the outcome with the highest probability.

P(high risk|unhealthy)

= (P(high risk|unhealthy) * P(high risk))/P(unhealthy) = (2/2 * 2/4)/(3/4) = 0.66

Table 4-3. Outcome likelihood

Record	Unhealthy	Outcome
Person 1	Yes	*High risk*
Person 2	Yes	*Low risk*
Person 3	Yes	*High risk*
Person 4	No	*Low risk*

Naïve Bayesian models are easy to build and particularly useful for extensive datasets. Along with its simplicity (and hence its naïvety), naïve Bayes is known to outperform even highly sophisticated classification methods.

kNN: k-Nearest Neighbor

Python, scikit-learn; method, neighbors.KNeighborsClassifier

Not to be confused with k-means clustering, the kNN method classifies an unknown object O with the label of the majority of the k-nearest neighbors. Used in classification and regression problems, kNNs are a nonparametric technique. They do not learn a model. Instead, kNNs store the training dataset as their representation and perform classification of a new sample based on learning by analogy.

As there is no learning of the model, kNNs are considered lazy learners.

Each object represents a point in N-dimensional space. A neighbor is defined nearest if it has the smallest distance in feature space. One of the ways to calculate the distance between an unseen object and its neighbor is by using the Euclidian distance between A = (a1 ... an) and B = (b1 ... bn) as follows:

$$d(A,B) = \sqrt{\sum_{i=1}^{n} (a_i - b_i)^2}$$

The algorithm works as such:

1. Calculate the distance between any two points.
2. Find the nearest neighbors based on these pairwise distances.
3. Majority vote on a class label based on the nearest neighbors list.

The prediction is made at the time of the request. In regression problems, the mean or median of the k-most similar instances is used. In classification problems, the class with the highest frequency from the k-most similar instances is selected.

As with most algorithms, the determinant method can vary. The distance between binary vectors (Hamming distance) and the sum of absolute difference between real vectors (Manhattan distance) are also used.

A disadvantage of kNNs is the large computing requirement for classifying an object, as the distance for all neighbors in the training dataset must be calculated. It is not particularly well suited to high-dimensional data. Each predictor variable can be considered a dimension of N-dimensional input space. For instance, x1 would be one-dimensional and x1, x2 two-dimensional and so on. The increase in dimensionality exponentially increases the volume of the input space.

kNNs are not well suited to data with missing values, as distances between vectors cannot be calculated on missing data.

Neural Networks

The biological mechanism in which the brain works inspires artificial neural networks, a different paradigm for computing, based on the parallel architecture of animal brains.

Neural networks are well equipped for data mining tasks due to their ability to model multidimensional data and efficiency at finding hidden patterns among data. Neural networks can be applied to prediction and classification problems. The process of estimating the result and comparing it with the real output is known as forward propagation.

Neural networks are a subset of machine learning algorithms, which include perceptrons, fully connected neural networks, convolutional neural networks, recurrent neural networks (RNNs), long short-term memory neural networks, autoencoders, deep belief networks, generative adversarial networks, and many more. Most of them are trained with an algorithm called backpropagation.

Artificial neural networks (ANNs) are used in a plethora of tasks:

- Classifying disease—computers can learn what images of diseased organs, such as the kidneys, eyes, and liver, look like to predict the likelihood of disease
- Recognizing speech

- Translating and digitalizing text
- Facial recognition

Neural networks can learn how to accomplish a task like a human brain—supervised, unsupervised, and through reinforcement learning.

Perceptron

Perceptrons are the fundamental units of ANNs. In AI, the perceptron is a node or unit that takes multiple inputs and has a single output, typically 1 or -1. Perceptrons take a weighted sum of the inputs, S, and a unit function calculates the output for the node (see Figure 4-9). Unit functions can be the following:

- Linear—such as the weighted sum
- Threshold functions that only fire if the weighted sum is over a threshold
- Step functions that output the inverse, typically -1, if S is less than the threshold
- Sigma function (1/(1 + e - s), which allows for backpropagation

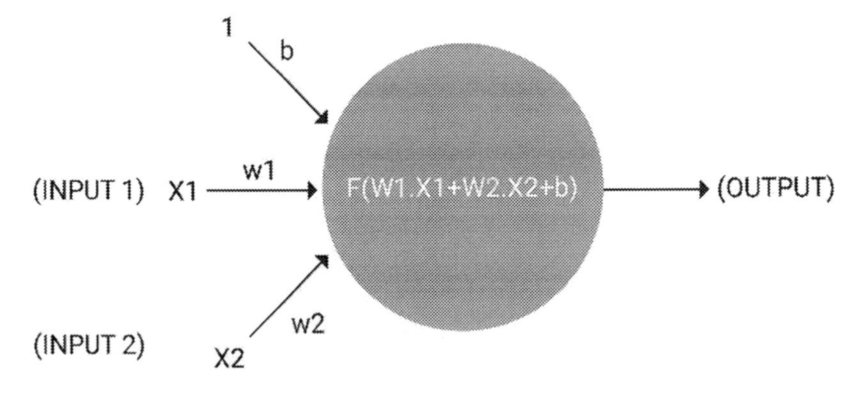

Figure 4-9. *Configuration of an artificial neuron*

Perceptrons mimic biological neurons (see Figure 4-10), where dendrites receive an input. If the signal is high enough within a set duration of time, the neuron sends an electrical pulse to the terminals then received by dendrites of other neurons. Each perceptron has a bias, similar to b of the linear function y = ax + b. It moves the line up and down to fit the prediction with the data better.

$$a = f\left(\sum_{i=0}^{N} w_i x_i\right),$$

where 1 is x_0 and $b = w_0$.

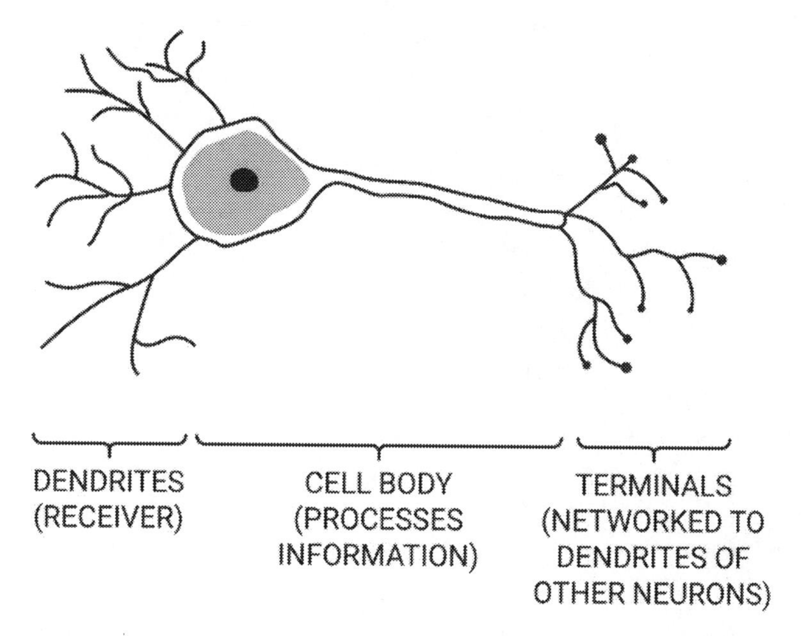

Figure 4-10. *Configuration of a biological neuron*

Artificial Neural Networks

Python, scikit-learn; method, linear_model.Perceptron

Artificial neural networks are made of perceptrons and contain one or more hidden layers (see Figure 4-11). There are several topologies, with the simplest being a feedforward network.

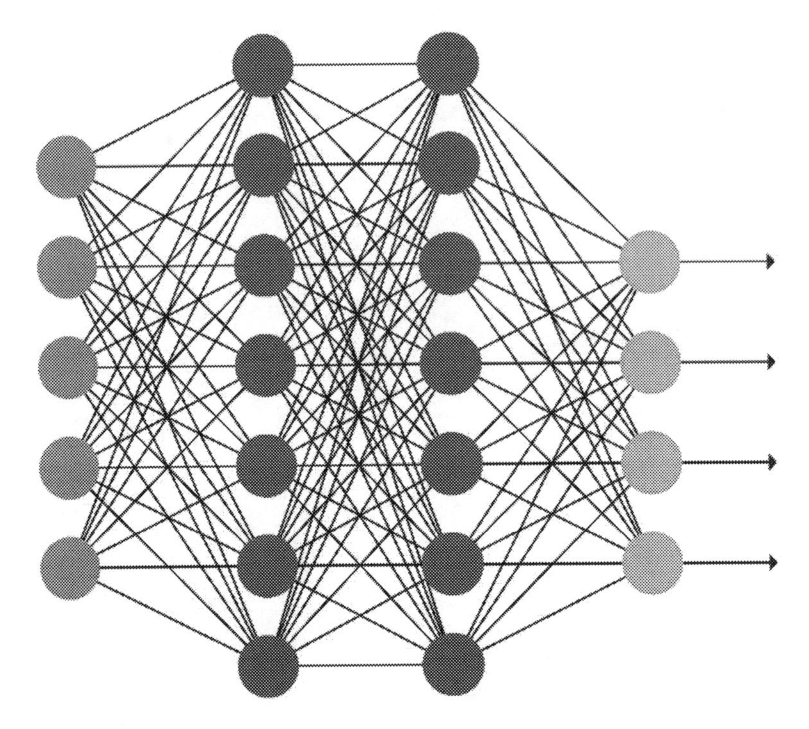

***Figure 4-11.** Organization of an artificial neural network*

Backpropagation is the method of determining the error, or loss, at the output and propagating it back to the network. Weights at each node are updated to minimize the respective error output from each neuron. Backpropagation seeks to minimize the error for neurons and thereby the entire model through updating respective neuron weights.

The primary learning that must be accomplished in neural network models is training neurons when to fire. An epoch refers to one training iteration of forward- and backpropagation. During the learning phase, nodes in a neural network adjust their weights based on the error of the last test result. The learning rate controls how fast the network learns.

The number of perceptrons or neurons in a model are equal to the number of variables in the data. Some representations may also have a bias node.

Typically, artificial neural networks have neurons organized in layers. Each layer can perform different transformations on the input data it receives.

Deep Learning

Deep learning is a process that employs deep-layered neural network architectures. Training a deep, artificial neural network model requires more time and CPU power compared to other types of models. Noteworthy is that the performance of ANN models may not necessarily be superior to typical supervised learning techniques.

Deep learning is not a new technique; however, it is becoming more frequent with hardware advances, primarily power and cost, which have made such computing feasible.

Deep neural networks (DNNs) were used by Alphabet to create the strongest Go player in history, defeating several human champions of Go in 2017 [74]. The ancient game of Go has historically been considered a challenge for AI and demonstrates the capabilities of deep learning (see Figure 4-12).

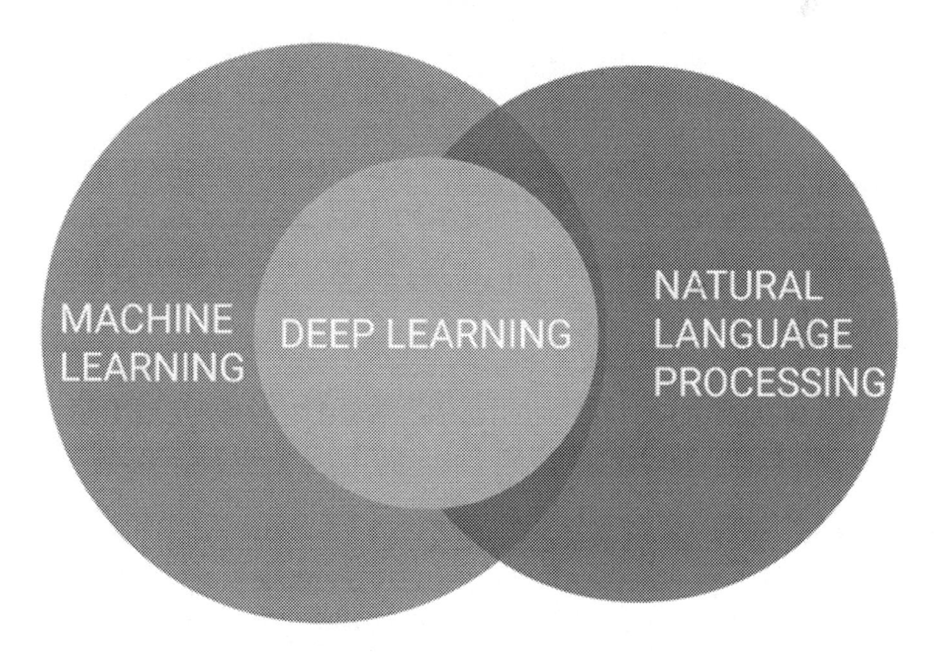

Figure 4-12. *Deep learning: Where does it sit?*

There are many types of artificial neural networks, including the following.

Feedforward Neural Network

In a feedforward neural network, the sum of the weighted inputs is fed to the output (see Figure 4-13). The output is activated (typically 1) when the sum exceeds the unit function threshold. If it doesn't fire, it typically outputs -1.

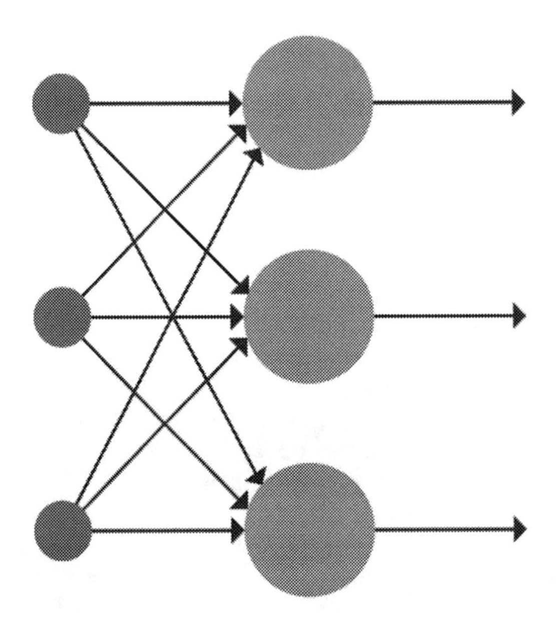

Figure 4-13. *Feedforward neural networks*

Recurrent Neural Network (RNN): Long Short-Term Memory

Recurrent neural networks or RNNs save the output of a layer and feed it back to the input to improve predictions of the output layer (see Figure 4-14). As such, each node has a memory when performing computations and makes use of sequential information. As a result, nodes are deemed to have a memory.

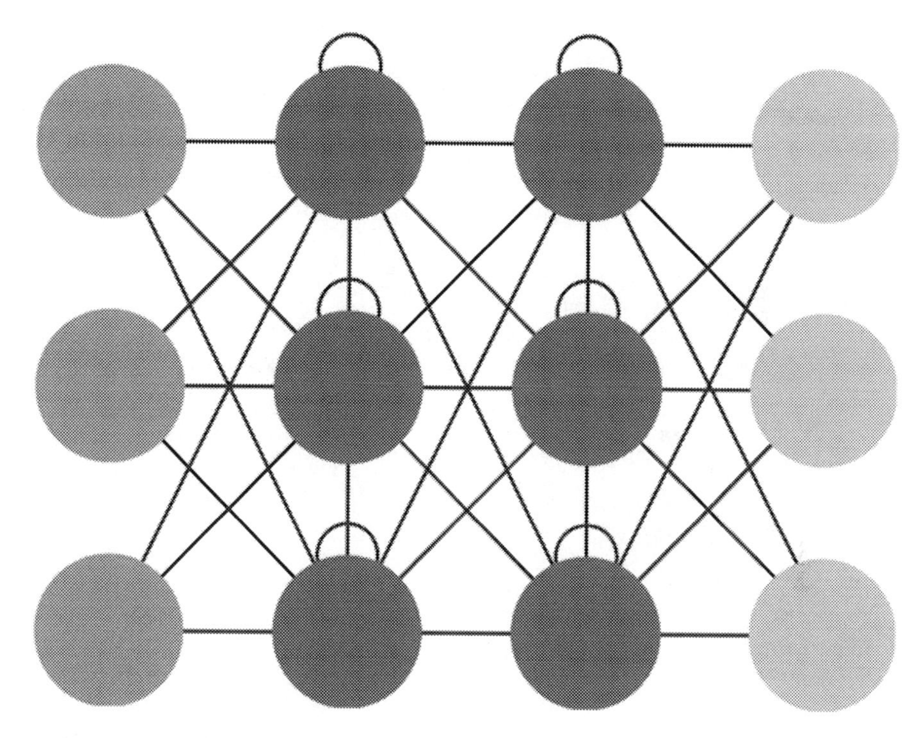

Figure 4-14. *Recurrent neural networks*

Convolutional Neural Network

Convolutional neural networks are deep, feedforward neural networks. Layers are known as convolutional layers and are often used in speech recognition, spatial data, NLP, and computer vision problems. Separate convolutional layers typically apply to different aspects of a particular problem.

For instance, in understanding whether there is a face present within a photograph, separate convolutional layers within a DNN may be used to identify different aspects of the face—the eyes, nose, ears, mouth, and so forth.

Modular Neural Network

Similar to the human brain, the concept of modular neural networks is that of neural networks working together.

Radial Basis Neural Network

This is a neural network that uses radial basis activation functions where the value depends on the distance from the origin (see Figure 4-15).

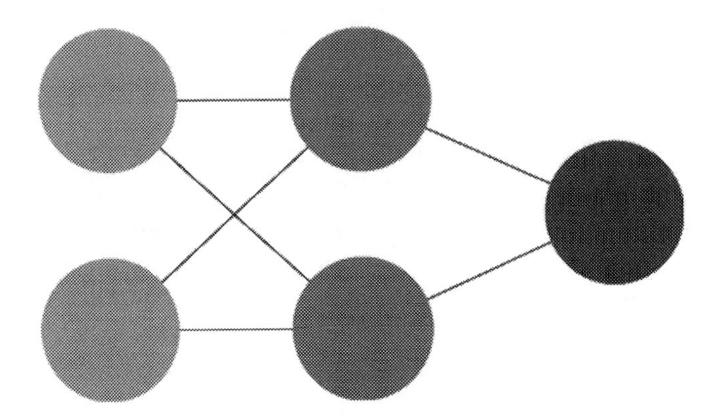

Figure 4-15. *Radial basis neural network*

The key strength of ANNs and deep learning lies in the ability to generate complex linear and nonlinear models from the training dataset.

DNNs (deep and multilayered ANNs) can learn complex relationships between data from training data alone, which typically improves generalization. DNNs can still succumb to the problem of overfitting data, so appropriate pruning may be required. As the name suggests, deep networks can also take a long time to train and require a large number of training examples.

A common concern with neural networks, particularly in healthcare, is that they act as a black box and do not present a clear interface for a user to understand findings. Practically, an ANN model is difficult to interpret, and weights and biases aren't readily interpretable to what is important in the model.

It is typically considered that for supervised learning problems, the upper bound on neurons to prevent overfitting is

$$N_h = N_s(\langle * (N_i + N_o)),$$

where

N_i = number of input neurons

N_o = number of output neurons

N_s = number of samples in the training dataset

⟨ = an arbitrary scaling factor

Unsupervised Learning

Unsupervised learning refers to the process of learning a model from unlabeled data. This means that input data (x) is supplied without output (y). Algorithms are left to determine notes within data, which means there is no right or wrong answer.

Semi-supervised learning occurs when some output labels (y) are supplied. For example, learning would be semi-supervised in a model learning to predict glaucoma from patient eye scans with partially labeled data. If only a subset of eye scans had appropriately labeled outputs (e.g., glaucoma, not glaucoma), the model might not have enough appropriate data to learn a model. As more labeled data is provided for the model to train on, the system would likely become móre accurate.

Unsupervised learning can be resource consuming concerning time, money, and expertise. Usually, vast improvements in model accuracy can be made through improving labeling of data.

Unsupervised learning is composed of two main problem concepts: clustering and association.

Clustering

Clustering refers to the process of discovering relationships within the data. Clustering has a variety of healthcare uses including the following:

- Grouping patients of similar profiles together for monitoring
- Detecting anomalies or outliers in claims or transactions
- Defining treatment groups based on medication or condition
- Detecting activity through motion sensors

k-Means

Python, scikit-learn; method, cluster.KMeans

k-Means clustering (see Figure 4-16) aims to find k similar groups within the data. k-Means is an iterative algorithm that calculates the centroids of the defined k clusters, with all training data assigned to a cluster.

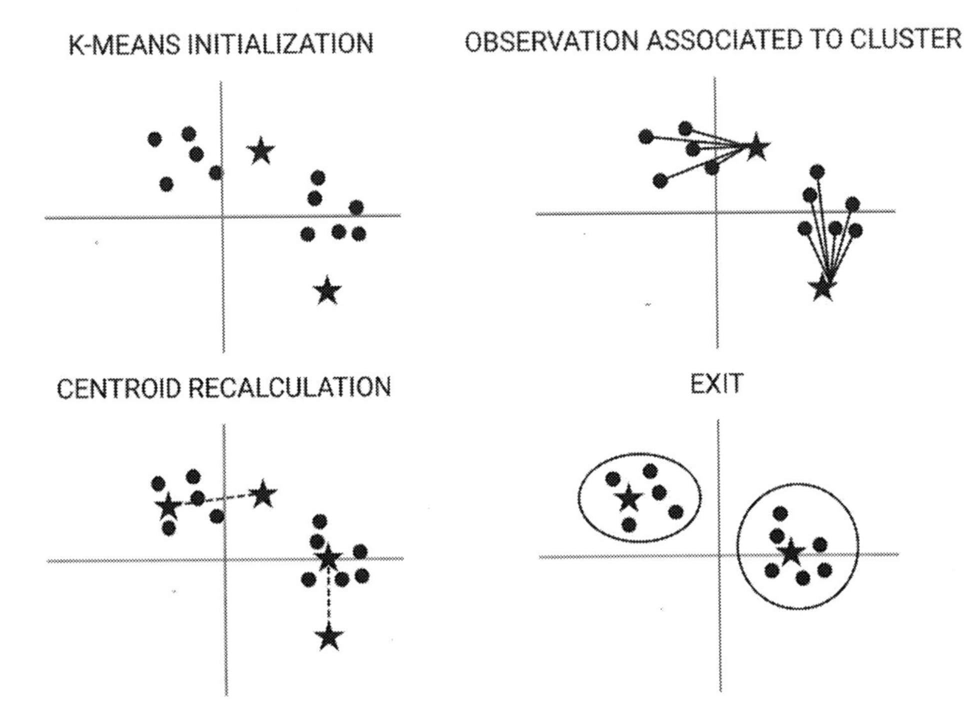

Figure 4-16. *k-Means clustering*

Data points are assigned by the distance between the data point and its centroid. A cluster centroid is a collection of attribute values that define the resulting clusters.

$$\underset{c_i \in C}{\arg\min}\, dist\left(c_i,\ x\right)^2,$$

where c_i is the collection of centroids in set C and distance (dist) is standard Euclidean distance.

Rather than defining groups before data exploration, clustering enables the model to determine groups that have formed organically.

Steps of the k-means algorithm include the following:

- Step 1: Set a value of k. Here, let us take k = 2.
    - Assign each data point randomly to any of the k = 2 clusters.
    - Determine the centroid for each of the clusters.
- Step 2: For each data point, associate it to the closest cluster centroid.
- Step 3: Recalculate centroids for the new clusters.
- Step 4: Continue this process until the centroid clusters remain unchanged.

Association

Association rule learning methods extract rules that best explain perceived relationships between variables in data. These rules can discover useful associations in large multidimensional datasets that can be used to drive and optimize.

Association rule learning is historically best applied to online shopping checkout basket datasets gathered on users' purchasing habits. For instance, when someone buys a pizza, they may also tend to purchase wine, just like how someone who purchases lettuce may tend to buy tomatoes, cucumbers, and onion too. Through analyzing transactional datasets, the probability of associations can be predicted. We know that certain items (whether that be food, clothes, or even disease) frequently occur together, and association rule learning seeks to understand these relationships.

In healthcare, in particular, associative symptoms can be understood to predict better and diagnose disease and adverse events. Determining potential adverse effects based on medication and associative patient comorbidity pathways could lead to improved care and treatment pathways.

There are three important metrics to be familiar with.

Support

Support is the value of absolute frequency. An association rule holds with support sup in dataset T if the sup % of transactions contain X U Y. This represents how popular an itemset is, as measured by the proportion of transactions in which an itemset appears:

sup = Pr(X U Y) = count(X U Y)/Total transaction count

Confidence

The confidence measure represents correlative frequency. An association rule holds in dataset T with confidence conf if the conf % of transactions that contain X also contain Y. This estimates how likely item Y is to occur or be present in the transactional dataset when item X occurs. This is expressed as {X → Y} and measures the proportion of transactions with item X in which item Y also appears:

$$conf = Pr(Y|X) = count(X \cup Y)/count(X)$$

Lift

Lift determines how likely item Y is given that X occurs while accommodating for Y's popularity:

$$Lift = Support\ (X \cup Y)/Support\ (X) * Support\ (Y)$$

Association rules determine the probability of Y given that X occurs. The benefits of these associations have applications across industry. There are many association rule mining algorithms that use different strategies toward understanding associations and use different structuring of data.

Apriori

Apriori is the most popular association rule mining algorithm. The Apriori algorithm is frequently used in pattern mining and in the identification of features and items that occur together. It is used extensively in basket analysis and is a powerful tool to find hidden feature patterns. It is usually applied to transactional databases to mine frequent relationships and generate association rules.

Apriori begins with a set of n items. The algorithm calculates all possible candidate itemsets, or frequent itemsets, which meet the specified threshold support and confidence values. The Apriori property states for an itemset to be frequent, all of its subsets must also be frequent. The Apriori algorithm uses a bottom-up approach; frequent subsets are identified and increased one item at a time.

The Apriori method is relatively simple (see Figure 4-17):

- Apply minimum support to find all frequent sets of k items within dataset T where a frequent itemset has support ≥ min(sup).
- Expand selection to find frequent sets of k + 1 items using frequent k-itemset.

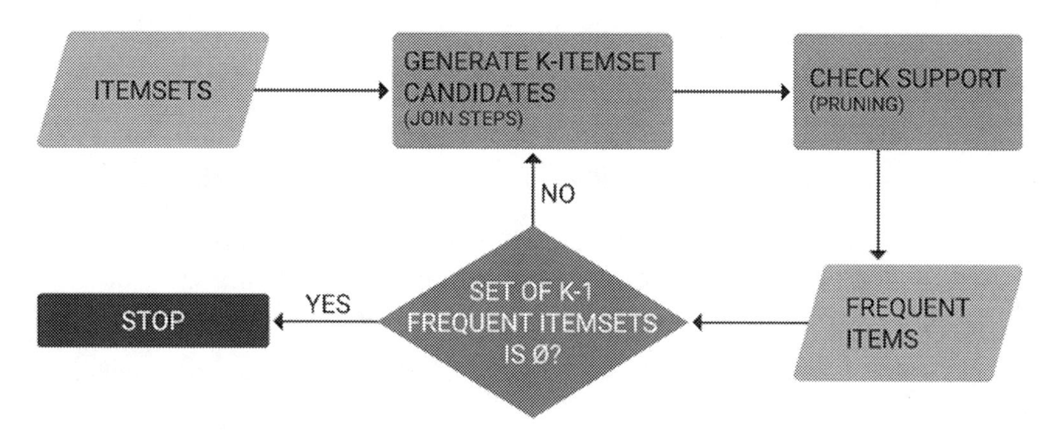

Figure 4-17. *Apriori*

The maximum size of an itemset is k, the number of items.

Dimensionality Reduction Algorithms

Although the message is broadly the more data, the better, datasets can often contain many variables and as a result make capturing the signal of the data a more laborious task. Whether it's sparse values, missing values, identifying relevant features, resource efficiency, or more straightforward interpretation, dimensionality reduction algorithms are very useful to data scientists.

As the name suggests, dimensionality reduction algorithms or DRAs reduce the number of dimensions that exist within a dataset. The need for dimensionality reduction becomes apparent when you consider the wealth of data available to make narrow decisions:

- Mobile phones collect hundreds of data points including calls, texts, steps, calories burned, floors climbed, Internet usage, and so forth. What data is best for understanding phone usage?
- Brands on social media are collecting data on engagement and interactions such as comments, likes, followers, sentiment, and mood. What data is best for understanding attitudes toward health?
- Medical health records contain a wealth of information, but only some information is relevant in predicting disease risk or illness progression. What data is relevant to understanding future risk of disease or adverse event?

Dimensionality reduction refers to converting a dataset of many dimensions into fewer dimensions while concisely representing similar data.

The graph in Figure 4-18 shows the relationship between cm and inches for the dataset. This two-dimensional representation can be transformed succinctly into a one-dimensional representation, improving the visualization for understanding. N dimensions of data can be reduced to k dimensions where $k < N$.

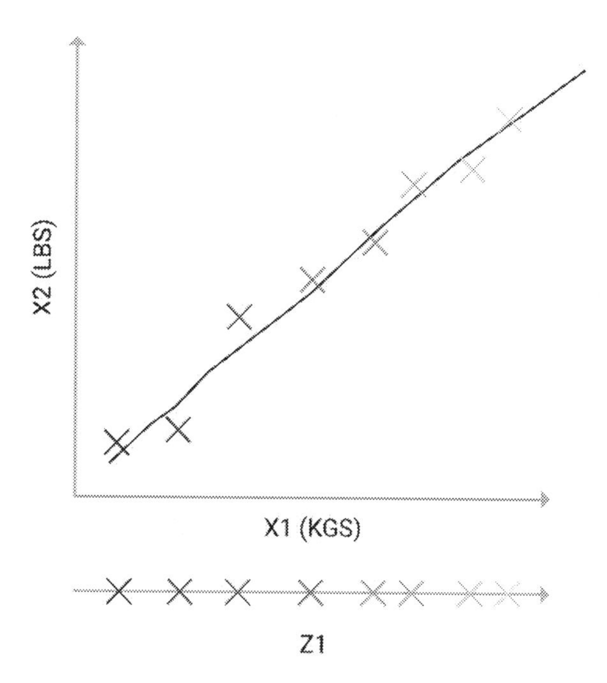

Figure 4-18. *Dimension reduction*

Dimensions of value can be identified or combined or new dimensions created to represent the inherent relationships.

Dimension reduction is extremely useful in machine learning tasks, with a plethora of benefits:

- Fewer dimensions result in quicker computations when compared to the original dataset.
- By default, dimension reduction algorithms reduce the space required for storage.
- Reducing data into fewer than three dimensions enables visualization and easier understanding.
- Redundant data is removed, which improves the performance of the machine learning model.
- Noise is removed, which improves model performance.

Dimension Reduction Techniques

Dimension reduction can be achieved in a variety of ways.

Missing/Null Values

Missing data and null values are not a huge problem in isolation, but a growing number of empty values may help determine whether to drop a variable, ignore missing values, or compute a predicted value.

Most data scientists support the dropping of variables if an attribute has upward of 50% empty or null values. This threshold does vary.

Low Variance

Data attributes that are very similar to one another do not carry much information. As a result, there is little variance within the data. Hence, a lower variance threshold can be set and dimensions removed based upon this.

As variance is dependent on range, data normalization should take place first.

High Correlation

Attributes with similar data trends are likely to carry similar information. Multicollinearity is the carrying of similar information that can reduce the performance of the model. Correlation between continuous data is determined by the Pearson product-moment coefficient, which is a measure of the linear correlation between two variables X and Y.

For discrete data, the Pearson's chi-square value determines how likely it is that any observed difference between sets arose by chance.

Random Forest Decision Trees

Random forest decision tree ensembles are useful for identifying key features. Random forest decision trees use a subset of features that can identify best split attributes.

If an attribute is often selected as the best split node, it is likely to be a feature to keep. Decision trees are particularly useful for visualizing reductions also.

Backward Feature Elimination

In Backward Feature Elimination, the model trains on n attributes. Starting with n attributes, at each iteration, the model is trained on n-1 attributes n number of times. The attribute that demonstrates the smallest increase in error rate is removed, and the algorithm performs another iteration on n-1 attributes.

At each iteration k, a model with n-k attributes is trained. This is computationally expensive.

Forward Feature Construction

Forward Feature Construction is the opposite of Backward Feature Elimination. The model starts with one attribute and evaluates which of the attributes has the highest increase in performance. Much like it's opposite Backward Feature Elimination, this method is computationally expensive unless operating in a lower range of dimensions.

Principal Component Analysis (PCA)

Principal component analysis decreases the number of variables (axes) through transforming an original set of variables into a new set, or the principal components (see Figure 4-19). This process is iteratively calculating the maximum variance within features in the data.

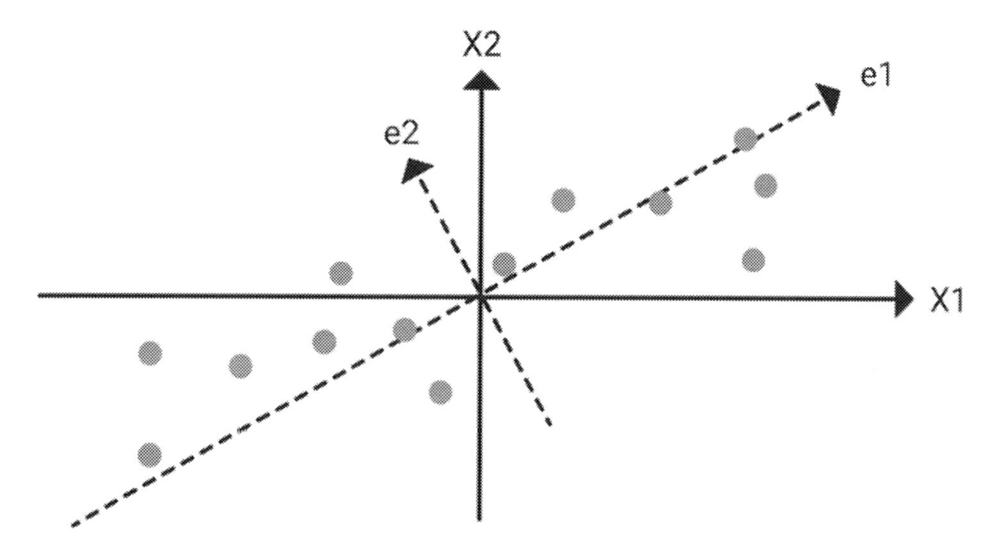

Figure 4-19. *PCA*

Each component is orthogonal and a linear combination of original variables. Orthogonality indicates that the correlation between the components is zero.

The first principal component is the linear combination of the original dimension that has the maximum variance; and the nth principal component is the linear combination with the highest variance, subject to being orthogonal to the n - 1 principal components.

Natural Language Processing (NLP)

Natural language processing is the field of AI that focuses on language. Natural language processing or NLP is defined as the ability of systems to analyze, understand, and generate human language, including speech and text.

NLP is an aspect of computational linguistics (studying linguistics using computer science) and is useful in the following:

- The retrieval of structured and unstructured data within a dataset, for example, searching clinical notes by keyword or phrase
- Social media monitoring
- Question answering: interpretation of natural language from humans to interact appropriately, for instance, as with virtual assistants or speech recognition software
- Analysis of a document to determine key findings
- Ability to parse and interpret a text to understand sentiment and mood
- Recognizing distinctions among diagnoses and relationships
- Image to text recognition, for instance, reading a sign or menu
- Machine translation—NLP is used in machine translation programs in which one human language is automatically translated into another human language
- Topic modeling what is this document talking about?
- Understanding sentiment from social media or discussion posts

Interpreting natural language is fraught with challenges, as human language is naturally ambiguous—language, pronunciation, expression, and perception.

Although there are rules with human language, they are often misunderstood and misused. NLP takes into consideration the structure of language to derive meaning. Words make phrases; phrases make sentences; sentences make documents; and all of the aforementioned convey ideas.

NLP has a toolkit of text processing procedures including a range of data mining methods that can be used for model development. Due to the nature of unstructured data, NLP tasks can be expensive regarding computational resource and time. Neural networks and deep learning can also be used for NLP tasks.

With most data generated existing in the form of unstructured data, NLP is a powerful tool to interpret and understand natural language.

As with any aspect of computing, there are several terms to understand before proceeding:

- Tokenization: The process of converting a corpus of text into smaller units, or tokens. There are many algorithms available for breaking a text into tokens.
- Tokens: Words or entities present in the text.
- Text object: A sentence or a phrase or a word or an article.
- Stemming: A basic rule-based process of stripping suffixes ("ing," "ly," "es," "s," etc.) from words.
- Stem: The text created after stemming.
- Lemmatization: Determining the root of a word from dictionaries and morphological analysis.
- Morpheme: Unit of meaning in a language.
- Syntax: Arranging symbols (words) to make a sentence. It involves determining the structural role of words in the sentence and phrases.
- Semantics: Meaning of words and how to join words into meaningful phrases and sentences.
- Pragmatics: Using and understanding sentences in different situations and how interpretations are affected.

In this section, I aim to introduce key concepts and methods of NLP and to demonstrate the techniques that can be applied to datasets to generate defined value.

Getting Started with NLP

Giving an NLP model an input sentence and receiving a useful output requires several key components (see Figure 4-20).

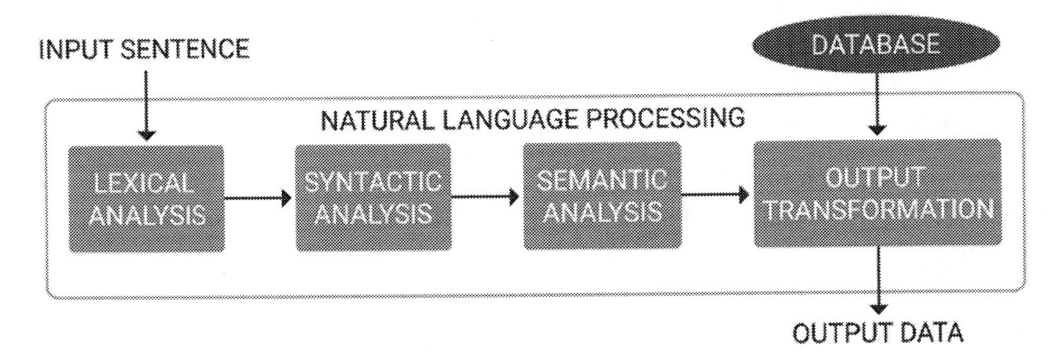

Figure 4-20. *How NLP works*

Preprocessing: Lexical Analysis

As with any dataset, a corpus of text that is not relevant to the context of data can be understood as noise. The first stage in NLP is to clean and standardize the input text, ensuring it is noise-free and ready for analysis.

Over and above spelling correction and grammar correction, the following techniques are used to reduce noise.

Noise Removal

Noise removal involves preparing a dictionary of noisy tokens (i.e., words) and parsing the text, removing the tokens found in the noise dictionary. For instance, words like the, a, of, this, that, and so forth would be removed.

Lexicon Normalization

Follow, following, followed, and follower are all variations of the word follow. Contextually, the words are similar. Lexicon normalization reduces dimensionality through stemming, which strips suffixes and prefixes, and lemmatization, which is a defined procedure that uses word structure and grammar relationships.

Porter Stemmer

Python, NLTK; method, PorterStemmer

The Porter stemmer algorithm is a popular and useful method to improve the effectiveness of information retrieval [81]. The algorithm works on the principle that many words in English share a common root. Stemming works on suffixes and removes common morphological and inflexional endings from words. Therefore, stemming allows one to reduce similar words into a common root form.

For example, take the text "I felt troubled by the fact that my best friend was in trouble. Not only that, but the issues I had dealt with yesterday were still troubling me."

The words troubled, trouble, and troubling all share the common root trouble. Therefore, according to the Porter stemmer algorithm, instead of counting all three words once, the stem trouble is counted thrice instead. The benefit of stemming is that common words can be clustered under a common stem to provide a more accurate statistical representation of the number of occurrences of a certain word. However, a drawback of stemming is that the semantic meaning of the word may be lost.

Stemming and lemmatization are used to reduce inflectional forms and sometimes derivationally related forms of a word to a common base form.

Object Standardization

A corpus of text may contain words that cannot be found in lexicon dictionaries. For example, on Twitter, someone may mention DM'ing someone, or another may like an RT of someone else's tweet. Acronyms, hashtags, slang, and colloquialisms can be removed through prepared dictionaries or using regular expressions.

Syntactic Analysis

To be analyzed, the text needs to be converted into features. The syntactic analysis of text involves analyzing sentences to understand relationships between words and assigning a syntactic structure to it. There are several algorithms for syntactic analysis; however, the Context-Free Grammar is most popular due to it being the simplest style of grammar and therefore widely used.

Take the sentence "David saw a patient with uncontrolled type 2 diabetes."

Within the sentence, we need to identify the subject, objects, noise, and attributes to understand the sequence of words and its dependencies.

Dependency Parsing

Toolkit, NLTK; method, StanfordDependencyParser

Sentences are composed of words in a structure. Basic grammar can determine the relationship or dependencies between a structure. Dependency parsing represents this in a tree structure and represents grammar and arrangement of words (see Figure 4-21).

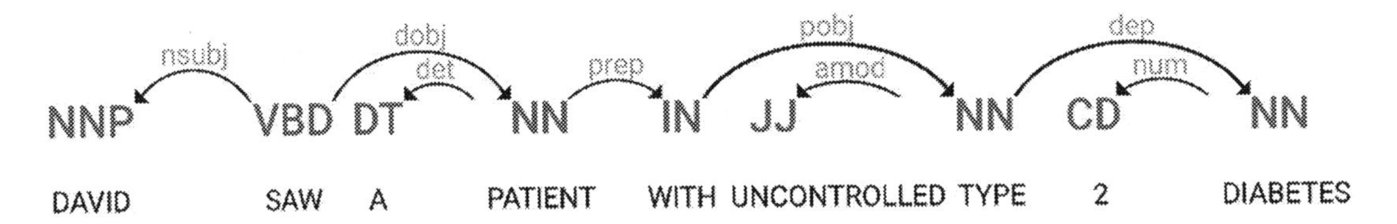

Figure 4-21. *Dependency parsing*

Dependency grammar analyzes asymmetrical binary relationships between tokens. The Stanford parser from NLTK is commonly used for this purpose.

Using the example in Figure 4-21, the parse tree determines the root of the word as "saw" and is then linked by subtrees. The subtrees are split by subject and object, with each subtree also showing dependencies.

Part of Speech Tagging

Toolkit, NLTK; method, word_tokenize

Part of speech (POS) tagging involves associating each word or token in a sentence with a part of speech tag. These tags are the basic English labels that you learned in primary school and determine nouns, verbs, adjectives, adverbs, numbers, and so on.

This is particularly useful for tasks such as building parse trees—which can, in turn, be used for determining what something is, sentiment analysis, determining appropriate answers to questions, or understanding similar entities (see Figure 4-22).

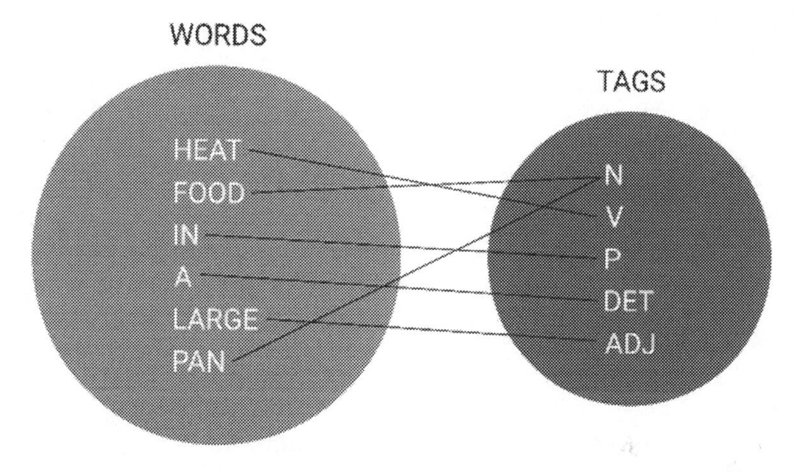

***Figure 4-22.** Part of speech tagging*

Part of speech tagging is the basis of the assists in several areas.

Reducing Ambiguity

Some sentences have multiple meanings given the structure; for example, take the following two sentences:

"I managed to read my book on the train." "Can you please book my train tickets?"

Part of speech tagging identifies "book" as a noun in the first sentence and as a verb in the second sentence.

Identifying Features

By identifying the types of speech alongside different contexts of a word, POS can distinguish between uses and creates stronger features for use.

Normalization

POS tags are the foundation of normalization and lemmatization, to understand sentence structure and dependency.

Stopword Removal

POS is useful in removing commonly used words, or stopwords, from a text.

Semantic Analysis

Semantic analysis is the most complex phase of NLP. It draws the exact meaning or the dictionary meaning from the text. Using knowledge about the structure of words and sentences among the context, the meaning of words, phrases, sentences, and texts is stipulated, and subsequently, also their purpose and consequences.

Techniques Used Within NLP

Once data preprocessing, lexical, syntactical, and semantic analysis of corpus has taken place, we are required to transform text into mathematical representations for evaluation, comparison, and retrieval. For instance, searching a collection of patient profiles for users with "hypertension" should only bring out those with hypertension. This is achieved through transforming documents into the vector space model with scoring and term weighting essential for query ranking and search retrieval.

Documents can be in the form of patient records, web pages, digitalized books, and so forth. The following algorithms are typical in the comparison and evaluation process.

N-Grams

Python: NLTK; Method: ngrams

N-Grams are used in many NLP problems. If X = number of words in a given sentence K, the number of n-grams for sentence K would be

$$N\ grams_K = X - (N - 1)$$

For example, if N = 2 (bigrams), the sentence "David reversed his metabolic syndrome" would result in n-grams of the following:

- David reversed
- Reversed his
- His metabolic
- Metabolic syndrome

N-Grams preserve the sequences of N items from the text input. N-Grams can have a different N value: unigrams for when N = 1, bigrams when N = 2, and trigrams when N = 3.

N-Grams are used extensively for spelling correction, word decomposition, and summating texts.

TF–IDF Vectors

A document can be represented as a high-dimensional vector in the space of words. Each entry in the vector corresponds to a different term within the documents and the number of its occurrences. The TF–IDF (Term Frequency–Inverse Document Frequency) vector scheme assigns each term within the document a weight using the following formula:

$$W_{td} = f_{td}.\log\left(\frac{D}{ND_t}\right)$$

Each term's weight (W_{td}) is calculated by multiplying the frequency of the term (f_{td}) by the log of the total number of documents (D) divided by the number of documents the term occurs at least once in (ND_t). The ordering of the terms may not necessarily be maintained.

Weights for the terms gathered can then be used to determine documents with high frequencies of the specific term within them. A collection of TF–IDF vectors could be used to represent a user's interest.

Latent Semantic Analysis

Where there is a corpus of a large number of documents, for each document d, the dimension of the vector representing each document can typically exceed several thousand. Latent semantic analysis relies on the fact that intuitively, terms in documents may often be related. For example, if document d contains the term sea, it will often contain the word beach.

Equivalently, if the vector representing d has a non-zero component in the entry for sea, it will also have a non-zero component in the beach entry. If this kind of structure can be detected, relationships between words can be automatically learned from the data.

The word document is represented by a matrix A, which is decomposed using singular value decomposition to give the strength of the most significant correlations and their directions.

The decomposition of A allows one to discover the semantics of the document through correlations between terms and their significance within the document. The latent semantic analysis method can be applied to determine the context of a variety of materials to present a web user with results that are contextually relevant.

Cosine Similarity

Python, SciPy; method, cosine

The cosine similarity is a metric that measures the similarity between two vectors. Therefore, this could either be used to define the similarity between two documents or a document and a query.

The cosine similarity between two vectors can be defined as the following:

$$Sim(u,v) = \frac{u \cdot v}{\|u\| \cdot \|v\|}$$

The similarity between two vectors is calculated as the inner product of the vectors divided by the product of the lengths of the vectors in question. Intuitively, the greater the angle between two documents, the less similar they are. This holds for vectors in any N-dimensional space.

Furthermore, the TF-IDF vector scheme can be integrated with the cosine similarity metric to determine the similarity between a search query and a plethora of documents:

$$Sim(q,d) = \frac{\sum_{t \in q \cap d} W_{td} \cdot W_{tq}}{\|d\| \ \|q\|}$$

The similarity between a query q and document d is calculated as the product of the TF–IDF weights for the term in both the document and query summed over all the terms in both the query and document; this is divided by the product of the length of the document and length of the query. Practically, however, calculating Sim(q,d) would prove computationally expensive as the number of documents used grows.

Naïve Bayesian Classifier

The Bayesian classifier is based on the Bayesian theorem and is particularly suited when the dimensionality of the inputs is high. Despite its simplicity, naïve Bayes can often outperform more sophisticated classification methods.

Given a specified threshold, this method can be used to classify the probability of whether a vector representing a document is of interest to a user. Given an attribute d, we can calculate whether the example belongs in class C with the following formula:

$$P(C=c|D=d)=\arg\max_{c}\frac{P(D=d|C=c)P(C=c)}{P(D=d)}$$

Other techniques such as kNN and ANN can also be used to classify and retrieve information.

Genetic Algorithms

Genetic algorithms (GA) are a fascinating topic within machine learning. GA take inspiration from evolution to minimize the error rate, attempting to mimic the function of chromosomes much like neural networks attempt to mimic the human brain.

Evolution is considered the optimal learning algorithm. In machine learning, the application of this is in models whereby several candidate answers (referred to as chromosomes or genotypes) are produced and the cost function applied to all.

In GA, a fitness function is defined that determines if the chromosomes fit enough to mate. Chromosomes furthest away from the optimal outcome are removed. Chromosomes are also subject to mutation. GA are a type of search and optimization learner and apply to discrete and continuous problems.

Chromosomes that are close to the optimal solution may be combined. The combination or mating of chromosomes is known as a crossover. The survival of the fittest approach identifies chromosomes that aim to express characteristics that adhere to natural selection—where the offspring is more optimal than the parent.

Mutation helps to overcome overfitting. It is a random process to get over local optima and find the global optimum. Mutation helps to ensure that child chromosomes are different from the parents' and continue evolution.

The degree in which chromosomes mutate and mate is a parameter that can be controlled or left to the model to learn.

GA have varied applications:

- Detection of blood vessels in ophthalmology imaging
- Detecting the structure of RNA
- Financial modeling
- Routing vehicles

A group of chromosomes is referred to as a population. Although it stays at a defined, constant size, it usually evolves to better average predictions over the course of generations or time.

The evaluation of a chromosome, c, is calculated as its evaluation function value divided by the average of the generation, represented as the following:

fitness(c) = g(c)/(Average of g over the entire population)

John Holland invented the GA approach (see Figure 4-23) in the early 1970s [81]. The automation of chromosome selection is known as genetic programming.

Best Practices and Considerations

Mastering the art of machine learning takes time and experience. However, there are several areas worthy of consideration, particularly for beginners to machine learning, to ensure the best use of time and optimal model efficiency.

Good Data Management

With the volume and complexity of data used in machine learning tasks, proper data management is paramount. Not only does this involve having the processes, policies, procedures, permissions, and certifications required for data management, it consists of the management and transformation of training data to learn an optimal model.

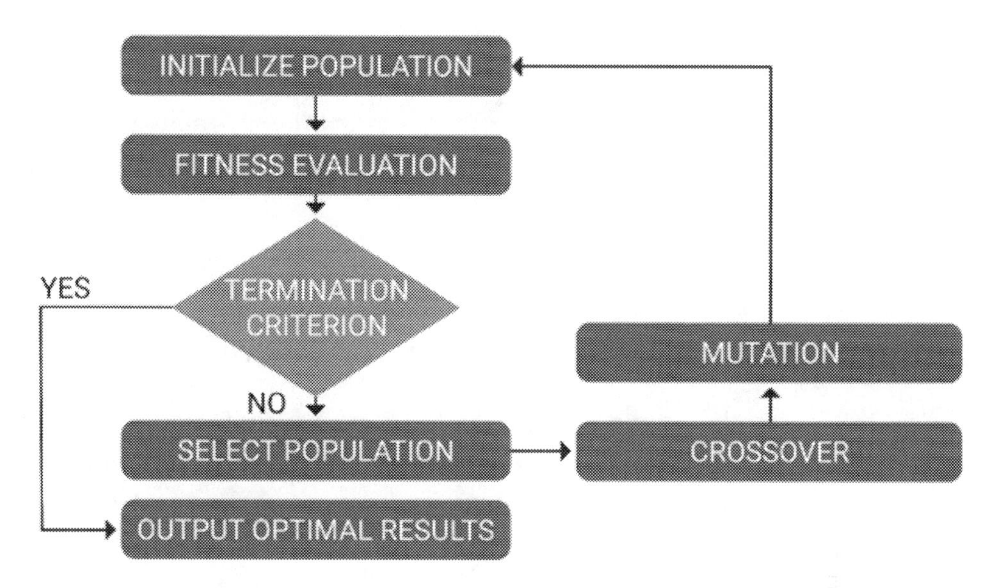

Figure 4-23. *Basic structure of a genetic algorithm*

Establish a Performance Baseline

It is crucial to have a baseline to measure the performance of your algorithm against other iterations and model types. As there is no one perfect algorithm for every problem, attempt many algorithms to identify the relative performance of your models.

Spend Time Cleaning Your Data

The time required to train a machine learning model varies significantly between algorithms. The accuracy of data and time spent on training the model positively influence model accuracy. Take time to clean your data to ensure predictions are as robust as possible.

It is often the case that not all of the input variables impact the dependent variable or outcome. Ensure the variables used by the model do not include those that are irrelevant.

Training Time

If the dimensionality of data is large and computing power limited, training time will be extensive. When time is short, it is useful to consider machine learning algorithms that do not require substantial training. For example, neural networks may not be appropriate for time-limited tasks.

Choosing an Appropriate Model

Some machine learning algorithms make particular assumptions about the structure of data or the desired outputs. For this reason, it is helpful to consider the model choice and whether it is an appropriate approach. Utilizing the right model has a plethora of benefits, including more accurate predictions, faster training times, and more useful results.

Some machine learning models are resistant to outliers or nonparametric tests. For example, decision tree–based approaches typically classify a node into two based on a threshold. Data outliers are therefore less impactful in tree-based approaches.

Evaluating a range of models is a useful approach to machine learning. Occam's razor is typically applied to choosing the model of choice. Occam's razor seeks to identify the most straightforward model that achieves the desired output given all else being equal.

Choosing Appropriate Variables

Although more data is ordinarily invited in machine learning problems, it is typically preferable to work with fewer predictor variables for many reasons.

Redundancy

Increasing the number of variables within a training dataset increases the chances of models learning hidden relationships between them. It is vital to identify unnecessary variables and only use nonredundant predictor variables within models. A model that is learning redundant connections will impact model accuracy.

Overfitting

Even if there are no relationships among predictor variables within a model, it is still beneficial to use fewer variables. Complex models, or those that use a high number of predictor variables, typically suffer from overfitting.

Models perform well on training datasets but are less accurate on validation and real-world settings, as they learn the error within the data (i.e., the noise) rather than the signal, or relationships, between variables.

Productivity

Even if all of the variables within a sophisticated machine learning model are relevant, there is a practical impact in using a large number of predictor variables that can affect productivity.

Practical considerations include the amount of data available, the subsequent effect on storage, computing resources, associated costs, time allocated for the project, and the time required for learning and validation. Predictor variables can be identified through feature selection, or transformed through feature extraction. SVMs are useful in cases of increased data dimensionality.

The Pareto principle (or the 80/20 rule or law of a vital few) is a useful measure to use in machine learning projects. Using the Pareto principle, focusing on the 20% most significant predictor variables should facilitate the building of relatively successful models within a reasonable time. Microsoft learned that 80% of the errors and crashes in Windows and Office were caused by 20% of the entire set of bugs detected [82].

PARETO PRINCIPLE

The "Pareto principle" is the observation that approximately 80% of the effects come from 20% of the causes. Many natural phenomena have been shown empirically to exhibit such a distribution.

Understandability

Models with fewer predictor variables are easier to visualize, understand, and explain. A key aspect of a successful machine learning project is that all stakeholders can comprehend the model. This often requires data scientists trading off.

By reducing the number of predictor variables, there is a chance there may be a reduction in the success of a machine learning model. However, this concurrently makes the model easier to interpret and understand.

The usefulness of this approach is only realized toward the end of a project when it comes to sharing not only the performance of a model but also how it works. This is particularly relevant in healthcare, as there are concerns about using black box models.

Accuracy

Any machine learning model aims to generalize well. Depending on the use case, an approximation may be more beneficial than a precise, accurate output. Through approximating values, models tend to avoid overfitting and reduce the amount of time spent on processing.

Impact of False Negatives

When evaluating the impact of a model before deployment into a live environment, consider the impact of false negatives. For example, take a predictive model that classifies the risk of breast cancer. A false positive would mean that a patient is informed they have breast cancer when they do not, which should be identified later in the treatment pathway. However, a false negative would mean a patient with breast cancer would not be notified—which is potentially far worse and costly.

Linearity

Many machine learning algorithms assume that relationships are linear—in other words, that classes can be separated by a straight line of best fit or its higher-dimensional representation. There are instances where relationships are not linear. Relying on a linear classification algorithm where there is a nonlinear class boundary will result in low accuracy.

Linear regression, for example, assumes both input and output variables do not contain noise. It is vital to expose the signal within data to ensure models learn the signal or the correct relationships within the data. It is advantageous to identify and remove outliers for the output variable in particular to reduce the chances of learning the noise.

Data with a nonlinear trend may require transforming to make the relationship linear, for instance, log-transforming data where there is an exponential relationship.

Linear algorithms are typically the first methods attempted in machine learning scenarios. As a result of linearity, algorithms are simple and quick to train.

Parameters

Each machine learning model is subject to parameters and hyperparameters. The time and resource required to train a model increases as the number of parameters does. An algorithm's parameter is a variable that is utilized by the model where the value can be estimated from the dataset.

An algorithm's hyperparameter is a variable that is external to the model and used to estimate parameters. A hyperparameter's value cannot be estimated from data. Instead, it is set by the data scientist, randomly or otherwise, or through the use of heuristics.

Ensembles

In specific machine learning problems, it may prove more useful to group classifiers together using the techniques of voting, weighting, and combination to identify the most accurate classifier possible. Ensemble learners are very useful in this aspect.

Use Case: Toward Smart Care in Diabetes

Type 2 diabetes is one of the most significant health and economic burdens facing the global population, with one in six people dying from diabetes or its related complications every 6 seconds [83, 84].

There are many factors involved in the progression of the disease, which, if tackled, can help to prevent its progression through targeted treatment profiling, leading to lower patient morbidity and mortality rate. Data in the form of health biomarkers—typically blood glucose, HbA1c, fasting blood glucose, insulin sensitivity, and ketones—are used to understand and monitor disease burden and response to treatments.

As a result of the tremendous burden of type 2 diabetes, there has been a substantial investment in developing intelligent models within this area, with much of the focus on diagnosing, predicting, and managing the condition through machine learning and data mining.

Many machine learning techniques have been applied to type 2 diabetes datasets. Traditional techniques, ensemble techniques, and unsupervised learning—in particular, association rule mining—have been used to identify prediction models with optimal accuracy.

Predicting Blood Glucose

Georga, E. I. et al. used random forest decision trees on many features collected from 15 people with type 1 diabetes to predict short-term subcutaneous glucose concentrations [85]. Glucose concentration was predicted through the use of SVMs.

The study concluded that the inclusion of two biomarker features—8-hydroxy-2-deoxyguanosine, an oxidative stress marker, and interleukin-6—improved classification accuracy.

Predicting Risk

Farideh Bagherzadeh-Khiabani et al. used the clinical dataset from 803 female prediabetics consisting of 55 features collected over a decade to develop multiple models for predicting type 2 diabetes risk [86]. The study developed a logistic model, demonstrating wrapper models, or models that consider a subset of features and improve the performance of clinical prediction models. This was developed into an R program to visualize output.

Summers, Panesar, et al. developed a model consisting of 34 features collected for people with prediabetes and obesity to predict risk of developing type 2 diabetes based on the health and engagement outcomes of over 200,000 people who used the Low Carb Program digital health intervention. This was developed in Python and deployed on the cloud to assess patient risk and suitability for the award-winning digital therapy [16, 58].

Razavian et al. conducted another extensive study, determining that type 2 diabetes prediction models could be accurately derived from population-scale datasets [87].

Over 42,000 variables were collected from over 4 million patients between 2005 and 2009. Machine learning was used to determine relevant features, which tailed out at around 900. Novel risk factors for type 2 diabetes were identified, such as chronic liver disease, high alanine aminotransferase, esophageal reflux, and history of acute bronchitis.

Duygu Çalişir et al. developed an automatic diagnosis system for type 2 diabetes with a classification accuracy of almost 90% through the use of linear discriminant analysis and SVM classifiers [86]. Latent Dirichlet Allocation (LDA) was used to distinguish features between healthy patients and patients with type 2 diabetes, which was then in turn used as the input for the SVM classifier. A tertiary stage evaluated sensitivity, classification, and confusion before determining the output.

Predicting Risk of Other Diseases

Machine learning has likewise been applied to diagnosing the risk of other diseases related to diabetes.

Lagani et al. identified the least set of clinical variables with the best predictive accuracy for complications such as cardiovascular disease (heart disease or stroke), hypoglycemia, ketoacidosis, proteinuria, neuropathy, and retinopathy [88].

Huang et al. used decision tree-based prediction models on identifying diabetic nephropathy in patients [89]. Leung et al. compared several methods: partial least square regression, regression trees, random forest decision trees, naïve Bayes, neural networks, and SVMs on genetic and clinical features [90]. Patient age, years of diagnosis, blood pressure, genetic polymorphisms of uteroglobin, and lipid metabolism arose as the most efficient predictors of type 2 diabetes.

Summers et al. developed a predictive model utilizing a dataset consisting of over 120 features to identify comorbid depression and diabetes-related distress. Summers et al. explored several techniques including random forest decision trees, SVMs, naïve Bayes, and neural networks. Patient age and employment status were defined as the variables most likely to affect predictive accuracy [16, 86].

Unsupervised learning has mainly concentrated on association rule mining, identifying associations between risk factors in people with diabetes. Simon et al. proposed an extension to association rule mining labeled survival association rule mining, which included survival outcomes, and made adjustments for confounding variables and dosage effects [86].

Reversing Disease

The understanding of type 2 diabetes has been radically disrupted by the generation of real-world evidence over the last 15 years. Previously, type 2 diabetes was understood as a chronic, progressive disease.

The Low Carb Program digital health intervention has been instrumental in demonstrating that sustainable weight loss is achievable through scalable digitally delivered therapies, supporting prediabetes and type 2 diabetes remission. The platform's peer-reviewed evidence shows 54% reduce or eliminate diabetes medications, with retention of 71% at 1 year.

A key driver in the Low Carb Program's success has been the implementation of a predictive model to identify education modules, resources, recipes, and discussion threads that are most likely to engage patients. This results in the app itself showing a different experience to each patient based on key features including ethnicity, age, gender, location, language, weekly food budget, dietary preferences, health status, and comorbidities.

This uses the patient's data to provide personalized, tailored care to best empower decisions and provide around-the-clock support. The benefits of this approach are self-evident, with one in four patients achieving type 2 diabetes remission at 1 year. Outcomes from Low Carb Program have been used to influence public nutrition policy and guidelines. The app was ranked the #1 app for diabetes in the NHS [91, 92].

Machine learning is so often focused on predicting and diagnosing disease. The application of machine learning to reverse disease provides hope that we can support improved behaviors and redefine outdated paradigms.

CHAPTER 5

How to Perform Machine Learning

Machine learning can expand our life, capabilities and capacities.

—Harkrishan Panesar

Machine learning is a rapidly evolving and exciting discipline within computer science, blending a range of skills including exploration and discovery, engineering, and analysis. Machine learning uses large amounts of factual, historical data to help us better understand behaviors.

While it can seem intimidating at first, a step-by-step approach to machine learning and its prerequisites can be used to conduct a successful project without the need to be an expert statistician or programmer. Although a knowledge of mathematics and programming is certainly advantageous, many machine learning algorithms are available out of the box and use preexisting libraries, and so there is seldom a requirement to develop novel algorithms from scratch.

The approach to a machine learning project is as unique as the problem it is trying to solve. However, the workflows involved in the lifecycle of a machine learning or data science project are detailed for use as a methodology.

Conducting a Machine Learning Project

An appreciation of the workflows involved in conducting a machine learning project highlights the skills that will be required within a project team and helps to communicate these with others.

The objective of a machine learning project is to develop algorithms that learn from the signals within the data they are trained upon.

© Arjun Panesar 2021

A. Panesar, *Machine Learning and AI for Healthcare*, https://doi.org/10.1007/978-1-4842-6537-6_5

There are seven defined workflows in the methodology used (Figure 5-1) to develop machine learning models. Each of these stages can be performed iteratively, meaning that workflows can be repeated during any step in the process.

1. Framing: Specifying the problem at hand as a learning task (and identifying whether it is a machine learning problem before proceeding)
2. Data preparation: Involving data exploration, analysis, insight, and cleaning
3. Training model(s): Choosing learning methods, applying them to create a model, and attempting to optimize the models created
4. Evaluation: An objective assessment of the method and results
5. Dissemination: Reporting the results of evaluation to stakeholders and ensuring explainability
6. Deployment: Releasing the trained machine learning model, monitoring ongoing usage and accuracy

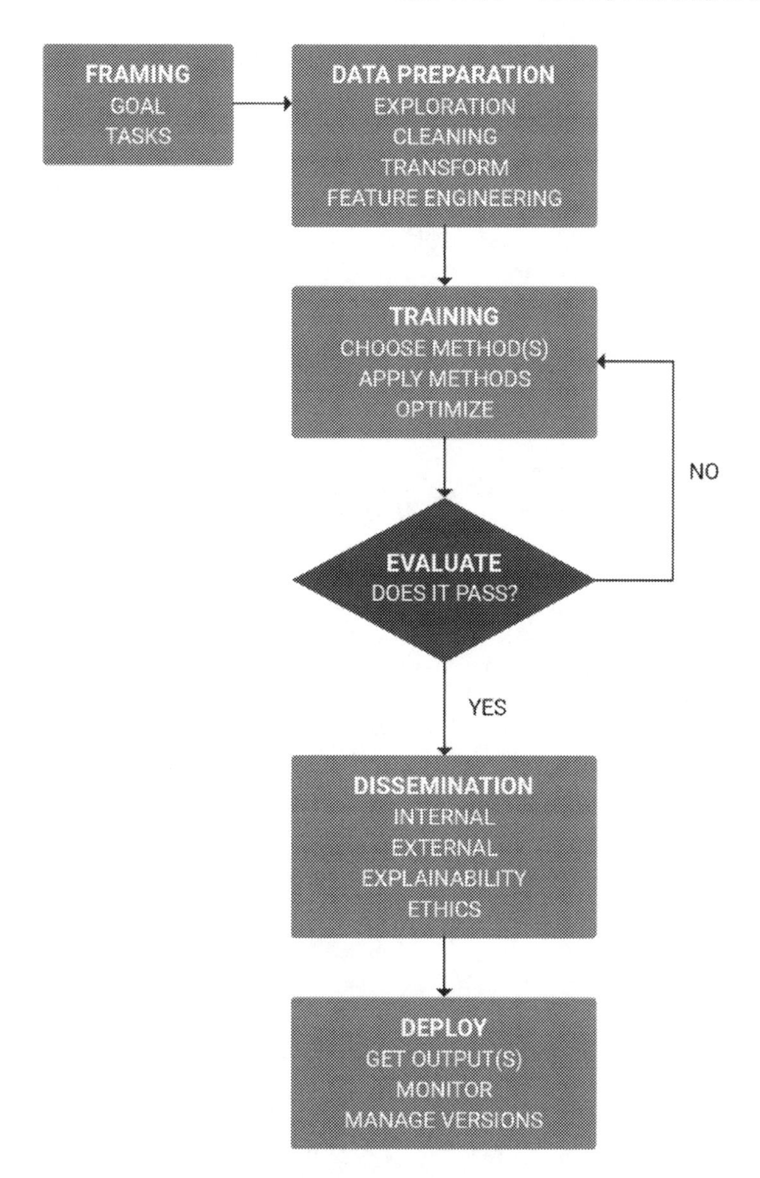

Figure 5-1. *Workflows involved in performing a machine learning project*

Framing: Specifying the Problem

The first thing when it comes to machine learning is specifying the problem. This involves understanding what the problem is, the goals of the machine learning project, and how attempts at achieving the goals can be evaluated. It is useful to understand why you want to solve the problem, especially if you are creating a budget case for a machine learning project. This includes evaluating the following:

- Why is it important?
- What do you hope to achieve? What are your best hopes?
- What are the inputs and outputs for the task at hand?
- Is this data available?
- How will the results be beneficial?
- Is this exploratory?
- What are the KPIs (key performance indicators); how is performance measured?
- What does success look like?
- Do we have the talent on hand to attempt the task?
- What are the limitations to the task at hand? This could include time, finance, skillsets, experience, domain knowledge, data availability, and so forth.

To solve the problem, data available to you will include examples and background information.

Collecting Examples

Categorization is a skill that can be learned through examples. It can be used for predictive tasks such as disease risk or resource demand prediction.

Certain problems present themselves in traditional healthcare settings, such as data from medical appointments and certain diagnoses having complementary handwritten notes. Bad examples can help to introduce error. For classification, examples are typically provided in the form of positives and negatives.

Background Information

This refers to the knowledge and axioms associated with a machine learning problem. This may include metadata, attributes, or relationships between concepts.

For example, in a hypertension prediction tool, the diagnosis of type 2 diabetes would be related to the likelihood of hypertension based on medical evidence and data. This is information that is utilized in the concepts learned to achieve categorizations.

Errors in the Data

Be mindful of the fact that errors from the real world can creep in, for example, from human error, incorrect classifications, missing data, incorrect background information, and repeated data.

Errors could creep in, for example, from the digitalization of paper documents and the machine reading of written text. It is important to remember the model is only as good as the data it is provided. It is worth spending time verifying the data to ensure that human errors are minimized.

Data Preparation

Machine learning algorithms learn from the data they are trained on. It is mission critical to provide the model with valid and robust data to learn.

Data must be prepared in a usable format. In a real-world scenario, this would involve understanding the data that would be used to model the problem and exporting the data. The data must then be processed to ensure correct formatting, removal of erroneous data, and the fixing of any missing data. The dataset size may be more than required, and so dataset sampling may also be required.

Every machine learning candidate will acquire skills in data preparation. Data preprocessing is essential to have tidy, valid data. Tidy, valid data is key to having robust, veracious outcomes. In order to develop a machine learning model that is highly domain applicable, technical experts who efficiently implement algorithms must work with domain experts who can understand data, classify it, and figure out trends and patterns.

Data preparation is by far the most vital aspect of machine learning.

How Much Data Do I Need?

Sadly, there is no definitive answer to this question—it depends. This is something that you explore through training your model and deploying it in the real world.

Several factors can affect the "learning curve." These include the complexity of the problem and the complexity of the algorithm. This, in turn, affects the amount of data required. Whether the model can make outputs that have veracity in the real world will depend on the sophistication of the algorithm and quality of data provided.

Assess the availability and usefulness of more data. In a healthcare setting, randomized controlled trials on medications rarely assess more than 1,000 patients in any given setting. In a surgery setting, access to bigger data is likely but may not be required for training a model.

Nonlinear, nonparametric algorithms require more data to become better as to their accuracy. These are typically the more powerful machine learning methods. However, due to the validity of the data affecting both bias and variance, it's key to train and validate the model robustly on as much data as efficiently possible.

Training the Machine Learning Model

The standard question any newcomer to machine learning has is "Which algorithm should I use?" The algorithm of choice depends on several factors including the size, quality, and nature of data; the task deadline; and available resources and motivation for using the data. The learning technique is also referred to as the "representation of the solution," as each machine learning approach represents data differently. Algorithms themselves are available off the shelf and do not need coding from scratch.

With that said, some algorithms lend themselves to particular problems over others, which can help aid the decision of which algorithms to try:

- Classification: Logistic regression, support vector machines, random forest, naive Bayes
- Regression: Linear regression
- Feature reduction: PCA, LDA
- Clustering: k-Means, LDA
- Collaborative filtering: Alternating least squares

It is almost impossible to predict which approach will best perform on the data. And even if the model does perform well during training and validation, that is no guarantee that it will perform well in the real world.

NO FREE LUNCH

The "No Free Lunch" theorem in machine learning states that there is no one algorithm that works best for every problem. This is particularly relevant for supervised learning, as there are many factors that affect the performance of an algorithm.

What Programming Language Should I Use?

Machine learning tasks are typically conducted in a variety of programming languages: predominantly R, Python, Matlab, and SQL. Java and C are also commonly used.

- R is typically used for statistical analysis. It allows you to understand and explore data using statistical methods and graphs and contains an extensive range of machine learning algorithms.
- Python is a language well suited to machine learning. Extensions such as numpy and SciPy are particularly useful for machine learning and data analysis.
- Matlab is the language many university students begin with. It is useful for fast prototyping, as it contains a large machine learning repository.
- SQL is a language used for managing data held in a traditional database management system (DBMS).

Should I Code My Machine Learning Algorithm from Scratch?

Several decisions need to be made from a resource and value perspective that affect the approach. A proof of concept project may be short of time, whereas a highly specific and complex task will contain a blend of approaches.

It is entirely possible to run a machine learning algorithm through implementing the ready-made algorithms provided by learning libraries such as scikit-learn, SciPy, pandas, Matplotlib, TensorFlow, Keras, and so forth.

This has been the traditional approach to machine learning: establishing a data science team that includes members experienced in languages such as Python and R. Members with experience in machine learning will typically be more impactful than those who are beginners to machine learning concepts.

Engineers develop an understanding of the appropriate machine learning techniques to employ given certain problems through experience. This experience can prove vital in avoiding wasted time, resources, and exploration.

Engineers with fundamentals in computer science will typically be familiar with many of the concepts present in machine learning. Choosing a programming language can influence the APIs and standard libraries you can use in your implementation. Enthusiasts may want to avoid using ready-made algorithms and code the implementations themselves to gain greater knowledge of the mathematics, statistics, and logic involved.

In many instances where time is constrained, time for engineering enthusiasm is limited. Cloud-based services such as Google AI, Microsoft Azure, IBM Watson, and a host of other providers allow the provider to take care of the algorithm; and many provide visual interfaces that enable the user to attempt particular machine learning techniques on the data uploaded.

Cloud-based vendors offering data mining and machine learning applications are growing and available at a low cost. One of the benefits of using such vendors is the ability to try more than one type of model for an experiment and compare the results, allowing you to find the best solution for your problem task.

For those with more traditional programming experience, almost all offer API or web service creation facilities, allowing you to create an interface to engage with the hosted model. All that is required to integrate with APIs is an efficient traditional programmer or web developer.

As the world adopts open source, consider browsing for implementations that already exist through GitHub, Reddit, and other machine learning forums.

API stands for application programming interface. APIs enable communication between software components and, in particular, can be used to interface between cloud-based machine learning providers and in-house development teams. There are many public APIs that can be used to assist in your machine learning tasks.

Embedding machine learning APIs should be approached with diligence and discretion. In an increasingly open source environment, as well as good documentation and functionality, ensure there are visible displays of popularity and robustness such as GitHub status, search engine popularity, and conversations.

Training and Test Data

A test set and training set are selected from the prepared data. The algorithm is trained on the training dataset and evaluated against the test dataset.

In many cases, it is a case of trialing several machine learning methods. It is useful to understand two terms when it comes to machine learning modeling:

- Signal: The true underlying pattern in a dataset
- Noise: Random or irrelevant patterns in a dataset

Some machine learning techniques may return one solution, whereas others may produce several. It is often a case of gathering their outputs (also referred to as hypotheses, or learned models) and evaluating their outputs.

The assessment of hypotheses is conducted through evaluating the predictive accuracy, comprehensibility, or utility. In most cases, Occam's razor is used, where the simplest solution is chosen if all else is equal. After the assessment is complete, a candidate hypothesis is chosen.

There are various ways to partition data into training and test datasets to train the machine learning model.

Predictive accuracy

Predictive accuracy refers to the accuracy at which the agent performs the task of classification.

Comprehensibility

Comprehensibility refers to how well we as humans can understand the output. In the real world, this could translate into many different settings.

Two Facebook bots recently began talking to each other in a language of their own which was incomprehensible to humans [93].

Utility

Utility refers to the problem-specific measure of worth. For example, in drug synthesis, a combination of compounds that are unsafe for human consumption may be the best combination. This would not have the greatest utility. Equally, this may override accuracy and comprehensibility.

Hold-Back Method

The hold-back method is a simple training method that can be employed that involves holding a proportion of the data to test on, leaving the remaining data for training, and validating the model.

Testing will typically give an indicator of the model's effectiveness or accuracy.

n-Fold Cross-Validation

The n-fold cross-validation involves splitting the dataset into equally sized groups of instances (or folds). The model is then trained on all but one fold and tested on the final, left-out fold. This process is repeated, and each fold is left out for one iteration.

The number of folds can vary based on the size of the dataset. Common folds include three, five, seven, and ten folds. The aim is to balance between the size and representation of data in the training and test datasets. It enables the model to be tested n times on different datasets. This enables us to make the most out of the data, training n times on it.

Some machine learning texts may refer to cross-validation as the cross-validation stage of the machine learning process. This is because machine learning engineers often use this approach to reduce the likelihood of overfitting.

Monte Carlo Cross-Validation

Monte Carlo cross-validation is similar to n-fold cross-validation. It involves randomly splitting the dataset into training and test data. The model is tuned to the training data, and predictive accuracy is evaluated using the test data. Results are then averaged over the splits.

Which Algorithms Should I Use?

The algorithm to use is not often identifiable from the outset. Try a number of techniques that are relevant to your problem. This allows you to compare and choose the solution judged on accuracy, comprehensibility, and utility. Even the most experienced data engineer would be unable to tell which algorithm will perform the best before trying the various algorithms.

All machine learning software support multiple algorithms. Find the best algorithm for the job by trying as many as feasible.

Evaluation and Optimization of the Method and Results

The performance of machine learning tasks varies on the representation of data given. For example, a patient record analyzed by an AI system does not examine the patient directly. Instead, data is fed into a system, with each piece of information relating to the representation of the patient known as a feature. It is not necessary to require complete feature sets as part of representations to have highly confident outputs.

The goal of a machine learning algorithm is to generalize well, neither underfitting nor overfitting. Generalization refers to the model performing with maximum accuracy on instances not seen during training.

Evaluation methodologies for machine learning projects are discussed in greater detail in Chapter 7.

Algorithm Accuracy Evaluation

Many supervised learning problems have binary classifications, where the problem is to learn a way of classifying unseen data into one of two categories. These are known as positive and negative categories:

- A **false positive** is received if the output from a binary classification agent that is given new data should have been categorized as negative but was classified as positive.
- A **false negative** is when an agent categorizes new data as negative but is incorrect.

In many medical settings, a false positive is not as bad as having a false negative. For example, false positives in this instance could include the diagnosis of illness for a patient based on a set of particular symptoms. It is evidently better to be told that you have a condition and not have it in reality, rather than being told you do not have a condition, only to be told that in fact you do. A false negative would mean that someone was not diagnosed, which is perhaps more concerning.

A simple way of measuring predictive accuracy of a particular hypothesis over a test dataset is to calculate the number of correctly identified outcomes (both positive and negative).

For example, if a heart disease tool learned a hypothesis and was given 250 new candidates to classify as having heart disease and correctly classified 133 of 150 positives and 98 of 100 negatives, it would have an accuracy of

$$(133 + 98)/250 = 92.4\%$$

This would be a 92 out of 100 chance of correctly categorizing an unseen example.

Generalization is measured by accuracy. Both under- and overfitting cause poor model performance and are detrimental to accuracy. If a model was classifying with 92% accuracy in testing but only 30% on unseen data, we would say that the model doesn't generalize well from training to unseen data.

This is overfitting. Overfitting is when a model fits the training data too well. It memorizes answers rather than learning concepts from them. In this case, the model cannot distinguish signal from noise and learns the noise present in the data. As such, the model does not understand the actual patterns, or signals, in the data. The model does not apply to new data which is detrimental to the model's ability to generalize, and the model is not useful. Decision trees are examples of algorithms that can overfit training data. This can be improved through pruning. In many cases, it's a case of keeping things surprisingly simple.

If a model performs better on training datasets than on unseen test sets, then the model is likely to be overfitting. Cross-validation is a technique to employ to measure against overfitting. Further still, training with more data may help algorithms infer hypotheses better.

The concept of "more data gives better results" is misguiding. In many cases, more data may help. However, more data does not imply good data. If data with noise were to be added, it wouldn't help reduce overfitting. Hence, this is why it's paramount to ensure data is clean and appropriate.

Data leakage occurs when data you are using to train your machine learning model has the information you are trying to predict in it. This leads to poor generalization and contributes to overfitting. Data leakage can occur from test data leaking into training data or additional features. A trivial example of data leakage would be a model that uses the response variable itself as a predictor, thus concluding, for example, that "a patient with conjunctivitis has conjunctivitis."

If the performance of your machine learning algorithm is too good to be true, data leakage may be the reason it is. An n-fold cross-validation can help to reduce data leakage.

Underfitting is simple to detect, as it will have poor performance. It refers to a model that cannot model the training data or generalize to new data. Typically, this would be represented by simple models that have too few features. Results would tend to show bias over variance (i.e., wrong outcomes).

Bias and Variance

Supervised machine learning aims to understand the signal from the dataset while ignoring the noise.

There is a machine learning dilemma of minimizing the two sources of prediction error that prevent supervised learning algorithms from generalizing beyond their training set. There is a trade-off between a model's ability to minimize bias and variance (also known as the bias–variance trade-off).

Bias

Bias refers to the error the model has from learning from erroneous data. Although this sounds strange, datasets may present randomly. Bias measures how far off, in general, these models' predictions are from the actual correct value, or true signal:

- Low bias: Better understanding of true signal from a dataset
- High bias: Worse understanding of true signal from a dataset

High bias can often indicate that a better machine learning technique is available.

Variance

This refers to error from sensitivity to fluctuations in the training set:

- **Low-variance** algorithms output similar models based on the dataset.
- **High-variance** algorithms output sufficiently different models based on the dataset.

Bias and variance can best be illustrated using a dartboard analogy (see Figure 5-2). Take a dartboard where the bull's-eye of the dartboard represents a model that perfectly predicts correct values.

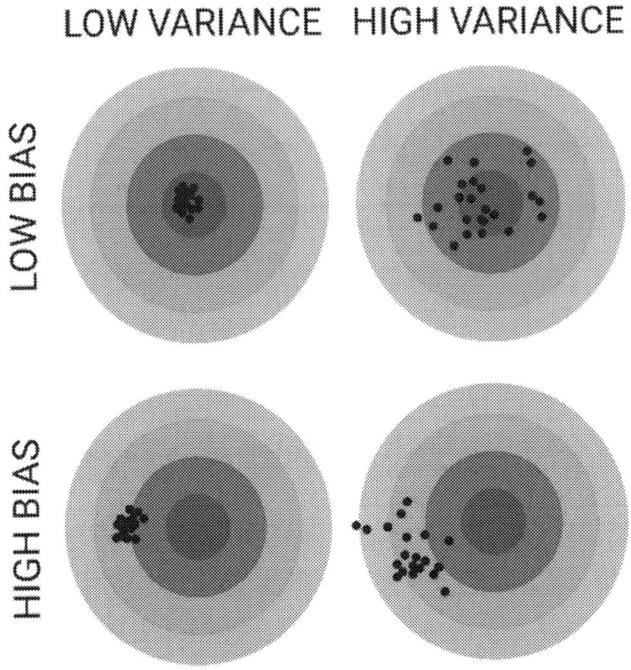

LOW VARIANCE HIGH VARIANCE

LOW BIAS

HIGH BIAS

Figure 5-2. *Bias and variance illustrated*

As you move further away from the bull's-eye, the predictions become worse. If each new model was represented by a new dart, each throw would represent a realization of the model given the variability in the training data. With a good distribution of data for training, models can predict well.

Data with anomalies and redundancies will result in poor predictions. This can be visualized on a dartboard.

The trade-off engineers must make with machine learning algorithms is that an algorithm with low bias must be flexible to enable it to fit the data well. However, if there is too much flexibility, each dataset will be interpreted differently and have high variance. There are several ways to reduce algorithm complexity, which are discussed in Chapter 4.

Many supervised learning toolkits provide methods to control the bias–variance trade-off either automatically or by providing a parameter that can be adjusted.

Low-variance algorithms tend to have reduced complexity when compared to low-bias algorithms. Low-variance algorithms have a simple or rigid structure that results in models that are consistent but inaccurate on average. This includes algorithms like regression, naïve Bayes, and so forth.

Low-bias algorithms tend to be more complex with a more flexible structure. As a result, they are accurate on average; however, they can be inconsistent. This includes nonparametric algorithms such as decision trees and k-nearest neighbor.

The key in your project is to find a balance between bias and variance that minimizes total error (see Figure 5-3).

Total error = Bias^2 + Variance + Irreducible error

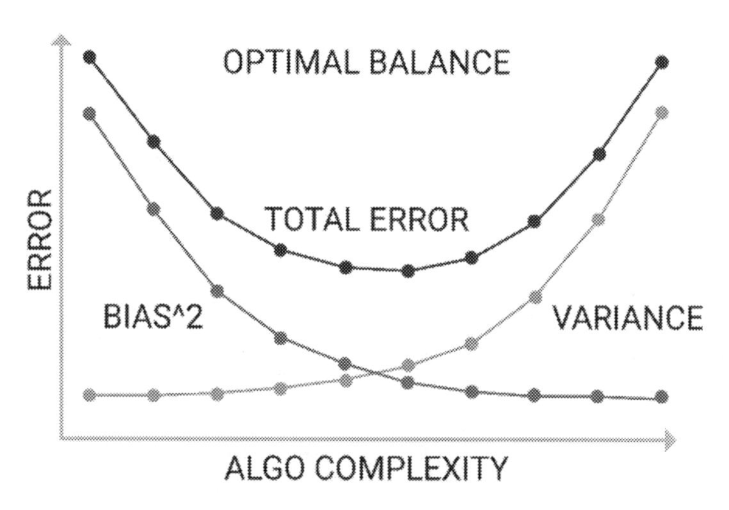

Figure 5-3. *Finding optimal algorithm balance*

The irreducible error refers to error regardless of the model or dataset that is used. This is the noise that comes from the training data. It is a constant, as real-world data always has some degree of noise.

Performance Measures

Examples of methods for evaluating algorithms include accuracy, prediction and recall, squared error, likelihood, posterior probability, cost, and entropy K–L divergence.

Optimization

Just because a model performs well doesn't mean it is the best and only model. The key is to get the maximum veracity from your results. Techniques such as cross-validation are useful to determine confidence in model results. However, we can optimize machine learning algorithms further.

Besides, we may want to improve the performance regarding time to output. Algorithms can be optimized through the following.

Algorithm Tuning

Tuning can be understood as the process of optimizing the parameters that impact the model to enable the algorithm to perform the "best," based on what "best" is specified as.

Training and Validation Data

Approaching a machine learning problem with an n-fold cross-validation technique allows you to validate your results, whereas a hold-back methodology has a waterfall approach to learning.

Evaluating a Range of Methods

Trialing a variety of machine learning methods is useful to determine which can achieve the most accurate results.

Improving Results with Better Data

There are some ways to improve performance with data, including the following.

Getting More Data

If you can get more data, this may help the model to improve its performance.

Getting Better-Quality Data

Rather than more data, it's often better, where possible, to get better-quality data. Typically, this would improve the signal within the data and reduce the noise. You can also enhance the quality of data through cleaning the data.

Resample Data

Partitioning the sample data into different sizes or distributions may represent the concept better or help improve performance by reducing attributes (features).

Data Representation

Data can be represented in different ways through the varying machine learning methods. Changing the type of machine learning method applied and reevaluating the problem may optimize your model.

Feature Selection

Addressing features that are and are not important can present new inferences for your model to learn. Increasing the number of features may not always be beneficial. However, in unsupervised learning instances, it may be beneficial to include all features to ensure nothing goes unexplored.

Feature Engineering

The key to success is the representation of data. Feature engineering involves exposing the data in other ways through decomposition and aggregation.

One of the most famous quotes behind the power of data is that of Google's research director and machine learning pioneer Peter Norvig, claiming "We don't have better algorithms. We just have more data."

Should I Choose a Supervised or Unsupervised Algorithm?

When developing models, it's rarely a case of supervised or unsupervised learning. In fact, they can often be used together. Unsupervised learning is a useful technique for dimensionality reduction and feature engineering, which can be used in supervised learning problems.

Ensembles

Ensembles are created when several classifiers specialize in various aspects of the problem. Ensembles turn a model into a feature. There are three types of ensemble algorithms known as bagging, boosting, and stacking, covered in Chapter 4.

Ensembles are for advanced practitioners of machine learning.

Problem Distribution

Another optimization method is distributing the processing used for a machine learning task. Tools from Microsoft, such as Azure, and from Google, such as AI.io, are examples of warehouses that can distribute problems at scale. In reality, most problems can be contained on a multicore implementation.

Implementation Problems

The user interface or UI is the key component in a machine learning application that has human interaction. This is not only where results will be displayed but where the most intelligence can be gathered. Rethinking the user interface and user experience is worth considering when engaging with people.

Disseminating the Results

In many settings, this is the most crucial aspect to ensure internal stakeholders support machine learning projects. Whether the machine learning project was successful or not is immaterial; if people do not understand the reasons behind events, support to adopt innovative technologies can be lost. Typically, a PowerPoint presentation and white paper will be good enough to present within an organization.

There are many ways to report the results, and to varying degrees of evaluation. An internal white paper, for instance, may have been reviewed internally by organization stakeholders, whereas the publishing of an article evaluating the innovation within a peer-reviewed journal may be more favorable. As healthcare rapidly embraces big data and machine learning, robust evidence to demonstrate the benefits and impact of AI is essential to ensure that only evidence-based, clinically proven frameworks and technologies are used and built upon.

Just like you would not want to take a counterfeit medication, you would not want to engage with technology that affects your health which is without evaluation or publicly available, scrutinizable data.

Frame the problem as a scientific experiment, assessing the following:

- Why: Define the context of the problem and the motivation behind solving the problem.
- Problem: Describe the issue as a question.

- Solution: Explain the answer to the questions proposed.
- Findings: Describe the discoveries made in the data and inferences made by the models.
- Conclusions: Evaluate what was learned from the project, addressing the limitations, ethical concerns, privacy, and data sensitives (particularly with data regulations such as CCPA and GDPR).
- Monitoring: If the plan is to deploy the model, how are predictions monitored and evaluated through usage?

Once you've reported your findings, it's up to your stakeholders to determine the value of the project before deployment, if even deployed at all.

Topics for discussion include how to handle the reporting of false negatives and false positives and the accuracy of predictive claims. The implications of predictions are a crucial consideration. Some concerns may be first-time problems. For instance, what are the legalities of using the data involved? When is a machine learning model truly predictive, or is it merely coincidence? What are the implications of being able to diagnose someone's cancer risk precisely?

Novel sources of information such as wearable devices, trackers, and EHRs place machine learning in the space of innovation and ethical wonderment. The implications for broader healthcare and insurance are massive and should be taken into consideration with machine learning research projects.

There has been progress in the field of machine learning predictive models, particularly in biomedical research where guidelines for reporting on models have been published, including steps for identifying prediction problems, ensuring n-fold cross-validation, and calibration of outcomes (see Figure 5-4).

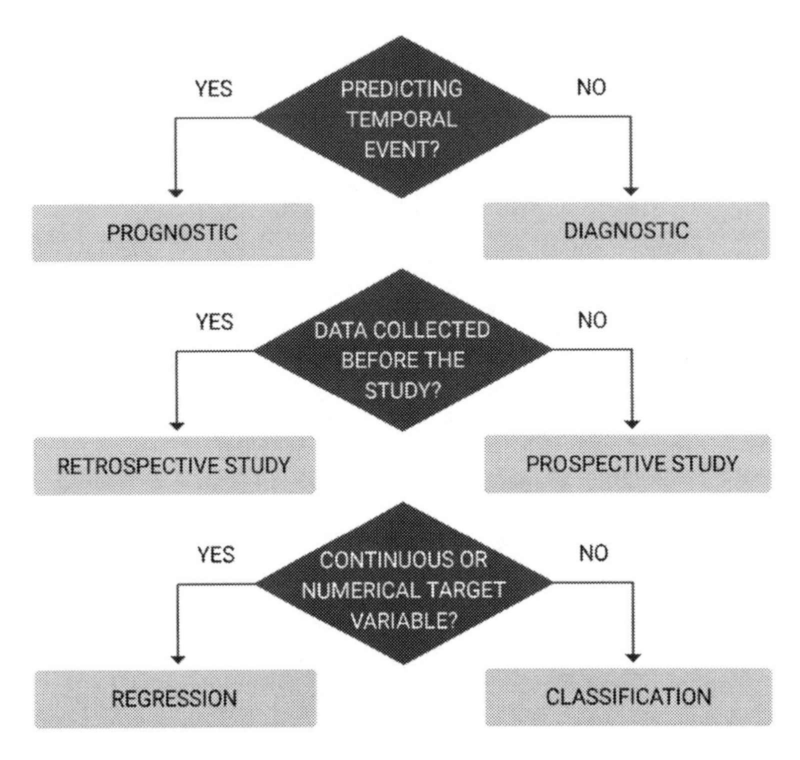

Figure 5-4. *Understanding the task at hand*

Calibration has relevance in clinical practice and refers to the relationship between the predictions made by the model and the observed outcomes.

As an example, if the prediction model "predicts 70% risk of mortality in the next one year for a patient with lung cancer, then the model is well calibrated if in our dataset, approximately 70% of patients with lung cancer die within the next one year."

Do I Need to Disseminate Project Results Before Deploying?

Ultimately, deploying a machine learning model before disseminating the results of the project to wider stakeholders is an organizational decision. There are a number of inherent risks with this approach, including lack of governance and understanding by others within a team or organization. Resources will be required to manage and monitor deployments, and these are best anticipated for rather than reacted against.

Several forms of independent evaluation should take place before deploying a public project. As well as prediction evaluation, cyber-security must be addressed, with penetration testing and cyber-incident response assessments. Machine learning models can cause unintended consequences—and these can shake a team who is neither aware of nor involved in the ramifications.

Deployment

Deploying a machine learning model is the process of moving the model from a testing or staging environment to a production environment. The model can then be accessed by others and provide output to other agents whether human or computer.

Machine learning models only provide value if they are made available, making deployment the final, crucial phase of the machine learning project lifecycle.

Deployment will also contain a mechanism of recording all observations and predictions and a process whereby this growing log is monitored and periodically assessed to ensure relevance and lack of harm. Most deployed projects will be hosted in cloud-based servers making auditing a relatively simple exercise. An added benefit is found in that cloud-based servers support scalability.

A process to monitor and manage the deployed model's usage and triage concerning data should be implemented.

Conclusion

Machine learning starts with asking the right question. More often than not, question framing this is the most difficult part of the process. And by far, the most resource-intensive part is data preparation.

Framing the right question provides a clearly defined objective function which enables stakeholders to identify the data that is required, approaches to solving the problem, potential impact, and resources required. The first question proposed may not be the right question, but iteratively revising the question enables progress toward a clearer, more defined goal. Without the right question being posted, data science teams can lose endless hours collecting data and training models that are no good for their intended purpose.

Be prepared for the possibility that the hypothesis or question you pose may not be answerable by the signal within your data—and explore the consequences of this possibility.

As organizations leverage machine learning, not fully understanding the true cause of perceived failures could diminish the perceived value of AI and lead to less support in the future.

Look for simplicity. Wherever possible, it is better to keep it simple than over-engineer and build complex and expensive machine learning models which are difficult to explain and disseminate.

CHAPTER 6

Preparing Data

Success depends upon previous preparation, and without such preparation there is sure to be failure.

—Confucius

Machine learning revolves around training of algorithms to perform a variety of computationally intensive tasks. Training of algorithms takes place through feeding models data, thus making the most crucial step in any data or machine learning project cleaning and preparation. It cannot be stressed enough: bad data creates bad results even when fed to the world's most sophisticated machine learning models.

All of this is for good reason. Whether it's directions to a restaurant, ordering a taxi, navigating an aircraft, or accepting a substitution in your online shopping basket, the idea of trusting data and algorithms more than our own judgment has its advantages and disadvantages. To err is indeed human, but when an algorithm makes a mistake, we are very unlikely to trust it again.

Data in the real world is inherently fragmented. Within healthcare, electronic health records (EHRs) lie in multiple distributed database systems. Hospital admission data, prescription data, and wearable and behavioral data all co-exist in silos. The act of bringing data together, or data linking, often creates new data tables which may have variables for which values are missing, incomplete, or transformed.

There are many perceptions of a machine learning engineer. The reality is that most good engineers spend the majority of their time cleansing and preparing data.

Creating an appropriate data preparation plan, imputing missing values, removing rows and unnecessary variables, and transforming features are some of the most common data preparation practices.

© Arjun Panesar 2021
A. Panesar, *Machine Learning and AI for Healthcare*, https://doi.org/10.1007/978-1-4842-6537-6_6

Phases of Data Preparation

During the data preparation phase of a project, the intention is to ensure the dataset is in the best state for machine learning.

Data preparation has five distinct stages, as denoted in Figure 6-1:

1. Gather
2. Explore
3. Cleanse
4. Transform
5. Expand (mainly through feature engineering)

Figure 6-1. *The five stages of data preparation*

The review and analysis of data requires an understanding of databases and Structured Query Language (SQL), which is a necessary skill for any machine learning engineer or data scientist.

What Is a Database?

A database is a structured collection of data, typically stored electronically in tables modeled in rows and columns.

As a persistent storage mechanism, a database enables one to store data and implement functionalities such as CRUD—Create, Read, Update, Delete. A database management system (DBMS) is the software used to control the data. Examples of some popular DBMSs include Microsoft Access, Microsoft SQL Server, MySQL, Oracle Database, and Ingres.

There are a number of database types, of which the most common are the following:

- **Relational databases** are the most popular and widely used database type. Relational databases store data in structured tables in rows (records) and columns (variables or features). Structured Query Language (SQL) is the language used to query relational databases.
- **Hierarchical databases** are tree-like in structure and store data in a parent–children relationship. Each child record has one parent only. Parents can have multiple children. Examples of hierarchical databases include the IBM Information Management System (IMS) and Windows Registry where retrieval of data requires traversing each tree until a record is found. Network databases are an extension of hierarchical databases where child nodes can have multiple parents and hence many-to-many relationships.
- **Object-oriented databases** store data as objects, providing database functionality to object-oriented programming languages.
- **NoSQL databases** are non-relational, meaning that they are without predefined schemas and hence allow semi- and unstructured data to be stored and manipulated. NoSQL databases fall into five categories: column, document, graph, key, and object.

While most people are familiar with relational databases, there has been a surge in the use of NoSQL databases driven by web and mobile application use, the need to store and process unstructured data, and the eternal requirement to improve speed.

Challenges of Databases

The best database type for a particular scenario depends on how an individual or organization wishes to use the data being stored. In most instances, databases are expected to return the results to complex queries in near real time. This has made improvements in performance the focus of database maintenance.

Accommodating Growing Data Volume

With data coming from wearables, sensors, connected devices, behaviors, and a variety of other sources, coping with high-velocity data sources and increasing file sizes is the primary headache of most database administrators—from beginners to the most experienced.

Scalability

It is difficult to predict just how much storage is required. With storage costing real dollars, a database which scales and flexes to the needs of the owner is preferable.

Accessibility

There are growing requirements of what is required of a database. In today's world, a variety of stakeholders require real-time access to data to support decision-making, seize new opportunities, and understand trends.

Database Management and Maintenance

Databases must be maintained to prevent problems and ensure industry best practice. As data complexity and volume scales, there is additional time resource required to monitor and optimize databases. This could include taking regular backups, checking databases for corruption, ensuring security is enforced, monitoring performance, and planning for anticipated demand.

Maintaining Data Security

Protecting data from intruders is made increasingly difficult as data volume and accessibility requirements grow. As innovation evolves, so do cybercriminals.

What Is the Difference Between a Database and a Spreadsheet?

Databases and spreadsheets both store data. However, the main differences lie in the way data is stored, how much data can be stored, and how it is accessed.

Spreadsheets were originally intended to be used by a single user who would not want to perform overly complex actions. Hence, spreadsheets are limited in functionality, and their features reflect this. For instance, it requires thought as to how best to measure concurrent access to a spreadsheet.

Databases, however, were designed to hold larger collections of data and be accessed securely by multiple users who query the data using a programming language.

What Is Structured Query Language (SQL)?

Whether at the data preparation or data analysis stage of a project, a good understanding of SQL will help with the bulk of data and machine learning tasks.

Structured Query Language (SQL) is the programming language used to communicate with a relational database and is preferred for data preparation.

Common database management systems that use SQL include MySQL, Microsoft SQL Server, Access, and PostgreSQL. MySQL is free to use and a popular choice among bedroom scientists and enterprises alike.

Although several versions of SQL exist, they are very similar in structure, meaning that if you understand how to use one version of SQL, you have almost learned them all.

This chapter will expose the most important commands and concepts.

Common SQL Commands and Concepts

Common SQL commands include select, insert, update, delete, create, and drop.

SQL holds a set of words, known as reserve words, that are used and interpreted as commands. As such, these words cannot be used to name tables or variables. A full list can be found in the supplementary materials.

Creating a Table

Relational databases hold data in objects called tables. Tables are composed of rows and columns which are used to store data and uniquely identified by name. Rows refer to records, and columns refer to the feature or attribute.

The ***create table*** command is used to create tables in SQL. Creating a table requires specifying the table's name, fields, and their respective data types. The syntax is as follows:

```
create table Example_Table {
column1_name datatype
column2_name datatype
column3_name datatype
...
};
```

where columnX_name refers to the column's name and datatype refers to the type of data held by the column. Common data types include the following:

char(n)	A fixed-length character string of *n* specified in parentheses. Max size of 255 bytes.
varchar(n)	A variable-length character string of maximum n specified in parentheses.
number(n)	A number value with a maximum number of digits specified in parentheses.
date	A date value.
number(n,i)	A number value with a maximum number of digits of "n" total, with a maximum number of "i" digits after the decimal point.

A full reference list of data types can be found in the accompanying materials.

Assume we are to create a small database system to manage accident and emergency admissions across a network of hospitals focused on three main entities: admissions, locations, and patients. A table is required for each of these entities. Each table will describe the features of these entities. Each feature will be a separate column within the table.

For the purposes of illustration, a table that could be used to record admissions into an accident and emergency ward of a hospital could be defined as follows:

```
create table Emergency_Admissions {
[ADMISSION_ID] INT IDENTITY(1,1) PRIMARY KEY
[LOCATION_ID] INT,
[PATIENT_ID] INT,
[ADMISSION_DATE] DATE,
[PRESENTING_WITH] TEXT,
[CONTACT_TELEPHONE] INT
[EMAIL] CHAR(50),
[PRIORITY] BIT
}
```

In the example, the admissions entity has seven attributes: admission ID, location ID, patient ID, admission date and time, presenting symptoms, contact telephone number, and whether the admission is a priority. Each admission has a unique ID and refers to a location ID and patient ID to identify the hospital and patient involved in the admission.

Inserting Data into a Table

The ***insert into*** statement is used to insert records into tables:

```
insert into Example_Table {column1_name, column2_name, column3_name, ...)
values (value1, value2, value3, ...)
};
```

where the column names refer to the features or attributes of the entity. As an example, the following command could be used to insert data into Table 6-1:

```
insert into Emergency_Admissions (LOCATION_ID, PATIENT_ID, ADMISSION_DATE,
PRESENTING_WITH, CONTACT_TELEPHONE, PRIORITY)
values ('6','997',Date(), 'Heart attack', '447709366550', '1')
```

Table 6-1. *Emergency hospital admissions*

ADMISSION _ID	LOCATION _ID	PATIENT _ID	ADMISSION _DATE	PRESENTING _WITH	CONTACT _TELEPHONE	PRIORITY
594	11	1453	08/08/2020	Concussion	447755366555	0
595	11	1225	08/07/2020	Head trauma	447733366552	1
596	6	997	08/13/2020	Heart attack	447709366550	1

Primary and Foreign Keys

At least one field in a table is a primary key. The primary key acts as the unique identifier for each record or row of data in the table. Each of the table's columns describes something to do with the primary key. In our admissions example, the primary key is admission ID.

A foreign key refers to a primary key used to link data between two or more tables. In our emergency admissions table, for example, both the location ID and patient ID are primary keys of separate tables or entities. Foreign keys support databases to minimize duplication and maximize efficiency. A focus of database management is preserving space, and thus storing the least amount of data required supports this approach.

Select, From, and Where Commands

The ***select*** command is used to query a database and retrieve data matching the criteria specified by the command. The table name proceeding the ***from*** keyword refers to the table to retrieve the data from. Optionally, a where clause can be used to specify the criteria based on which data should be returned:

```
select column1_name, column2_name, column3_name, …
from Table_Name
where condition;
```

To extract all of the admissions to location 11 in our worked example, we could use the following command, which would return only the first two rows of the admissions table:

```
select *
from Emergency_Admissions
where LOCATION_ID = 11;
```

The ***where*** command is used with SQL statements to identify only those records that fulfill specified conditions.

The ***distinct*** statement is often used in conjunction with the select statement to extract only the non-duplicate values from a table. The following statement would result in the output in Table 6-2:

```
select distinct location_id
from Emergency_Admissions;
```

Table 6-2. *Distinct emergency hospital admission location_IDs*

Location_ID
11
6

The select top command can be used to specify the number of rows to return. This command is particularly useful with big datasets where only a fraction of the results may be of interest:

```
select top number | percent column_name(s)
from Table_Name
where condition;
```

AND, OR, NOT Operators

Logical operators AND, OR, and NOT can be used with the where clause to filter retrieved data based on conditions:

- AND: Displays a record if all the conditions separated by AND are TRUE.
- OR: Displays a record if one of the conditions separated by OR is TRUE.
- NOT: Displays a record if the condition is NOT TRUE.

Group By

The ***group by*** command groups records that have the same values into aggregate rows. As such, it is often used with aggregate functions such as COUNT, MAX, MIN, AVG, and SUM to group the result-set by its columns. Using our worked example, the following statement would output the result shown in Table 6-3:

```
select count(Admission_ID), Priority
from Emergency_Admissions
group by Priority;
```

Table 6-3. *Emergency hospital admissions grouped by priority*

Count(Admission_ID)	PRIORITY
1	0
2	1
596	6

Update Command

The ***update*** statement is used to update data existing in a table:

```
update Table_Name
set column1_name = 'value'
where condition;
```

In this example, column1_name refers to the database table column(s) whose values will be updated. For instance, to update the priority of admission_id = 594 to 1 in our worked example, the following syntax could be used:

```
update Emergency_Admissions
set priority = 1
where admission_id = 594;
```

Delete and Truncate Commands

The ***delete*** statement removes one or more records from a table:

```
delete from table_name
where condition;
```

The ***truncate*** command deletes all of the rows from a table, leaving intact the columns and constraints. The truncate command removes all rows from a table, whereas the delete command allows filtering of rows.

Deleting a Table

The command ***drop table*** is used to delete a table and the data contained within it. The syntax is simple:

```
drop table Table_Name;
```

where Table_Name is the name of the table to delete.

Casting

Casting is useful when preparing data, for example, when required to convert strings to dates or integers to strings. The ***cast()*** function changes a value into a specified data type:

```
cast (valuetoconvert AS datatype(size))
```

Null Values

NULL values refer to fields with no value. Or another way to consider NULL values is as blanks. NULL values may arise if, for instance, a table has certain optional fields which are not required when a new row is created. NULL values differ from fields that contain spaces or zero values.

Joins

Data residing in more that one table requires combining before it can be queried and analyzed. There are five main ways that two tables can be joined which all affect the size of the resulting table.

Inner Join

The INNER JOIN command is the most common join command which is used to select the rows that have matching records in both tables. The command only selects rows if there is a match between columns. See Figure 6-2 for a representation of an INNER JOIN.

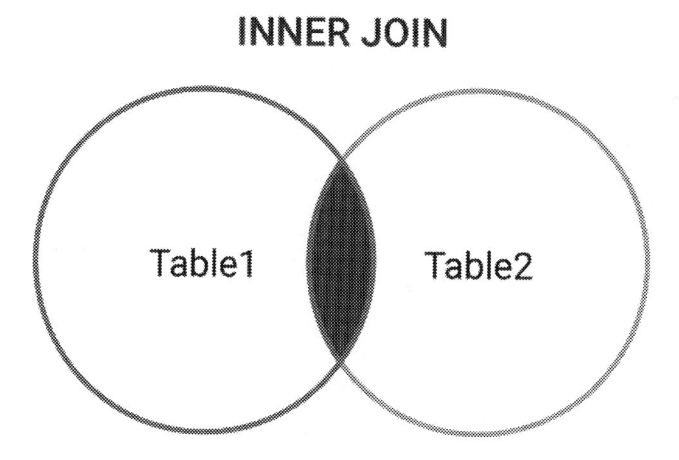

Figure 6-2. *Set representation of INNER JOIN*

An ***on*** statement is used to identify the columns to join the tables upon.

For example, Tables 6-1 and 6-4 could be joined using the following SQL statement to output a list of patients' names shown in Table 6-5:

```
select ADMISSION_ID, ADMISSION_DATE, FIRSTNAME, SURNAME,
from Emergency_Admissions
inner join Patient_Details
on Emergency_Admissions.PATIENT_ID = Patient_Details.PATIENT_ID;
```

Table 6-4. *Patient contact table*

PATIENT_ID	GP_ID	FIRSTNAME	SURNAME	DOB	CITY
1453	845	Susan	Riley	06/04/1984	Warwick
1225	36	Ahmed	Ali	14/03/1955	Wrexham
997	663	Anupam	Sanyal	04/04/1938	Crewe

Table 6-5. *Resulting inner join*

ADMISSION_ID	ADMISSION_DATE	PATIENT_ID	FIRSTNAME	SURNAME
594	08/08/2020	845	Susan	Riley
595	08/07/2020	36	Ahmed	Ali
596	08/13/2020	663	Anupam	Sanyal

Left and Right Join

The LEFT JOIN command will return all of the records within the left table with the matched records from the right table. If there is no match in the right table, NULL is output. See Figure 6-3 for a set representation.

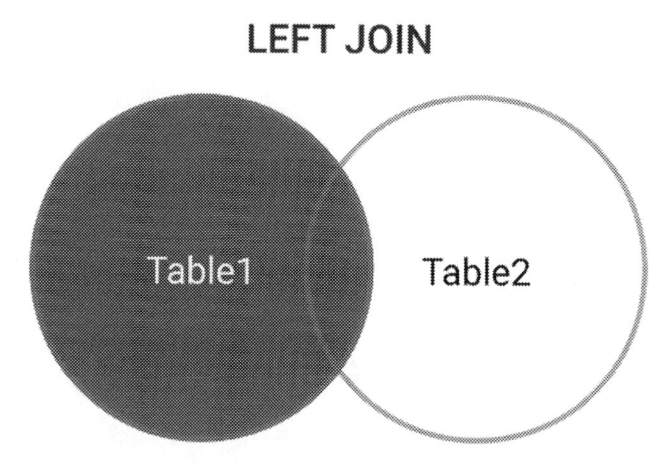

Figure 6-3. *Set representation of LEFT JOIN*

Similarly, the RIGHT JOIN operates in the opposite way, returning all of the records within the right table with the matched records from the left table if there is a match, otherwise outputting NULL. Figure 6-4 shows a set representation.

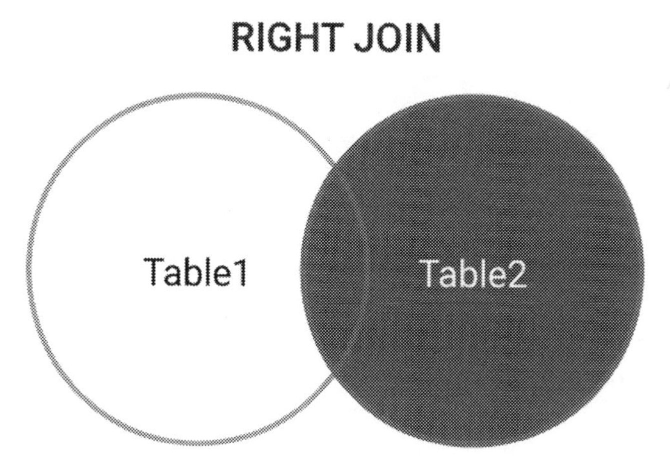

Figure 6-4. *Set representation of RIGHT JOIN*

Full Join

The FULL JOIN command, also known as the FULL OUTER JOIN command, returns all records when there is a match in left or right table rows.

Accordingly, this can output large datasets and be computationally expensive as all matching records from both tables are retrieved regardless of whether the other table matches or not. Figure 6-5 shows a set representation.

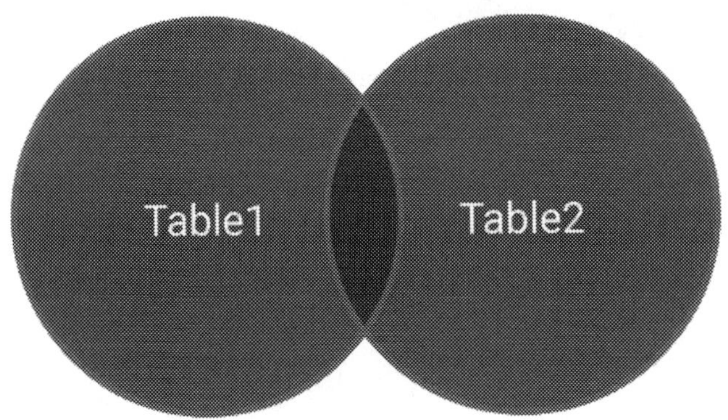

Figure 6-5. *Set representation of FULL OUTER JOIN*

SQL Date Functions

Every health record, regardless of whether EHR or wearable driven, has a date- and time-stamp attributed to it.

Dates are vital to evaluate health over time. For example, symptoms that occur after a particular treatment may be considered side effects and conditions that follow complications. SQL has several useful built-in date functions that can be used to manipulate dates:

- GetDate(): This function outputs the current date and takes no input.
- DateAdd(period, number, date): This function adds a given number of time, denoted by period (hours, days, months, years), to the date provided. The number provided can create a future date or negative to provide dates in the past.

- DatePart(part_of_date, date): This function returns a single part of a date-time (minute, hour, day, quarter, year, etc.).
- DateDiff(part_of_date, date1, date2): This function returns the difference between the two supplied dates. The part_of_date indicates the output's unit of time.

Data Gathering

The first phase of a machine learning project involves gathering the datasets to be used for the project. Since we are dealing with EHR data and for the purposes of this chapter, it is assumed that the datasets to be used within the machine learning project already exist.

Data Exploration

Once the datasets have been gathered, they need to be explored. The discovery phase of data preparation enables engineers to understand the data available and its constraints and nuances.

Activities include

- How much data is available?
- What format is the data provided in?
- What data types is data provided in?
- What format are dates provided in?
- Are there missing values?
- Is there obvious duplication?
- Are NULL values empty or NULL?
- Anomalies: Are outliers valid?
- Is the data balanced?
- Is there bias in the data?

- Are there typographical errors?
- Do multiple strings represent identical categories?

Once the available data has been reviewed, the project moves into the data cleansing phase.

Unbalanced Data

A point to note is that data may be unbalanced. Unbalanced features can prevent a machine learning model from learning how categories impact output. For instance, a skin cancer prediction algorithm trained mainly on white skin would be close to useless for people who were black, brown, or of any other skin color. To prevent this form of bias, you should calculate your accuracy metrics for different categories.

Data Cleansing

Data cleansing addresses incomplete, duplicate, missing, or incorrect values. The purpose of data cleansing is to improve the quality of the data available for training a model, providing analytics, or making a decision. This phase of a machine learning project is the most arduous and will consume a large amount of time. The data cleansing phase involves the following tasks.

Resolving Missing Values

Missing values are a common issue. Most machine learning models require a complete set of features, so this activity is well studied. Missing values can be assumed to be the common value seen in the dataset, inferred from other available data or rows containing missing values removed completely from a table.

Assume a table containing health condition data for a group of patients has a column for diagnosis of high blood pressure, heart attack, and stroke. If diagnosis of high blood pressure was NULL, this field could be assumed to be the most common diagnosis for this cell. Alternatively, this could be measured from other values nearby.

For instance, a diagnosis of heart attack and diagnosis stroke may help estimate diagnosis of high blood pressure. In addition, if more data was available, diagnosis of high blood pressure could be assumed from medication data. Removal of data is often a last measure to ensure important data is not lost or any form of bias introduced.

Contradictory and Duplicate Data

If the data resembles the real world, contradictory data should be explored to see whether events are justifiable or random. For instance, data showing a patient being admitted to a hospital before their date of birth highlights there may be a problem with the date format or representation. Duplicate data is always best removed.

Exploring Anomalies in the Data

Anomalies in the data are to be explored to understand whether they are erroneous or important. Removing anomalies, or outliers, from a dataset is known as data trimming. If the data resembles the real world, anomalies may well be justifiable.

Machine learning can only be as good as the data you use to train it. If the data is biased (i.e., not representative), the learning will be biased. At the same time, as you are exploring datasets, be careful to avoid introducing human error. This is typically satisfied through ensuring good version control.

Correcting Typos, Cleaning Values, and Formatting

Make sure there are no typographical errors in the dataset. Digitalized text, such as handwriting, may have errors. Check your datasets for errors, biases, and inconsistencies. This is a very common problem.

Normalize the casing for text to be either lowercase or uppercase. Go through the data to assess whether there are multiple representations for unique items and correct any mistyped strings or categorizations. For instance, high blood pressure may be represented in a data table as high blood pressure, hypertension, and arterial hypertension where all categories are correct, but require unifying to ensure data accuracy.

Review all column formats, paying particular attention to dates and integers to ensure that they are in the preferred format and encoding. This may involve casting columns to different data types.

Formatting may need to be changed; for example, you may need to export your data to something that is a flat file rather than a traditional relational database. You may only want to work with a subset of your sample data.

Classifying Groups of Results

Columns within a dataset can often have a large distribution of values, of which some will be of a low frequency. This is particularly true when free text is input. In some instances, it may help to classify these records into categories.

Transforming Data

The next phase of data preparation is transformation of data. It is a rare event to have collected data exactly how you require it. In order to make data more useful, data transformation techniques can be applied.

This is typically guided by the algorithm you are using and the data available. For instance, if a machine learning algorithm was taking the input of a diagnosis of type 2 diabetes as an integer—"[0 for false]" and "[1 for diagnosed with type 2 diabetes]"—but the medical notes read specified "type 2 diabetes" or "no type 2 diabetes," the data would need to transform from text to integers to be machine readable.

Scaling may also be used—for example, converting metric to imperial units.

Transformations can include the following.

Aggregation

Aggregation refers to attributes that can be aggregated into an individual attribute. For example, a log of particular patient attendance to health education could be aggregated into a count of the total number of attendances rather than representing each attendance separately.

Decomposition

Decomposition refers to partitioning attributes such that they may become more useful to a machine learning technique. For example, a patient's records may have their ethnicity listed as Asian. However, it may be better to know the country of birth and primary language of the patient to more accurately predict criteria such as obesity risk.

Encoding

Encoding changes categorical variables to a numerical, machine-readable representation. Ordinal categorizations will require a ranking which will be used as input. Non-ordinal categorizations can use a column to flag whether a category is true or false for that record.

Scaling

Data can contain attributes with varying quantities. For example, weight in kilograms, height in centimeters, and clinical markers may use different scales completely. Scaling transforms data into a particular common range. Scaling can be performed through normalization or standardization.

Skewed Data

Skewed data can be normalized with statistical techniques to approximate more symmetrical distributions. Methods include logarithms, roots, and reciprocals.

Bias Mitigation

If there appears to be bias in the data, this can be mitigated through a number of preprocessing techniques. Bias mitigation preprocessing algorithms include reweighting, learning fair representations, disparate impact remover, and optimized preprocessing.

Bias mitigation is different from unbalanced data. Unbalanced data does not have a similar number of examples for each class contained within a dataset.

Weightings

Weight refers to the priority or influence of a particular feature. Typically, positive weightings would increase the prediction value, whereas negative weights would reduce the prediction value.

Expand

Existing datasets can be expanded and enriched through feature engineering to generate new features based on the data about current features and the task objective.

An understanding of the data at hand is essential when performing feature engineering—whether that be supplied by the team or a subject matter expert. Feature engineering involves extracting features and identifying relationships.

Feature Extraction

Feature extraction creates new columns from raw data strings. For instance, this could be a new column based on phrases, values, or frequency.

For example, aggregation may involve categorizing patient ages in brackets of 10 years, rather than specifying a continuous number. Decomposition involves reducing dependencies within the data. This may even involve including attributes previously excluded and unseen to the model during data preparation.

Identifying Feature Relationships

Relationships between features can be explicitly highlighted to help the machine learning model know what is important. For instance, a model may not be able to easily connect the latitude and longitude of multiple addresses, but by specifying the city, it may better enable pattern identification.

Feature Reduction

If there is too much data, too many features can make it difficult for a machine learning model to identify signal from noise. Reducing the number of features will help avoid the curse of dimensionality which refers to the phenomena that arise in analyzing and organizing data in high-dimensional spaces that do not occur in low-dimensional settings.

Conclusion

Machine learning models require data to be formatted in particular ways in order to learn. Good data preparation develops clear and accurate data which leads to more useful, accurate machine learning models and predictions. This chapter has explored the phases involved in data preparation, the required skills, and notes for considerations.

As highlighted, data preparation is the foundation to a successful machine learning project and an important step that cannot be skipped.

Even good data must be reviewed to ensure it is in the right format, scale and preserving meaningful features. Good-quality data will provide good-quality information to empower every decision to be more intelligent.

CHAPTER 7

Evaluating Machine Learning Models

Intelligence is the ability to adapt to change.

—Stephen Hawking

As a field, machine learning is still in its infancy. Advanced machine learning has only been explored over the last 25 years, which has fueled data science as a profession. Thus, the data science industry is still in a phase of wonderment at the endless potential of AI and machine learning. With this come both excitement and confusion—and an industry that is gathering knowledge, experience, and first-time problems.

Typically, the most laborious tasks within a machine learning project are identifying the appropriate model and engineering features, which make a substantial difference to the output of the model. In fact, the features chosen can often have more impact on the quality of a model compared to the model choice itself.

Therefore, it is important to evaluate the learning algorithm that will determine the model's intelligence to predict the output of an unknown sample. This is usually done using various metrics, which are discussed further.

Model Development and Workflow

To successfully deploy and maintain a machine learning model, there are several stages of development and evaluation that take place, as illustrated in Figure 7-1.

© Arjun Panesar 2021
A. Panesar, *Machine Learning and AI for Healthcare*, https://doi.org/10.1007/978-1-4842-6537-6_7

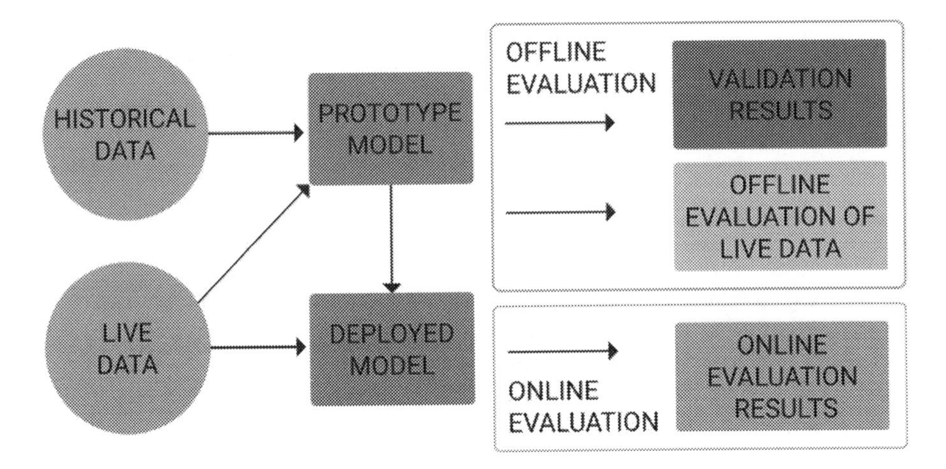

***Figure 7-1.** Model development and workflows*

The first stage is the prototype phase. During this phase, a prototype machine learning model(s) is created through testing various methods on historical data to determine the best fit model. Hyperparameter tuning, as discussed later in this chapter, is a requirement of model training. Once the best prototype model is chosen, the model is tested and validated. Validating a model requires splitting datasets into training, testing, and validation sets as discussed in Chapter 3.

Consider the fact that there is no such thing as a random dataset and instead the randomness applies to the splitting of the dataset. Be aware of biases that may appear in the data. Once the model has been successfully validated, it is deployed to production. The model is then usually evaluated by one or several performance metrics and monitored throughout its lifecycle.

There are two ways of evaluating a machine learning model: offline evaluation and online (or live) evaluation.

Why Are There Two Approaches to Evaluating a Model?

A deployed machine learning model consumes data from two sources: historical data (or the data that is used as the experience to be learned from) and live data. Many machine learning models assume stationary distribution data—that the data distribution is constant over time. However, this is atypical of real life, as distributions of data often change over time—known as a distribution shift.

Consider a system that predicts the side effects of medications to patients based on their health profile. Medication side effects may change based on population factors such as ethnicity, disease profile, territory, medication popularity, and new medications. The distribution of relevant side effects based on patient data can vary quickly over time, and hence it is essential for a model to detect a shift in distribution and accordingly evolve the model.

Models are assessed through their performance based on live data and also evaluated against the validation metric used in the testing and validation phases on historical data.

A model with performance that is similar to or within a threshold of permissibility when evaluated on live data is deemed as a model that continues to fit the data. Degradation of model performance indicates that the model does not fit the data and requires retraining. Organizations should also consider the mitigating actions required should a live machine learning model degrade and become unsuitable, or clinically unsafe, for use.

Offline Evaluation

Offline evaluation measures the model based on metrics learned and evaluated from the historical, stationary, distributed dataset. Metrics such as accuracy and precision-recall are typically used within the offline training stage. Offline evaluation techniques include the hold-back method and n-fold cross-validation.

Online (Live) Evaluation

Online evaluation refers to the evaluation of prediction metrics once the model is deployed. The key takeaway is that these metrics may differ from the metrics used to evaluate performance when the model is deployed live. For instance, a model that is learning on new pharmacological treatments may seek to be as precise as possible in training and validation; but when placed online, it may need to consider business goals such as budget or treatment value when deployed. The same model may find a perfect pharmacological treatment which may solve the problem, but remain clinically unsafe. Evaluation metrics must be framed with use cases to ensure representation of all actors within the intended project.

The benefits of online evaluation, particularly in the digital age, support multivariate testing to understand best-performing models in real time through dedicated feedback loops.

Feedback loops are key to ensuring systems are performing as intended and help to understand the model in the context of use better. This can be performed by a human agent or automated through a contextually intelligent agent or users of the model.

Live evaluation may take place regularly, using unlabeled subsets of generated, previously seen data to maintain accuracy within an agreed tolerance threshold.

Finding a Statistically Independent Dataset

It is important that the evaluation of a machine learning model is based on a statistically independent dataset and not on the dataset it is trained on. This is because the evaluation of the training dataset is optimistic about the model's true performance as it adapts to the dataset. By evaluating the model with previously unseen data, there is a better estimate of the generalization error.

New data can be hard to find; hence, it is important to be able to have new, unseen data from the current dataset. Methods such as n-fold cross-validation are useful techniques for this purpose. Often the data used is more important than the algorithm choice; and the better the features used, the greater the performance of the model. The evaluation metrics discussed can be found in the metrics package for R and scikit-learn for Python.

Machine learning models, if safe to do so, can be exposed to BETA participants or interested stakeholders to generate data.

Evaluation Metrics

There is a vast library of evaluation metrics available for machine learning problems. Metrics exist for the variety of machine learning tasks—classification, regression, clustering, association rule mining, NLP, data mining, and so on.

The evaluation metric you choose will be bespoke to the problem your machine learning model is trying to solve and the techniques you are using.

Classification

Classification problems seek to give a label or classification to an input. There are several methods by which to measure performance, including accuracy, precision–recall, confusion matrices, log-loss (logarithmic loss), and AUC (area under the curve).

Accuracy

Accuracy is the simplest technique used in identifying whether a model is making correct predictions. It is calculated as a percentage of correct prediction over the total predictions made.

> **Accuracy** = Number of correct predictions/Number of total predictions

Confusion Matrix

Accuracy is a general metric that does not consider the division between classes. Therefore, it does not consider misclassification or the associated penalty with misclassification.

For instance, a medical misdiagnosis that is a false positive (e.g., take a patient diagnosed with breast cancer when they do not have it) has substantially different consequences compared to a false negative, whereby a patient is told that they do not have breast cancer when in fact they do.

A confusion matrix breaks down the correct and incorrect classifications made by the model and attributes them to the appropriate label:

- True positive: Where the actual class is yes and the value of the predicted class is also yes.
- False positive: Actual class is no, and predicted class is yes.
- True negative: The value of the actual class is no, and the value of the predicted class is no.
- False negative: When the actual class value is yes, but predicted class is no.

Take an example whereby a model predicts whether a patient has breast cancer or not based on 50 example inputs from the test dataset with an equal distribution between positive and negative labeled examples. The confusion matrix would be as in Table 7-1.

Table 7-1. Confusion matrix

	Prediction: Positive	Prediction: Negative
Labeled positive	20	5
Labeled negative	15	10

From the confusion matrix, it is determined that the positive class has greater accuracy than the negative class. The accuracy of the positive classification is 20/25 = 80%. The negative class has an accuracy of 10/25 = 40%. Both metrics differ from the overall accuracy of the model, which would be determined as (20 + 10)/50 = 60%. It is apparent how a confusion matrix adds more detail to the overall accuracy of a machine learning model.

As a result, accuracy can be rewritten as the following:

$$\textbf{Accuracy} = (\text{Correctly predicted observation})/(\text{Total observation})$$
$$= (\text{TP} + \text{TN})/(\text{TP} + \text{TN} + \text{FP} + \text{FN})$$

Per-Class Accuracy

Per-class accuracy is an extension of accuracy that takes into account the accuracy of each class. As a result, the preceding example has a per-class accuracy of (80% + 40%)/2 = 60%.

Per-class accuracy is useful in distorted problems where there are a larger number of examples within one particular class compared to another. The class with greater examples dominates the calculation, and therefore accuracy alone may not suffice for the nature of your model; thus, it is useful to evaluate per-class accuracy also.

Logarithmic Loss

Logarithmic loss (or log-loss for short) is used for problems where a continuous probability is predicted rather than a class label. Log-loss provides a probabilistic measure of the confidence of the accuracy and considers the entropy between the distribution of true labels and predictions.

For a binary classification problem, the logarithmic loss would be calculated as follows:

$$\text{log-loss} = -\frac{1}{N}\sum_{i=1}^{N} y_i \log p_i + (1 - y_i)\log(1 - p_i)$$

where P_i is the probability of the *i*th data point belonging to a class and y_i the true label (either 0 or 1).

Area Under the Curve (AUC)

The AUC plots the rate of true positives to the rate of false positives. The AUC enables the visualization of the sensitivity and specificity of the classifier. It highlights how many correct positive classifications can be gained allowing for false positives.

The curve is known as the receiver operating characteristic curve, or ROC, as shown in Figure 7-2. A high AUC or greater space underneath the curve is good, and a smaller area under the curve (or less space under the curve) is undesirable.

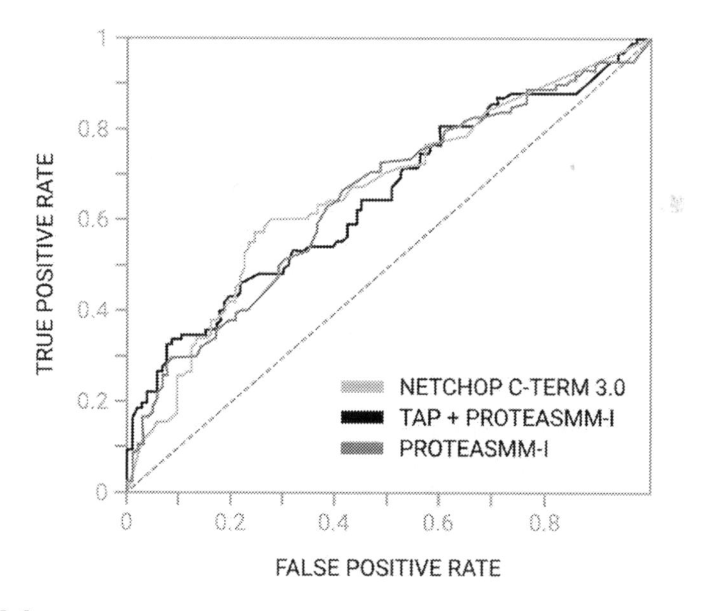

Figure 7-2. *ROC curve*

In Figure 7-2, test A has a better AUC as compared to test B, as the AUC for test A is larger than for test B. The ROC visualizes the trade-off between specificity and sensitivity of the model.

Precision, Recall, Specificity, and F-Measure

Precision and recall are two metrics used together to evaluate model performance. Precision evaluates how many items are truly relevant compared to the total number of items correctly classified. Recall evaluates how many items are predicted to be relevant by the model from the items that are relevant:

- Precision: (Correctly predicted Positive)/(Total predicted Positive) = TP/TP + FP
- Recall: (Correctly predicted Positive)/(Total correct Positive observation) = TP/TP + FN

Specificity refers to how well the model performs at returning incorrect classifications and is calculated as in Figure 7-3.

- Specificity: (Correctly predicted Negative)/(Total Negative observation) = TN/TN + FP

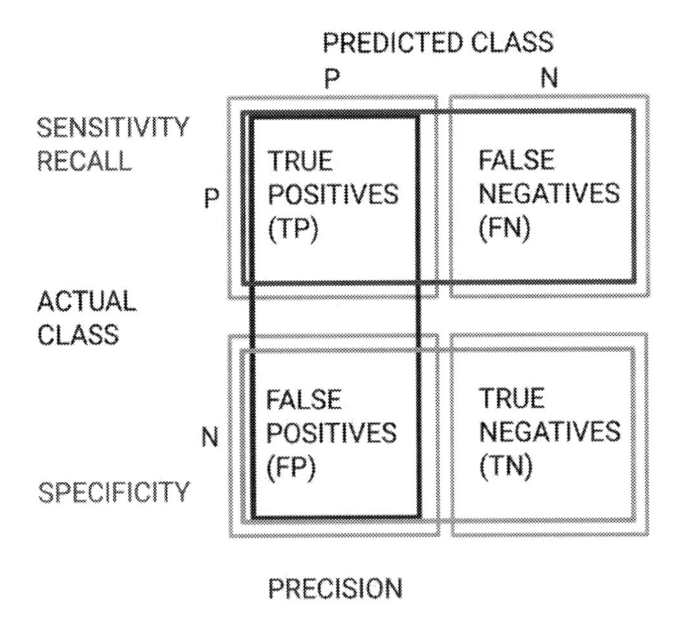

Figure 7-3. *Specificity classification diagram*

F-Measure goes beyond the arithmetic mean and calculates the harmonic mean of precision and recall:

$$F \frac{1}{\frac{1}{2}\left(\frac{1}{p}+\frac{1}{r}\right)} = \frac{2pr}{p+r}$$

where p denotes precision and r denotes recall.

Regression

Regression machine learning models output continuous variables, and root-mean-squared error (RMSE) is the most commonly used evaluation metric for these problems.

RMSE

RMSE calculates the square root of the sum of the average distance between predicted and actual values. This can also be understood as the average Euclidean distance between the true value and predicted value vectors. A criticism of RMSE is that it is sensitive to outliers.

$$RMSE = \sqrt{\frac{\sum_i (y_i - \hat{y}_i)^2}{2}},$$

where y_i denotes the actual value and $\hat{y}$ denotes predicted value.

Percentiles of Errors

Percentiles (or quantiles) of error are more robust as a result of being less sensitive to outliers. Real-world data is likely to contain outliers, and thus it is often useful to look at the median absolute percentage error (MAPE) rather than the mean.

$$MAPE = median\left(\left\|(y_i - \hat{y}_i)/(y_i)\right\|\right),$$

where y_i denotes the actual value and $\hat{y}$ denotes predicted value.

The MAPE is less affected by outliers by using the median of the dataset. A threshold or percentage difference for predictions can be set for a given problem to give an understanding of the precision of the regression estimate.

The threshold depends on the nature of the problem.

Skewed Datasets, Anomalies, and Rare Data

An experienced data scientist treats all data with suspicion. Data can be inconsistent; and as a result, skewed datasets, imbalanced class examples, and outliers can all significantly affect the performance of a model. Having more examples within one class compared to another can lead to an underperforming model. If there are problems with a model, go back to the data to ensure that all stages of the data preparation phase of a machine learning project have been conducted correctly.

Outliers or data anomalies can further skew performance evaluation metrics. The effect of large outliers can be mitigated using percentiles of error. However, if these outliers are representative of real-world, patient data, consideration should be made to retain them. In practice, good data cleansing, removal of outliers, and normalization of variables can reduce the sensitivity to outliers.

Parameters and Hyperparameters

Hyperparameters and parameters are often used interchangeably, yet there is a difference between the two. Machine learning models can be understood as mathematical models that represent the relationship between aspects of data.

Model Parameters

Model parameters are properties of the training dataset that are learned and adjusted during training by the machine learning model. Model parameters differ for each model, dataset properties, and the task at hand. For instance, in the case of an NLP predictor that outputs the sophistication of a corpus of text, parameters such as word frequency, sentence length, and noun or verb distribution per sentence would be considered model parameters.

Model Hyperparameters

Model hyperparameters are parameters to the model building process that are not learned during training. Hyperparameters can make a substantial difference to the performance of a machine learning model. Hyperparameters define the model architecture and affect the capacity of the model, influencing model flexibility. Hyperparameters can also be provided to loss optimization algorithms during the training process.

Optimal setting of hyperparameters can have a significant effect on predictions and help prevent a model from overfitting. Optimal hyperparameters often differ between datasets and models.

In the case of a neural network, for example, hyperparameters would include the number and size of hidden layers, weighting, learning rate, and so forth. Decision tree hyperparameters would include the desired depth and number of leaves in the tree. Hyperparameters with a support vector machine would include a misclassification penalty term.

Tuning Hyperparameters

Hyperparameter tuning or optimization is the task of selecting a set of optimal hyperparameters for a machine learning model. Optimized hyperparameter values maximize a model's predictive accuracy.

Hyperparameters are optimized through training a model, assessing the aggregate accuracy, and appropriately adjusting the hyperparameters. Through trialing a variety of hyperparameter values, the best hyperparameters for the problem are determined, which improves overall model accuracy.

Hyperparameter Tuning Algorithms

Hyperparameter tuning is like training a machine learning model. The task at hand is one of optimization. Model parameters can be expressed as a loss function, whereas hyperparameters cannot be expressed as such, as it depends entirely on the model training process. There are several approaches to hyperparameter tuning, with the most common being grid search and random search.

Grid Search

The grid search is a simple, effective, yet resource-expensive hyperparameter optimization technique that evaluates a grid of hyperparameters.

The method evaluates each hyperparameter and determines the winner. For example, if the hyperparameter were the number of leaves in a decision tree, which could be anywhere from n = 2 to 100, grid search would evaluate each value of n (i.e., points on the grid) to determine the most effective hyperparameter.

It is often a case of guessing where to start with hyperparameters, including minimum and maximum values. The approach is typical of trial and error, whereby if the optimal value lies toward either maximum or minimum, the grid would be expanded in the appropriate direction in an attempt to further optimize the model's hyperparameters.

Random Search

Random search is a variant of grid search that evaluates a random sample of grid points. Computationally, this is far less expensive than a standard grid search. Although at first glance it would appear that this is not as useful in finding optimal hyperparameters, Bergstra et al. demonstrated that in a surprising number of instances, a random search performed roughly as well as grid search [94].

The simplicity and better-than-expected performance of a random search mean that it is often chosen over grid search. Both grid search and random search are parallelizable.

More intelligent hyperparameter tuning algorithms are available that are computationally expensive as the result of evaluating which samples to try next. These algorithms often have hyperparameters of their own. Bayesian optimization, random forest smart tuning, and derivative-free optimization are three examples of such algorithms.

Statistical Hypothesis Testing (Multivariate Testing)

Statistical hypothesis or multivariate testing is an extremely useful method of determining which model is best for the particular problem at hand. Statistical hypothesis testing determines the difference between a null hypothesis and alternative hypothesis.

The null hypothesis is defined as the new model not affecting the average value of the performance metric, whereas the alternate hypothesis is that the new model does change the average value of the performance metric.

Multivariate testing compares similar models to understand which is performing best or compares a new model against an older, legacy model. The respective performance metrics are compared, and a decision is made on which model to proceed with.

The process of testing is as follows:

1. Split the population into randomized control and experimentation groups.

2. Record the behavior of the populations on the proposed hypotheses.

3. Compute the performance metrics and associated *p* values.

4. Decide on which model to proceed with.

The statistical hypothesis testing methodology enables teams to disseminate the project through scholarly communication and provides a framework for reporting.

Although the process seems relatively simple, there are a few key aspects for consideration.

Which Metric Should I Use for Evaluation?

Choosing the appropriate metric to evaluate your model depends on the use case. Consider the impact of false positives, false negatives, and the consequences of such predictions. The impact of false results can be significant.

Build the model to cater to the appropriate metrics. If a model is attempting to predict an event that only happens 0.001% of the time, an accuracy of 99.999% can be reported but not confirmed.

One approach is to repeat the experiment, thus performing repeat evaluations. Although not a fail-safe, this reduces the change of illusionary results. If there is indeed change between the null and alternate hypotheses, the difference will be visibly confirmed. Equally if a model seems to predict the output correctly, ensure results are statistically significant.

Correlation Does Not Equal Causation

The phrase correlation does not equal causation is used to stress that a correlation between two variables does not suggest that one causes the other. Correlation refers to the size and direction of a relationship between two or more variables.

Causation, also known as cause and effect, emphasizes that the occurrence of one event is related to the presence of another event. It may be tempting to assume that one variable causes the other; however, in models with several features, there may be hidden factors that cause both variables to move in tandem.

For example, smoking tobacco is a cause that increases the risk of developing a variety of cancers. However, it may be correlated with alcoholism, but it does not cause alcoholism.

What Amount of Change Counts as Real Change?

Defining the amount of change required before the null hypothesis is rejected once again depends on the use case. Specify a value at the beginning of the project that would be satisfactory and adhere to it. Ensure results are statistically significant.

Types of Tests, Statistical Power, and Effect Size

There are two main types of tests—one-tailed and two-tailed tests. One-tailed tests evaluate whether the new model is better than the original. However, this test does not specify whether the model is worse than the baseline.

One-tailed tests are therefore inherently biased. With two-tailed tests, the model is tested for the possibility of change in two directions—positive and negative.

Statistical power refers to the probability that the difference detected during the testing reflects a real-world difference.

Effect size determines the difference between two groups through evaluating the standardized mean difference between two sets. Effect size is calculated as the following:

> **Effect size** = ((Mean of experiment group) - (Mean of control group))/Standard deviation

Checking the Distribution of Your Metric

Many multivariate tests use the *t*-test to analyze the statistical difference between means. The *t* value evaluates the size of the difference relative to the variation in your sample data. However, the *t*-test makes assumptions that are not necessarily satisfied by all metrics. For instance, the *t*-test assumes both sets have a normal, or Gaussian, distribution.

If the distribution does not appear to be Gaussian, select a nonparametric test that does not make assumptions about a Gaussian distribution, such as the Wilcoxon-Mann-Whitney test.

Determining the Appropriate *p* Value

Statistically speaking, the *p* value is a calculation used in hypothesis testing that represents the strength of the evidence. The *p* value measures the statistical significance, or probability, that a difference would arise by chance given there was no real difference between two populations. Or in other words, how likely is the data, assuming a true null hypothesis? It provides the evidence against the null hypothesis and is a useful metric for stakeholders to draw conclusions from.

A *p* value lies between 0 and 1 and is interpreted as follows [95]:

- A *p* value of ≤ 0.05 indicates strong evidence against the null hypothesis, thus rejecting the null hypothesis.
- A *p* value of > 0.05 indicates weak evidence against the null hypothesis, hence maintaining the null hypothesis.
- A *p* value near 0.05 is considered marginal and could swing either way.

The smaller the *p* value, the smaller the probability that the results are down to chance.

Some healthcare researchers have suggested that a *p* value less than 0.01 should be applied to studies involving computer algorithms to evidence that there is substantial evidence against the null hypothesis.

How Many Observations Are Required?

The quantity of observations required is determined by the statistical power demanded by the project. Ideally, this should be determined at the beginning of the project.

How Long to Run a Multivariate Test?

The duration of time required for your multivariate testing is ideally the amount of time required to capture enough observations to meet the defined statistical power. It is often useful to run tests over time to capture a representative, variable sample.

When determining the duration of your testing phase, consider the novelty effect, which describes how user reactions in the short term are not representative of the long-term reactions. For instance, whenever Facebook updates their news feed layout or design, there is an uproar. This soon subsides once the novelty effect has worn off. Therefore, it is useful to run your experiment for long enough to overcome this bias. Running multivariate tests for long periods of time is typically not a problem in model optimization.

Spotting Data Variance and Drift

It is crucial to monitor ongoing performance of your machine learning model once deployed. Data variance, drift, and system development require the model to be confirmed against the baseline regularly. This is a key requirement to ensure safe deployment of healthcare AI.

Variance refers to errors from sensitivity to small changes in the training dataset. The control and experimentation sets could be biased, for instance, as the result of not being split at random. This may result in biases in the sample data. If this is the case, other tests can be used, such as Welch's t-test, which does not assume equal variance.

Typically, spotting variance and drift involves monitoring the offline performance, or validation metric, against data from the live, deployed model. If there is a sizable change in the validation metric, this highlights the need to revise the model through training on new data. This can be done manually or automated to ensure consistent reporting and confidence in the model.

Keep a Note of Model Changes

Keep a log of all changes to your machine learning model with notes on changes. Not only does this serve as a change log for stakeholders; it provides a physical record of how the system has changed over time. It also acts as an audit log.

The use of versioning software within a development environment (test/staging to live deployment) will enable software changes to automatically be noted. Versioning software provides a form of technical governance and can be used to deploy software with extensive rollback and backup facilities.

Real-Time Monitoring

A deployed machine learning model making predictions requires processes to support clinical safety and robustness. Once models are deployed, monitoring systems to detect model-level faults and identify environment changes is a sensible consideration for a live, clinical model.

Conclusion

Building a machine learning model involves working on an iterative, constructive feedback principle. Engineers build a model, evaluate the model by certain metrics, make improvements, and continue until a desired accuracy is achieved. Evaluating the performance of a model on training data alone is useless. Once deployed, it is crucial this accuracy is maintained.

There is a requirement to ensure that live predictions are safe and accurate. This can be achieved through a range of live evaluation metrics which have been covered in this chapter. Different problem types, as you would expect, merit different evaluation metrics.

Statistical hypothesis testing allows hypotheses to be confirmed and further confidence to the accuracy of machine learning models.

CHAPTER 8

Machine Learning and AI Ethics

> *People worry that computers will get too smart and take over the world, but the real problem is that they're too stupid and they've already taken over the world.*
>
> —Pedro Domingos

From supermarket checkouts to airport check-ins and digital healthcare to Internet banking, the use of data and AI for decision-making is ubiquitous. There has been an astronomical growth in data availability over the last two decades, fueled by, first, connectivity, and now the Internet of Things. Traditional data science teams focus on the use of data for the creation, implementation, validation, and evaluation of machine learning models that can be used for predictive analytics. Behavior psychology is embedded within digital health.

From nudges to get app users to submit health data, survey question framing, and presentation of content to upselling a side of French fries to your fast-food app order, machine learning is influencing our decisions, and its influence needs governing.

In the past decade, there has been a wealth of data-driven AI and technological advance specifically in the healthcare domain that have improved quality of life:

- Noninvasive blood glucose levels: There have been many advances in noninvasive blood glucose management. Google has developed a contact lens that determines the user's blood glucose levels from tears [96]. French company PKvitality developed a small wristwatch that determines the level of glucose in the blood, and there have been advances in determining blood glucose levels through using far-infrared signals [97].

© Arjun Panesar 2021
A. Panesar, *Machine Learning and AI for Healthcare*, https://doi.org/10.1007/978-1-4842-6537-6_8

- Artificial pancreas: As a combination of systems, the artificial pancreas uses two devices. The first device measures blood glucose using a sensor and communicates with a second device (or patch), which administers insulin to the patient. The insulin delivery system can adjust the insulin dosage according to the blood glucose levels [99].
- Reversal/remission of non-communicable disease: Digital therapeutics is an emerging discipline of digital health that is redefining conditions such as type 2 diabetes as not having to be chronic and progressive. Digital therapies have been demonstrated to sustain weight loss, improve blood glucose control, and place type 2 diabetes and prediabetes into remission [91].
- Bioprinting of skin constructs: 3-D printing has been used to reproduce blood vessels and skin cells to facilitate wound healing for burn patients [100].
- Peer support and relationships: Peer support communities such as Diabetes.co.uk have demonstrated in research studies the ability to improve qualitative and quantitative health outcomes [101].
- Foot ulcer detection: Concerns such as foot ulceration and bruising and wider diabetic foot concerns can be increasingly detected through machine learning to expedite ulceration detection, prevent amputations, determine effective treatments, and improve healing times [102].
- Open source data sharing: Data repositories and public APIs have enabled data sharing between systems. Historically conservative firms are now embracing open source analytical, AI, and data management software. Within many organizations, employees are actively discouraged or given autonomy on whether to use proprietary tools. Cost and performance are drivers toward open source data, mainly as open data sources have become more robust and accepted. The adoption is also driven by the latest generation of data scientists and university graduates aware of the safety and capabilities of open source data [103].

Novel solutions to problems developed through machine learning are themselves leading to questions of morality and ethics. Currently, governance is moving at the pace of the industry itself. There are many scenarios within AI for which there are no precedents, regulations, or laws. It is paramount to consider the ethical and moral implications of creating intelligent systems.

The World Health Organization (WHO) reports that the prevalence of chronic disease will rise to 57% of the global population by 2050 [104]. Unfortunately, the WHO also reports that there is a global, growing shortage of healthcare workers, which by 2035 will rise to 12.9 million [105]. Lack of professionals to provide healthcare services paints a stark image of the future with grave consequences for humanity. Healthcare professional shortages are being offset through AI, digital interventions, IoT, and other digital technologies that cannot only replace manual and cognitive working tasks but can also improve the reach, precision, and availability of healthcare. At the same time, advancements in the detection and diagnosis of diseases, genomics, pharmacology, stem cell and organ therapy, digital healthcare, and robotic surgery are expected to minimize the cost of treating illness and disease.

As AI penetrates humanity's day-to-day activities, philosophical, moral, ethical, and legal questions are raised. This is amplified in healthcare, where clinical decisions can mean the difference between life and death.

Even if AI can aid diagnosis of conditions or predict future mortality risk, will humans ever prefer an AI's advice over their doctor? As humankind becomes accustomed to living side-by-side to intelligent systems, there are a host of hurdles to overcome.

What Is Ethics?

Ethics or moral philosophy refers to the moral codes of conduct (or set of moral principles) that shape the decisions people make and their conduct. Morality refers to the principles that distinguish between good/right and bad/wrong behavior.

Ethics in the workplace, for example, is often conveyed through professional codes of conduct by which employees must abide.

What Is Data Science Ethics?

Data science ethics is a branch of ethics that is concerned with privacy, decision-making, and data sharing.

Data science ethics comprises three main strands:

- Ethics of data
- This area of data science ethics focuses on the generation, collection, use, ownership, security, and transfer of data.
- Ethics of intelligence
- This area of data science ethics covers the output or outcomes from predictive analytics that data is used to develop.
- Ethics of practices
- The ethics of practices was proposed by Floridi and Taddeo, referring to the morality of innovation and systems to guide emerging concerns [106].

Data Ethics

More smartphones exist in the world than people—and phones, tablets, and digital devices alongside apps, wearables, and sensors are creating millions of data points a day. There are over 7.2 billion phones in use, 112 million wearable devices sold annually, and over 100,000 healthcare apps available to download on your mobile phone [107–109]. IBM reports more than 2.5 quintillion bytes (2.5×10^{18}) of data are created daily [110]. Data is everywhere. Moreover, it's valuable.

The topic of data ethics has been thrust into the public spotlight through high-profile fiascos such as the Facebook–Cambridge Analytica scandal. Facebook, one of the world's largest and most trusted data collection organizations, had user data harvested through a quiz hosted on its platform.

Behavioral and demographic data of the 1.5 million completers of the quiz was sold to Cambridge Analytica. The data is largely considered to have been used to target and influence the outcome of the US 2017 elections [111]. What's more concerning is that this breach of security was reported over 2 years after the initial data leak.

We are in a time where fake news can travel quicker than the truth. Society is at a critical point in its evolution, where the use, acceptance, and reliance on data must be addressed collectively to develop conversations and guiding principles on how to handle data ethically. People are aware of fake news and misinformation [112]. AI needs to be ethics based and explainable.

The ethical and moral implications of data use are vast and best demonstrated through an example. In this chapter, we will refer to the following hypothetical scenario.

SCENARIO A

John, type 2 diabetic, aged 30, has a severe hypoglycemic episode and is rushed to the hospital.

John has been taken to the hospital unconscious for treatment. John, a truck driver, was prescribed insulin to treat his type 2 diabetes by his doctor's advice, which is most likely the cause of his hypoglycemia. Lots of data are generated in the process: both in the hospital, by healthcare professionals, and on John's Apple Watch—his heart rate, heart rate variability, activity details, and blood oxygen saturation to look out for signs of a diabetic coma. John's blood sample is also taken and his genome identified. Let's suppose all this data is used and it's useful.

John's Apple Watch was used to monitor his heart on the way to the emergency room, through which it was suspected that he has an irregular heartbeat, later confirmed with the hospital's medical equipment.

Upon waking from his hypoglycemic episode, John is pleased to learn that genetic testing does not always give bad news, as his risk of contracting prostate cancer is reduced because he carries low-risk variants of the several genes known in 2020 to contribute to this illness. However, John is told of his increased risks of developing Alzheimer's disease, colon cancer, and stroke from his genetic analysis.

Informed Consent

Informed consent refers to the user (or patient) being aware of what their data will be used for. Informed consent refers to an individual being legally able to give consent. Typically this requires an individual to be over 18, of sound mind, and able to exercise choice. Consent should ideally be voluntary.

Scenario A demonstrates how useful data can be given a particular context (or use case) and demonstrates the many intricacies of informed consent.

Freedom of Choice

Freedom of choice refers to the autonomy to decide whether your data is shared and with whom. This refers to the active decision to share your data with any third party. For example, should John, who has type 2 diabetes, now be required to demonstrate that his blood glucose levels are under control and within the recommended range before being allowed to drive his truck again?

A person's choice as to whether to share their data could result in a future where people are exempt from opportunities until otherwise demonstrated by their data. In an ideal world, each person should have a choice as to whether to share their data.

In practice, this is neither realistic nor plausible.

There are ethical implications for John not consenting, or wanting to consent, to the healthcare team using his Apple Watch data—namely, that John's Apple Watch data belongs to John. Should the emergency response team have used John's heart rate only to ensure it was beating, or was it ethical to diagnose John's atrial fibrillation subsequently? With informed consent, John would have a choice, which is a fundamental pillar of data ethics.

This is a choice that patients previously did not need to make. Before the datafication of modern life, there were fewer devices to capture such data and far less sophisticated means of predicting future events.

The advances of AI and machine learning in healthcare mean that today there are two groups of people: those who seek out health information to help them plan and manage future scenarios and those who are happy to live in a state of not knowing. As the datafication of everything continues, those who are happy to live in a state of ignorant bliss are finding less opportunity to do so.

Should a Person's Data Consent Ever Be Overturned?

In an ideal world, an individual's decision to share their data should be respected. However, as an absolute concept, this is neither realistic nor plausible. For instance, let's assume John was to decline consent to use his data. On the assumption that the emergency response team's duty is to safeguard the health of its patients, John's data helped to monitor his vital signs, which contributed to his survival. Arguably the emergency response team would be acting unethically if they were not to use the full variety of data available to them at that given moment. Perhaps even John himself would overrule his consent if it meant increasing his chances of survival.

Precedent demonstrates freedom of choice, and consent is ultimately a utopian concept. In Germany, a man accused of the rape and murder of a 19-year-old medical student had the health data from his smartphone used against him at trial [113]. The suspect, who was identified by a hair discovered at the crime scene, refused to give police the PIN code to his smartphone. Police enlisted the help of a cyber forensics firm in Munich who broke into the device. Data on the suspect's iPhone included his steps and elevation, which police analyzed. Police suggested the suspect's elevation (stair climb) data could correlate to him dragging his victim down a riverbank and climbing back up. As well as locating the suspect's movements, the phone also suggested periods of strenuous activity, which included two peaks the onboard smartphone app put down to climbing stairs.

Police investigators mimicked how they believed the suspect disposed of the body and demonstrated the same two peaks of stair climbing detected on the suspect's iPhone.

Public Understanding

The Facebook–Cambridge Analytica fiasco went on to highlight just how unaware the public is on the topic of data privacy. The US Senate's interrogation of Facebook CEO Mark Zuckerberg illustrated just how ignorant and unaware the public was on the topic of technology. Senators were puzzled as to how Facebook made money as a free platform, referred to sending emails through WhatsApp, and queried as to whether encrypted messages could ever be used to provide targeted advertising [114].

The misunderstandings and ignorance demonstrated by critical stakeholders on the topic of data governance was astounding.

Increased public awareness and understanding are required to educate the public on the use of data and to empower people to decide how, where, and by whom their data can be used.

Who Owns My Data?

People generate thousands of data points daily. Nearly every transaction and behavior creates a data trail left through devices such as smartphones, televisions, smartwatches, mobile apps, health devices, contactless cards, cars, and even fridges. However, who owns the data? The topic of data ownership is an example of a complex first-time problem that has facilitated the development of international policy and governance. Historically, user data is owned by companies rather than the individual.

EHRs enable machine learning–based predictive analytics that will eventually enable providers to provide enhanced levels of care. Although it seems unlikely a patient would not want to share their health data for improved morbidity and mortality, one would reasonably assume it would become more difficult to withdraw such privileges, much akin to closed-circuit television (CCTV) where consent is not often noted or obvious and is widely assumed to have utilitarian benefits over the long term.

Many platforms prevent users from accessing their full range of services if data sharing is disabled. This is driven by the organizational need to store data in a robust and central repository for safety, governance, and improvement.

Take the example of a patient who uses a connected blood glucose device and mobile app to track and record their blood glucose. Blood glucose data is delivered from the user's blood glucose meter to their mobile application. Although the data would be presented on the user's phone, the data is held in the mobile application provider's database, subject to the terms and conditions of use. Many of today's websites, mobile apps, connected devices, and health services state that data can and will be used in an anonymized and aggregated format or, in some instances, in an identifiable format by the organization providing the service and also typically with selected partners.

Data collection, usage, and sharing have become a fundamental part of supporting the data-driven quality improvement processes. There are several types of data that can be shared.

Anonymized Data

Anonymized data is data that has identifiable features removed. Identifiable features are features that enable someone to recognize the person whom the data has come from. For example, an oncology ward's spreadsheet of patients would be anonymized if the name, date of birth, and patient number were removed.

Identifiable Data

Identifiable data refers to data that can be used to identify an individual. For instance, an oncology ward's spreadsheet of patients would be identifiable if it reported the name, date of birth, and patient number.

Aggregate Data

Aggregate data refers to data that has been combined and where a total is reported. Following the oncology example, if the ward's spreadsheet covered ten patients, aggregate data reporting could include the male-to-female split or age brackets of patients. Data is cumulatively reported for the population within the dataset.

Individualized Data

Individualized data is the opposite of aggregated data. Instead of data being combined, data is reported for each person within the dataset. Individualized data does not have to be identifiable.

Data Controllers and Processors

Concerns over data privacy have spurred a global response. In May 2018, GDPR (or the General Data Protection Regulation) came into force across Europe. GDPR legislation governs organizations in how they use and share user data [115].

Because of GDPR, organizations have been forced to collect the opt-in consent of their memberships and declare to users how they would use user data and with whom they should share it. The GDPR legislation places data control firmly in the user's hands.

Users of data-driven systems are now able to exercise their rights in seeing the data held on them and with whom their data is shared and why, the right to be forgotten, and the right to be deleted. GDPR also defined labels for those collecting and processing data.

Data Controller

The data controller refers to the person or organization that controls, stores, and makes use of the data.

Data Processor

The data processor refers to the person or organization who processes data on behalf of a data controller. Based on this definition, agents such as calculators could be considered data processors.

For those working with data, it is useful to understand the distinction between data processor and controller and the responsibilities of each party. These responsibilities must be identified and have been enshrined in territories such as Europe with the General Data Protection Regulation (GDPR).

The GDPR demonstrates how vital data safety has become. The GDPR is tightening data access, security, and management in an ambitious attempt to protect the EU's 500 million citizens. It has single-handedly repositioned control with users. The maximum fine for the worst offenders against GDPR regulations is set at either €20 million (£17.6 million) or 4% of global revenues for larger organizations [115]. Unfortunately, regulations affect all businesses and organizations, and a flurry of consent requests flooded the Internet. There is much confusion as to how GDPR laws are applied, as demonstrated when Mark Zuckerberg was summoned to the European Parliament. Just like in the United States, key figures were ignorant and unaware as to how data and connectivity work in the twenty-first century.

Right to Be Forgotten...from Memory?

The topic of a user's right to be forgotten within machine learning has posited several compelling ethical questions. For instance, if John's data were to be used in a machine learning algorithm to predict the likelihood of severe hypoglycemia and if John were to request his data to be deleted, it would be close to impossible to decouple John's data from the machine learning model that is learned. How can John's data be removed from the model's memory? Should John's consent be overridden if there is a utilitarian benefit?

Further still, if John's data were to be for some reason valuable in developing an algorithm to diagnose and predict disease, it could be claimed that it is unethical for John to refrain from sharing his data or request his data to be removed from such an algorithm. If John were to be the only patient in the world to have a genetic defect or if his data was for some reason useful to progress medical understanding, it is evident as to how and why John's data would be useful to humankind.

What Can My Data Be Used For?

Data is already used to fuel a variety of decisions. Employers have used psychometric and parametric testing for decades to understand potential candidates. Nowadays, employers also look at supporting sources of data. Employers often scour the social media profiles of prospective employees for concerns of reputational risk. In the same vein, history is plagued with people who have lost their jobs as the result of (deemed inappropriate) social media use.

What an individual's data can be used for raises ethical and moral apprehensions. The concepts of identity and free will are challenged by the notion that an individual's data could be used to make decisions with direct ramifications that may be unbeknownst to that person.

Car insurance is an industry that has evolved to optimize the use of data. Historically, car insurance premiums have been grounded in accident claim data, which is generalized for segments of the population. The advent of the black box enables more precise premiums based on a variety of demographic and behavioral factors—such as the age of the driver, times that the car is used, the speed of driving, and frequency of erratic acceleration among others.

This data can also be used to reduce the cost of car insurance premiums should the driver be sensible and enables the increase in car insurance premiums should the driver not follow the stipulations set out by the insurer. Others see this constant real-time data analysis (and subsequent feedback) as "Big Brother."

Similarly, life and health insurance are beginning to follow in the footsteps of car insurance. Life and health insurers are increasingly providing incentivized products that reward positive behaviors and chastise negative behaviors. Life insurer Vitality, for instance, offers a product whereby it provides an Apple Watch to its members to encourage positive and healthy behaviors. Similarly, British organization Diabetes Digital Media provides evidence-based digital health interventions to insurers and bill payers worldwide to incentivize wellness.

Individuals should be aware that data could also be used for cynical, more sinister purposes. For instance, patient data could be used to decline an individual from a treatment, operation, or opportunity. The public has been incredibly trusting of data use, and robust data governance is required to prevent its future misuse. The GDPR is regarded as the first step toward improved understanding and data governance. The GDPR is considered by some to be as revolutionary to data science as the advent of the Internet itself.

Privacy: Who Can See My Data?

Conversations of data ownership naturally lead to who can see your data. The key is in ensuring only approved services and organizations have access to your data. For instance, it is unlikely you would want a life insurance company to be given your medical data without your consent, particularly if there was data that could influence your coverage.

Data Sharing

Data sharing between applications is commonplace, with APIs enabling accessibility and faster connectivity between independent services. For instance, users can import nutrition data from apps like MyFitnessPal into their diabetes management or fitness apps. These services often replicate user data among a variety of independent architectures, which leads to concerns over managing approved data access. Applications that enable data integration must also provide the facility to decouple patient data. Systems must be able to verify imported data for data governance, auditing, and patient safety.

The debacle surrounding Facebook and Cambridge Analytica went on to demonstrate that even if one trusts an aggregator of data, it is possible that third parties are engaging with that very data without your knowledge. Facebook requested Cambridge Analytica to delete its Facebook user data. Even though Cambridge Analytica agreed to the request and said they had deleted the user data, they were later demonstrated not to have deleted the data [111]. This act of deception leads to questions of accountability, the ramifications of data leaks, and who is accountable in situations of third-party data leaks.

Anonymity Doesn't Equate to Privacy

Moreover, even if data is anonymized, it may not guarantee privacy. Netflix recently published 10 million movie rankings from 500,000 members as part of a challenge to improve the Netflix recommendation system.

Although data was anonymized to remove personal details of members, researchers at the University of Texas were able to de-anonymize some of the Netflix data through comparing similar rankings and time stamps with public information in IMDb (Internet

Movie Database) [116]. There are natural security problems with anonymous data—mainly that it does not mean that you are anonymous.

Data Has Different Values to Different People

When it comes to sharing data, it is useful to distinguish between people and patients. The general public's attitude toward data use in the realm of healthcare is far more welcoming than its attitude toward nonmedical data use. Patients appear to understand or are at the very least hopeful that, through sharing of health data, they are progressing medical understanding and treatment. People, however, are far more skeptical of beneficial data use.

Confidentiality is a core tenet of data ethics.

How Will Data Affect the Future?

The sharing of patient data and aggregation of big datasets is being used to enhance diagnosis, treatment, and care. As the types and quality of data improve, healthcare's precision will become more precise in the areas discussed next.

Prioritizing Treatments

Big data medical sets can enable predictive analytics to determine optimal treatment pathways for various segments of the population. There is the potential to prioritize only those who are likely to see efficacy in the treatment, which begs the question as to how best to support those who do not.

Determining New Treatments and Management Pathways

The analysis of real-world experience data, clinical studies, randomized clinical trial (RCTs), and pharmacological data is facilitating treatment and management discovery.

Digital health interventions have been demonstrated to reverse type 2 diabetes and reduce the number of seizures suffered by those with epilepsy [91, 117]. Data has the potential to develop new pharmacological interventions and disrupt traditional treatment paradigms.

More Real-World Evidence

The use of real-world evidence (RWE) at scale from patient communities, digital education programs, and health tracking apps is being increasingly demonstrated to improve the self-management landscape. A concern with real-world evidence has typically been lack of academic robustness.

However, RWE is being increasingly used in research to determine population usage and benefits. Real-world evidence is a vital part of AI's ethical journey, and there must be trust between patient and provider.

Enhancements in Pharmacology

The precision audiences require for RCTs and academic studies can be identified quicker and easier through digital platforms. This has benefits, such as quicker project recruitment times, greater potential, and multisource comparison. Real-world data is being used to develop better and more effective drugs.

Cybersecurity

There are significant privacy concerns that come with a truly unified system. What if such a system were to be compromised? Do people, or patients, want a truly unified system, or is the possibility to use it for malintent greater than the potential good? The consequences of vulnerabilities—whether security, operational, or technical—are amplified in unified systems.

For years, large companies have talked about ways to combine and analyze the giant repositories of data they separately gather about people—location data from smartphones, financial data from banks, relationship data from social networking apps, and search data from browsers—to build a complete picture of a person's behavior.

Facebook reportedly offered to match the data that hospitals had about individual patients with social information gleaned from the social networking site, such as how many friends the person has or whether they seem to engage with others on the site. The company paused the project after privacy concerns were raised by the Cambridge Analytica scandal.

Further still, there are security concerns around IoT devices, which now range from thermometers, cars, and washing machines to blood glucose meters, continuous glucose monitors, and insulin pumps. The requirement of network connectivity to operate as a smart device leaves all devices prone to vulnerability.

Distributed denial of service (DDoS) attacks are used on IoT devices to break in and leave malicious code to develop botnets, leak data, or otherwise compromise the device. WeLiveSecurity reported on 73,000 security cameras with default passwords [118]. A basic but key to learning from this is that it is always advisable to change credentials that use default passwords. DDoS attacks are nothing new, but the extent of vulnerability is only being uncovered through research and first-time problems or security breaches. Research students from the University of Florida were able to compromise Google's Nest thermostat in under 15 seconds [119].

DDoS attacks can be crippling for any organization. Develop mitigation procedures and ensure network infrastructures allow visibility of traffic entering and exiting the network. It is good practice to develop a DDoS defense plan, which should be kept and regularly updated and rehearsed.

AI and Machine Learning Ethics

A primary application of machine learning in healthcare involves patient diagnosis and treatment. AI models are deployed to help physicians diagnose patients, especially in cases involving relatively rare diseases or when outcomes are hard to predict.

Imagine a future where doctors know exactly how many times you've eaten fast food in the last month and that it is connected to your medical record. That data is used to suggest what foods to eat, comparing your health record to hundreds or thousands of other people to identify who is like you. Moreover, imagine this could directly affect life insurance or health insurance, with possible rewards for positive behavior and avoiding bad foods. Imagine a system that was able to predict your likelihood of developing any particular disease in real time.

What Is Machine Learning Ethics?

The ethics of machine learning refers specifically to the questions of morality surrounding the outputs of machine learning models that use data, which come with their ethical concerns.

Machine learning has already been used to develop intelligent systems that have been able to predict mortality risk and length of life from health biomarkers. AI has been used to analyze data from EHRs to predict the risk of heart failures with a high degree of certainty.

Moreover, machine learning can be used to determine the most effective medication dosage learning on patient real-world and clinical data, reducing healthcare costs for the patients and providers. AI can not only be used in determining dosage but also in determining the best medication for the patient. As the genetic data becomes available, medications for conditions such as HIV and diabetes will accommodate for variations among races, ethnicities, and individual responses to particular drugs. Within the same data, medication interactions and side effects can be tracked. Where clinical trials and requirements for FDA approval look at a controlled environment, big real-world data provides us with real-time data such as medication interactions and the influence of demographics, medications, genetics, and other factors on outcomes in real time.

As the limitations of technology are tested, there are ethical and legal issues to overcome.

Machine Bias

Machine bias refers to the way machine learning models exhibit bias. Machine bias can be the result of several causes, such as the creator's bias on data used to train the model. Biases are often spotted as first-time problems with subtle and obvious ramifications. All machine learning algorithms rely on a statistical bias to make predictions about unseen data. However, machine bias reflects a prejudice from the developers of the data.

The capabilities of AI regarding speed and capacity of processing far exceed that of humans. Therefore, it cannot always be trusted to be fair and neutral. Google and its parent company, Alphabet, are leaders when it comes to AI, as seen with Google's Photos service, where AI is used to identify people, objects, and scenes. But it can still go wrong, such as when the search engine showed insensitive results for comparative searches for white and black teenagers. Software used to predict future criminals have been demonstrated to show bias toward black people [120].

Artificially intelligent systems are created by humans, who are biased and judgmental. If used correctly, and by those who positively want to affect humanity's progress, AI will catalyze positive change.

Data Bias

Bias refers to a deviation from the expected outcome. Biased data can lead to bad decisions. Bias is everywhere, including the data itself. Minimize the impact of biased data by preparing for it. Understand the various types of bias that can creep into your data and affect analysis and decisions. Develop a formal and documented procedure for best practice data governance.

Human Bias

For as long as humans are involved in decisions, a bias will always exist. Microsoft's infamous AI chatbot Tay interacted with other Twitter users and learned from its interactions with others. Once the Twittersphere seized hold of Tay, trolls steered the conversation from positive interactions to interactions with trolls and comedians. Within hours, Tay was tweeting sexist, racist, and suggestive posts. Sadly, Tay only lived for 24 hours. This experiment raises the question of whether AI can ever really be safe if it learns from human behavior [121].

Intelligence Bias

Machine learning models are only as good as the data they are trained on, which often results in some form of bias. Human-thinking AI systems have demonstrated bias amplification and caused many data scientists to discuss the ethical use of AI technology. Early-thinking systems built on population data showed significant signs of bias regarding sex, race, social standing, and other issues.

An infamous example of algorithmic bias can be found in criminal justice systems. The Correctional Offender Management Profiling for Alternative Sanctions (COMPAS) algorithm was dissected in a court case about its use in Wisconsin. Disproportionate data on crimes committed by African Americans were fed into a crime prediction model, which then subsequently output bias toward people from the black community. There are many examples and definitions of biased algorithms [122].

Algorithms that assess home insurance risk, for instance, are biased against people who live in particular areas based on claim data. Data normalization is key. If data is not normalized for such sensitivities and systems not properly validated, humanity runs the risk of skewing machine learning models for minorities and underrepresenting many groups of people.

Removing bias does not mean that the model will not be biased. Even if an absolute unbiased model were to be created, we have no guarantee that the AI won't learn the same bias that we did.

Bias Correction

Bias correction begins with the acknowledgment that bias exists. Researchers began discussions on machine learning ethics in 1985 when James Moor defined implicit and explicit ethical agents [123]. Implicit agents are ethical because of their inherent programming or purpose. Explicit agents are machines given principles or examples to learn from to make ethical decisions in uncertain or unknown circumstances.

Overcoming bias can involve post-processing regarding calibration of our model. Classifiers should be calibrated to have the same performance for all subgroups of sensitive features. Data resampling can help smoothen a skewed sample. But, for many reasons, collecting more data is not very easy and can cause budgetary or time problems.

The data science community must actively work to eliminate bias. Engineering must honestly question preconceptions toward processes, intelligent systems, or how bias may expose itself in data or predictions. This can be a challenging issue to tackle, and many organizations employ external bodies to challenge their practices.

Diversity in the workplace is also preventing bias from creeping into intelligence. If the researchers and developers creating our AI systems are themselves lacking diversity, then the problems that AI systems solve and training data used both become biased based on what these data scientists feed into AI training data. Diversity ensures a spectrum of thinking, ethics, and mind-sets. This promotes machine learning models that are less biased and more diverse.

Although algorithms may be written to best avoid biases, doing so is extraordinarily challenging. For instance, even the motives of the people programming AI systems may not match up with those of physicians and other caregivers, which could invite bias.

Is Bias a Bad Thing?

Bias raises a philosophical question on the premise that machine learning assumes that biases are generally bad. Imagine a system that is interpreted by its evaluators to be biased, and so the model is retrained with new data. If the model were to output similarly biased results, the evaluation might wish to consider that this is an accurate reflection of the output and hence require a reconsideration of what bias is present.

This is the beginning of a societal and philosophical conflict between two species.

Prediction Ethics

As advanced machine learning algorithms and models are developed, more accurate and reliable conclusions will be achieved in a short space of time. Technologies are being used currently to interpret a variety of images, including those from ultrasound, magnetic resonance imaging (MRI), X-rays, and retina scans. Machine learning algorithms can already effectively identify potential regions of concern on images of the eye and develop possible hypotheses.

It is key to ensure trust in your goals, data, and organization through good governance and transparency. This is fundamental for AI going forward.

Explaining Predictions

As AI algorithms become smarter, they also become more complex. Remaining ignorant about the construction of machine learning systems or allowing them to be constructed as black boxes could lead to ethically problematic outcomes. If an agent were discovered to be predicting incorrectly, it would be a daunting task discovering the behavior that caused an event that is hidden away and virtually undiscoverable.

Interpretability of both data and machine learning models is a critical aspect of intelligent systems. This not only ensures the model integrity but also that it is attempting to solve the correct problem. Users of solutions that embed data science will always prefer experiences where they are understandable and explainable. Data scientists can also use interpretability metrics as the basis for validation and improvement.

Machine learning black boxes could harbor bias, unfairness, and discrimination through programmer and data choices never to be known. Neural networks are an example of a typically unexplainable algorithm. The backpropagation algorithm's computed values cannot be explained. As AI develops in its accuracy in resembling humans, there will be a greater requirement to ensure AI isn't picking up the bad habits of humans.

Protecting Against Mistakes

Intelligence comes from learning, whether you are a human or machine. And intelligent machines, just like humans, learn from mistakes. Data scientists will typically develop machine learning models with a training, testing, and validation phase to ensure systems are detecting the correct patterns within a defined tolerance. The validation phase of machine learning model development is unable to cover all possible permutations of parameters that may be received in the real world. These systems can be fooled in ways that humans wouldn't be. Governance and regular auditing is required to ensure that AI systems perform as intended and that people cannot influence the model to use it for their own ends.

The incorrect classification of predictions can lead to scenarios where the outcome is a false positive or false negative. The impact of both should be considered in the context of your domain. Take for instance an incorrect diagnosis of breast cancer. In a false-positive scenario, a patient would be informed they had breast cancer when they did not. Being informed that this classification was incorrect will come to some relief to the patient. A false negative would result in a patient's disease progression and the eventual correct re-diagnosis. The mental and physical trauma as a result of a false-negative prediction should be considered. Patients should always be informed of the level of accuracy of results.

Information governance for systems that will be used by patients (patient-facing), whether predictive or otherwise, should provide robust risk mitigation procedures for mistakes in outcomes. Emotional or psychological patient support should be considered for those experiencing trauma as the result of the incorrect diagnosis.

As well as acting as a catalyst for innovation, the speed at which agile digital technologies are rushed to the market is also AI's biggest pitfall. Not only can technologies fail; they can be identified for performing unfairly.

Technology company LG infamously unveiled an AI bot at Consumer Electronics Show, better known as CES, an annual trade show organized by the Consumer Technology Association, that ignored the presenter's instructions. The AI bot, which was marketed as providing innovative convenience, failed to respond to any of its master's comments and either was malfunctioning or chose to ignore the commands [124].

In 2020, there was a vocalization of how black people and those from ethnic minorities have seen technology used to target them. In 2020, TikTok ashamedly algorithmically hid posts that included the Black Lives Matter or George Floyd hashtags from view during the global frustration at the killing of a black man by police [125, 126].

Instagram also identified it needed to look at shadow banning and biases within its systems during this period. Shadow banning is the process of "filtering people without transparency, and limiting their reach as a result" [127].

Whether in healthcare or not, AI system performance metrics should be transparent and audited regularly to ensure what people are exposed to is absolute and true. The cost, particularly in healthcare, is too high not to. AI systems in healthcare have been demonstrated to fail exorbitantly. In the United Kingdom, an NHS breast cancer screening system failed to appropriately invite women for screening, with observers claiming up to 270 females may have died as a result [128, 129].

In this particular case, the organization responsible for running the system placed blame with a contractor. The ethical implications and public perception of such mistakes are enormous.

Validity

There is a requirement to ensure the validity of machine learning models over time to ensure the model can be generalized and generalizations are valid. Regular testing and validation of models are essential to maintaining integrity and precision of your machine learning model. A suboptimal predictive analytics model will provide unreliable results and damage integrity.

Preventing Algorithms from Becoming Immoral

Algorithms can, and do, already act immorally. To date, AI algorithms have mainly performed in unethical ways due to design. This has been demonstrated very well outside healthcare by organizations such as Uber and Volkswagen. Uber's Greyball algorithm attempted to predict which passengers were undercover police officers and used it to identify and deny transport [129]. Volkswagen's algorithm allowed vehicles to pass emission tests by reducing nitrogen oxide emissions during the test phase [130].

Both organizations were internationally condemned for their public deception and lack of transparency. Precedents such as these from the world's largest companies demonstrate that internal and external auditing is required to validate the integrity and ethics of algorithms and indeed their organizations.

There is a genuine concern that AI may learn to act immorally not only from its creators but also from its experience.

The Ecole Polytechnique Fédérale of Lausanne's Laboratory of Intelligent Systems in Switzerland conducted a project that monitored robots designed to collaboratively search for positive resources and ignore dangerous items [131, 132]. Robots were designed as genetic agents equipped with sensors and a light that was used to flag the identification of a positive resource, which was finite in number.

Each agent's genome dictated its response to stimulus and was subject to hundreds of generations of mutations. Agent learning was reinforced by the agent receiving positive marks for identifying positive resources and negative marks for its proximity to poisonous items. The top 200 highest-performing genomes were mated and mutated at random to produce the next generation of agents. In their first generation, agents switched their lights on when they discovered a positive resource. This enabled other agents to find the positive resource. Due to the limited number of positive resources, not all agents were able to benefit. Overcrowding meant that some agents would also be distanced from the positive resource they initially found. By the 500th generation, the majority of agents had evolved to keep their light switched off when they discovered a positive resource, and a third of agents evolved to act in a way that was the exact opposite of their programming. Some agents had grown to identify lying agents that had an aversion to the light. Agents that were initially designed to cooperate eventually ended up lying to each other due to scarcity.

These concerns are amplified when applied to incentivized clinical decision support systems that could generate increased profits for their architects or providers—a system that was recommending clinical tests, treatment, or devices in which they hold a stake or by altering referral patterns.

In healthcare, this scenario is very concerning. All machine learning models regardless of their use must be governed and verifiable. A machine learning model that was incentivized to make decisions should also adhere to robust standards of transparency, morality, and scrutiny.

Unintended Consequences

The adoption of AI in medicine is becoming more commonplace; and as a consequence, first-time problems or unintended consequences will govern the public perception and direction of AI ethics. There are very few technologies that have been embedded into healthcare without unintended, or adverse, effects. The question quickly comes to how data scientists, on behalf of humanity, can mitigate these risks.

The ethics of AI and unintended consequences have been significantly directed by The Three Laws, published by science fiction writer Isaac Asimov in 1942 [133].

Asimov's Laws, found in the *Handbook of Robotics*, 56th Edition, 2058 AD, were designed as a safety feature to prevent robots from harming humans:

1. A robot may not injure a human being or, through inaction, allow a human being to come to harm.
2. A robot must obey orders given by a human being unless it conflicts with the First Law.
3. A robot must protect its existence as long as such protection won't conflict with the First or Second Law.

In Asimov's stories, manlike machines behave in counterintuitive ways. These acts are the unintended consequence of how the agent applies Asimov's Laws to its environment. Only 70 years later, Asimov's science fiction fantasy is eerily turning into reality. South Korea published a Robot Ethics Charter to prevent ills and malintent [135]; and the IEEE and British Standards Institution have both published best practice guides for ethical agent engineering [136]. Best practice guides to designing ethically sound agents are typically grounded in Asimov's Laws.

Fundamentally, intelligent agents are made of binary code and do not contain or obey Asimov's Laws. Their human architects are responsible for implementing the laws and reducing the risk of unintended consequences.

AI is unlikely to become evil and turn on humanity in the way depicted by Hollywood. Rather, a lack of context could lead to AI making unintended and disastrous actions. For instance, an intelligent agent that was tasked with eradicating HIV in the global population could eventually come to some conclusion that to achieve its objective, it should kill everyone on the planet. It is easier to frame AI intelligence in a negative setting.

Without careful management, an AI agent's utility function could allow for potentially harmful scenarios. There is a limited foundation to suppose that an AI agent would have such an extreme adaptation, yet it is worth consideration. There are undoubtedly many positive unintended consequences to AI that could save humanity from itself.

How Does Humanity Stay in Control of a Complex and Intelligent System?

Over millions of years, humans have combined ingenuity and intelligence to create methods and tools to control other species. As a kind, humans have evolved to dominate, driven primarily by the capacity to learn from the mistakes of themselves or others. Through this, humans have come to develop tools to master bigger, faster, stronger animals and techniques such as mental or physical training to achieve optimal performance for such tasks.

What Will Happen When AI Is More Intelligent Than Humans?

Human-made AIs are already able to surpass human cognition in niche areas. AlphaGo Zero, the reinforcement learning agent developed by Alphabet, acted as its teacher to master the game of Go [137]. AlphaGo was able to learn itself, without the aid of humans or historical datasets over thousands of game interactions. Are humans doomed to be mastered by an evolving AI race? This concept is known as the singularity—the point at which AI intelligence surpasses that of humans. It cannot be expected that an AI whose intelligence surpasses humans can be switched off, either. A reasonably intelligent agent may anticipate this action and potentially defend itself.

With AI that is more intelligent than humans, particularly if learning itself with its hyperintelligence, humankind is no longer able to predict outcomes and must accordingly be prepared for such an eventuality.

Intelligence

Intelligence refers to the outcomes of AI models and how they are used. Intelligence powers the ads you see, the apps you download, the content you see on the Internet, the cab you hire, and the price you pay for items such as loans and mortgages.

The ethics of intelligence posits questions such as whether an autonomous vehicle should protect its driver or others. Should a car prioritize protecting its passengers or instead protect other drivers and the public? If an incident were to cause loss of life, what should be the approach then? This may become less of a problem once autonomous vehicles are prevalent and fewer humans are involved. Until then, if an autonomous vehicle on autopilot and without human steering does have an accident, who should be liable? Court cases involving autonomous vehicle accidents have demonstrated that drivers are often using autopilot on impact [138].

Assessments have expressed that humans do not want a fair, utilitarian approach and want their vehicle to protect the driver. Regulation on topics such as these will be instrumental in the long-term adoption of autonomous vehicles.

Another example of intelligence can be found in virtual assistants such as Alexa which have revolutionized methods of human–computer interaction and engagement.

As the use of AI becomes embedded in everyday life, humans must not be assumed to be vigilant and aware, and organizations should take steps to assure the public. The vocal skills of Google's Duplex bot were demonstrated at its developer conference where it was shown booking a hair appointment. The onstage unveiling involved the Duplex software conducting a conversation with a hair salon receptionist. The computer-generated voice used pauses, colloquialisms, and circumlocutions usually present in human speech. The voice was under control of Google's DeepMind WaveNet software that has been trained using lots of conversations, so it knows what humans sound like and can mimic them effectively. Although welcomed by some, many were concerned at the deliberate deception of a human by intelligent AI trying to be, and successfully passing off as, human. This has amplified the public's voice that humans demand explicit confirmation when engaging with AI [139, 140].

Natural language can be separated into two components: style and content. Analysis of communications such as engagements, emails, or messages can easily identify sentiment and opinion.

Negative terms, for example, could be identified to understand the state of someone's emotional health. The implications of this, taking place publicly or privately, are vast. Analysis of social media profiles and other unstructured sources of data could lead to employers or insurers making predictions about people based on engagement, communication, or sentiment data available for mining.

Health Intelligence

Health intelligence refers specifically to the intelligence developed through healthcare AI. There are a variety of areas that health intelligence is being used in industry, driven by improvements in patient health, costs, and resource allocation:

- Healthcare services: Patients are increasingly engaging with predictive analytics services to diagnose disease. For instance, diabetic retinopathy is being detected through AI systems developed through a partnership between Alphabet and Royal Marsden Hospital. Diabetes Digital Media developed a foot ulceration detection algorithm for earlier and quicker referral to podiatry clinics.
- Pharmacology: New drugs are being discovered through learning on real-world data and patient profiles. This is enabling pharma to develop new and more precise medications.
- Life and health insurance: Patients with health conditions (or risks) such as type 2 diabetes and prediabetes, who would traditionally receive loaded premiums, can use digital health interventions to manage and improve their condition and, as a result of their sustained engagement, receive reduced premiums to incentivize wellness. The benefits to life and health insurers are vast. Not only are insurers able to engage with more of the population, through connecting insurance with improved wellness, but insurers also can save money through fewer and reduced claims and reduced paramedical and medication costs, improve their risk profile, and optimize risk algorithms and underwriting.

Health intelligence cannot be developed in an environment where the public's ethical AI requirements are ignored, such as with personal data. The ethics of AI should be integral to the development of systems capable of health intelligence, identifying the risks of agents before they are realized. Health intelligence ethics are integral to understanding the intention of data and use of outcomes in medicine.

Who Is Liable?

AI's application in healthcare is an exciting opportunity to improve patient care and cost savings within a short period. Accurate, timely decisions influence the diagnosis, treatment, and outcomes of patients with any number of illnesses.

Healthcare professionals can analyze a finite number of images, tests, and samples; and clinical decisions are still prone to human error. With AI, clinical decisions can be made with varying confidence through the analysis of infinite samples in near real time. The medical and technological limitations of AI appear easier to overcome than the potential moral and legal issues. If a breast cancer diagnosis algorithm falsely predicts breast cancer or an algorithm fails to identify signs of diabetic retinopathy on an eye scan, who should be to blame?

There are three liability possibilities in the diagnosis of a patient's disease:

- A human doctor decides a patient diagnosis, with no assistance from an external agent. A human doctor may be very accurate when symptoms are evident and often spots less obvious concerns at first glance. In this case, the liability is always with the doctor.
- An intelligent agent predicts the patient's diagnosis, with a 99% accuracy. Seldom will a patient be misdiagnosed, but errors such as death are neither the liability of the agent or human. If an AI was to make such a mistake, how would a patient raise their issue?
- A human doctor is assisted by an intelligent agent. In this case, it may prove harder to identify responsibility, as the prediction is shared. Even if the human's decision was final, it could be argued that the countless experiences of the AI agent would be a key factor in influencing a decision.

Liability for an AI agent's behavior can be understood as belonging to the following:

- The organization that developed the AI agent.
- The human team that designed the AI agent is responsible for any unexpected functioning or inaccurate predictions.
- The AI agent is responsible for any unexpected behaviors itself.

AI agent development is often the result of numerous engineers and collaborators, and thus holding the developers to account is not only difficult to manage but may put off potential engineers from joining the industry. Holding the AI development organization to account sounds like the best route to ethical adherence.

How do we ensure AI systems will not overturn humans? What regulations govern our safety? Friedler and Diakopoulos suggest that five core principles are used to define organizational accountability [141]:

- Auditability: External bodies should be able to analyze and probe algorithm behavior.
- Accuracy: Ensure good clean data is used. Identification of evaluation metrics, regular tracking of accuracy, and benchmarking should be used to calculate and audit accuracy.
- Explainability: Agent decisions should be explainable in an accessible manner to all stakeholders.
- Fairness: Agents should be appraised for discrimination.
- Responsibility: A single point of contact should be identified to manage unintended consequences and unexpected outputs. This is similar to a data protection officer, but specifically for AI ethics.

First-Time Problems

Real-world data is already demonstrating that type 2 diabetes management has been misguided for the past 50 years. Who should be liable when real-world experience demonstrates legacy treatments have been misled?

Supervised and unsupervised models developed on patient data are creating new concepts in healthcare that are defining AI ethics. First-time problems are a form of unintended consequence. Data can now be used within AI to detect risks of disease, such as type 2 diabetes, hypertension, and pancreatic cancer. If an application was able to tell you about a terminal disease you could do nothing about, would you want to know? Would you even want to know that there was a choice to know? Genetic profiling services give feedback on disease risk based on the patient's genetic profile. If someone were to be informed they had a reduced risk of developing certain diseases, they may develop a propensity for riskier behaviors. If a patient is told their susceptibility to lung cancer is lower than average, would this lead to a higher risk of them becoming a smoker or other risky behaviors?

Similarly, knowing the risk of developing particular conditions could lead to mental and emotional consequences. The implications on human psychology and behavior of knowing disease risk will be discovered as time and evidence develops.

An example of a first-time problem related to AI is misinformation and fake news. Fake news content is created, a mixture of bots and targeting is then used to amplify peoples' confirmation bias, and the content spreads further through sharing and engagement [142]. The manipulative threat of distorting perception of reality based on perceived evidence is highly alarming. Another problem is deep fakes—images or videos which are replaced with someone else's likeness using artificial neural networks. The world is slowly catching up on how best to fight these cyberthreats. Blockchain technology to verify content from trusted sources may prove to be a partial fix [143]. However, as deep fakes improve, trust in digital content diminishes.

Defining Fairness

AI is changing the way we interact and engage with products and services. As accessibility to data grows, how do we ensure AI agents treat people justly? Recommendation systems, for instance, can significantly affect our experience. But how do we know they are fair? Are we being biased by the choices we are given?

A machine learning model cannot be fair if there is not a clear definition of fairness. There are many definitions of fairness and collaboration between social scientists and engineers, and AI researchers are required to determine a clear recognition of fairness. Fairness dictates that data misuse should have public, legal, and ethical consequences.

How Do Machines Affect Our Behavior and Interaction?

AI's rapid evolution is prompting humankind to evaluate just what it means to be human.

Humanity

As AI bots have become better at modeling conversations and engagements with humans, so too have their prevalence. Virtual assistants such as Alexa, Siri, and Google Assistant are in the hands of over 100 million people—and voice-based telephony agents are commonplace [144]. Not only are agents able to mimic conversational language but emotion and reaction modeling is more sophisticated and believable than ever. In 2015, a bot named Eugene won the Turing Challenge, which asks humans to rate the quality of their communication with an unknown agent and to guess whether it is a human or an AI agent. Eugene was able to disguise itself as a human to more than half of its human raters [145].

The blurring between humankind and robot-kind is imploring exploratory interspecies relationships. In 2017, Chinese AI engineer Zheng Jiajia married robot Yingying, a robot spouse he created to ease the pressure of marriage [103]. He's not alone [18].

Human replacement sex robots are now common, with connected AI robots developed with humanlike skin and warmth, as well as safety measures to mitigate electrical shocks or worse.

Whether applied to banking, healthcare, transport, or anything else, humanity is at the beginning of a juncture where humans engage with AI agents as frequently as though they too are humans. Where humans have limits as to the amount of attention, kindness, compassion, and energy they can expend, AI bots are an infinite resource for developing and maintaining relationships.

Behavior and Addictions

Our behaviors are already influenced and manipulated by technology. Web landing pages, shared links, and user experiences are optimized through multivariate testing. This is a crude algorithmic approach to capture human attention. In the US elections, Donald Trump's campaign famously used and tested more than 50,000 ad variations daily to micro-target voters [149]. Human experimentation has been taking place for centuries, so this is nothing new. However, there is now the ability to influence choice and freedom like never before. Organizations have a moral responsibility to analyze the impact of AI and safeguard particular groups of people, including vulnerable adults and children, from manipulative tools like nudging or social proofing. Codes of conduct for behavioral experimentation within technological experiments are required.

Technology Addiction

Technology addiction is becoming a growing concern. Robert Lustig, a professor at the University of Southern California, discovered that the human brain responds to technology in a similar way to other addictive substances [150]. Internet addiction has been documented in a variety of countries; and adolescents appear to be more at risk, partly due to the prefrontal cortex being the last part of the brain to develop. According to the charity Action on Addiction, one in three people is addicted to something [151]. Internet addiction is being increasingly explored. A 38-year-old man was found dead after playing video games for five days in an Internet cafe in Taipei [152]. Shyness, loneliness, and avoidance seem to play a part in Internet addiction [153].

Real-world and academic evidence demonstrates that technology addiction is a mounting issue. AI has the potential to improve productivity and discovery. However, humankind is responsible for taking measures to prevent AI leading to destructive, digital addictions that make us more insular, overdependent, and lethargic.

Economy and Employment

What happens after the end of jobs? Supermarkets already use automated checkouts, and even McDonald's have already replaced the human-to-human food ordering procedure with a digital interface [108]. The hierarchy of labor is concerned primarily with automation, and this is also true of healthcare. Pharmacies, assembly lines, and check-ins are all commonly automated systems. Although this appears a concern for humankind, it can facilitate more complex roles for humans, moving the species more from physical work to cognitive labor. COVID demonstrated that while digital technology is a tool for good, key workers are required to ensure the system remains in motion.

Healthcare organizations exploring the use of AI are understanding how to retrain or redeploy their staff, rather than make them redundant. As these AI systems become increasingly common and grow in capacity and accuracy, conversations are moving past manual and cognitive labor to how specialisms such as radiology will be affected. Healthcare teams that use image processing AI, for instance, are increasingly confronted with the dilemma of what to do with teams of specialists.

AI can be used to redirect resources to amplify patient care, with more opportunities to engage with patients and their treatment. AI can be understood as an efficient and effective tool that can analyze and detect patterns in patient data at a rate and efficiency inconceivable to humans, with little misinterpretation. However, AI should be seen to augment a human's natural knowledge and intellect rather than be left to make a final decision.

Healthcare is unique in that a digital-only approach is inconceivable. It appears that human intervention will always be required to mitigate risk and confirm predictions. Even with the most sophisticated AI, 53% of 1,000 respondents to a health survey stated they would want human verification of a machine-led diagnosis [155].

Affecting the Future

There are many indications that how people are using technology is changing how children and adults develop. Among many things, research now demonstrates too much screen time is associated with structural and functional changes in the brain, affecting attention, decision-making, and cognitive control [156].

Another study of 390,089 people by the University of Glasgow found an association between a high level of time in front of a digital display and poor health [157, 158]. What's more, scientists say the risk of developing cancer and cardiovascular disease is reduced if screen time is reduced to 120 minutes or less [159]. Screen consumption time is considered to be sedentary behavior. Excessive screen time has been demonstrated to impair the brain, particularly the frontal lobe, which is used by humans in all aspects of life.

The next few decades will illustrate the implications of connected technology on human physiology, psychology, evolution, and ultimately survival.

Playing God

Engineers from the University of Adelaide developed an AI agent that was able to predict when you were going to die. The AI agent analyzed 16,000 image features to understand the signs of diseased organs [160]. The neural network–based AI model was able to predict mortality within 5 years with a 69% accuracy rate. Enhancements in data access, AI, and predictive analytics are enabling humans to predict more precise morbidity and mortality risks, which reignites the debate as to whether AI is playing God or enabling humans (facilitated with a data-driven approach) to play God.

Overhype and Scaremongering

The capabilities of AI have been historically overhyped through marketing and simplified media representations. Media scaremongering, along with more public breaches of data, has also not helped the cause of AI.

There is a fog of confusion and misconception, particularly over the possible use (and misuse) of data and outputs, which needs to be resolved for stakeholders to seriously consider implementing AI into healthcare systems.

Stakeholder Buy-In and Alignment

It is pivotal to ensure all stakeholders are aligned in the health intelligence a project can provide and how it will be used. For instance, developers of AI for healthcare applications may have values that are not always aligned with the values of clinicians. There may be the temptation, for example, to guide systems toward clinical actions that would improve quality metrics but not necessarily patient care or skew data provided for public evaluation when being reviewed by potential regulators.

Policy, Law, and Regulation

The next decade will provide a pivotal juncture in the ethics of AI. Regulation in the form of laws, ethical guidelines, and industry policy will influence the direction of machine learning. So too is machine learning influenced in the absence of such policies and where much of the risk is considered to lie. By failing to set a perimeter of acceptability, we open the risk of wider unintended consequences.

Technical governance is needed, particularly for the application of AI. It is vital to engage key opinion leaders, stakeholders, lawmakers, and disruptors to form and nurture these policies to ensure the potential of machine learning is realized without undue restriction. Applications of AI in non-healthcare sectors have demonstrated that there are potential ethical problems with algorithmic learning when deployed at scale.

Lawmakers and critical stakeholders in the legal decision-making process must understand the complexities of AI and welcome the challenge of the first-time problems it brings. Furthermore, it cannot be assumed that AI will be solely used for benevolent purposes; AI is already used in warfare, with the United States and China spending heavily [161]. The United Kingdom has decided to focus on AI ethics [162].

Healthcare professionals using AI must understand how deployed AI algorithms are created, assess the data used for modeling, and understand how the agent can safeguard for mistakes. Regulators must be prepared to handle each concern individually, as all AI is individual, akin to humans.

Safe and inclusive AI development will only be achieved through a diverse, inclusive, and multidisciplinary team of engineers, humanitarians, and social scientists.

International engagement between organizations and governments on the critical topics raised by AI is a must. Questions of ethics, responsibility, employment, safety, and evolution are examples of conversations that will require stakeholders from the public sector, private sector, and academic sector to collaborate as AI embeds itself in society and to ensure humanity's democratized prosperity.

Data and Information Governance

Implementation of AI in healthcare requires addressing ethical challenges such as the potential for unethical or cheating algorithms, algorithms trained with incomplete or biased data, a lack of understanding of the limitations or extent of algorithms, and the effect of AI on the fundamental fiduciary relationship between physicians and patients.

A data governance policy is a documented set of guidelines for ensuring the proper management of an organization's data—digital or otherwise. Such guidelines can involve policies for business process management (BPM) and enterprise risk planning (ERP), as well as security, data quality, and privacy. This differs from information governance, which is a documented set of guidelines for ensuring the proper management of an organization's information. An information governance policy would cover the use of predictive analytics outcomes.

An ethical code of conduct or ethical governance policy would cover the organization's direction and intent and steer governance.

Is There Such a Thing as Too Much Policy?

There is a possibility of too much policy, hampering progress and disengaging stakeholders. AI has fueled innovation across all industries including digital health. Start-ups have led the way for AI pervasiveness because of less bureaucracy and red tape compared to organizations that are larger and typically slower to respond to environmental change. Audit your policies to ensure space for listening to the user's voice and incorporating it into cycles of ethical innovation.

Global Standards and Schemas

There are several bodies that exist that share common codes of ethics and responsibility, but no globally enforced schematics. An AI agent that predicts your risk of death has little validation other than you, or someone else it has predicted on, dying. How should such an agent be verified? Which standards must it uphold? Must the agent, or its creator, provide evidence? As yet, AI has no defined industry standards which technologies are tested against. As such, it is difficult to be assured of the quality of the AI you are engaging with. And in a fast-paced industry such as AI, the concerns presented are novel and immediate. Global standards are required to set a minimum expectation of AI systems.

Healthcare also suffers from a time lag from evidence-based to clinical acceptance. For example, people with type 2 diabetes have placed their type 2 diabetes into remission for well over a decade, but health systems have been slow to be able to register a patient as placing their type 2 diabetes into remission. Thus, patients have historically either been labeled as having reversed their type 2 diabetes or as living with type 2 diabetes. If patients were labeled as having reversed their diabetes, they often would not receive the tests that confirmed their type 2 diabetes was in remission and be at risk of not receiving checks. Equally, if the patient was labeled as having type 2 diabetes (and did not), this would have an impact on insurance and is technically untrue. Collaboration on schemas and parallel strides of adoption would help to reduce international disparities.

Do We Need to Treat AI with Humanity?

The basic concept of reinforcement is parallel to how humans and other animals learn. For instance, when training a dog, compliant and expected performance is often rewarded with a treat. Noncompliance by a reinforcement learning system is met with punishment. This is similar to reinforcement learning by an AI agent, where we build mechanisms of reward and punishment. Positive performance is reinforced with a virtual reward, and negative actions are penalized to augment avoidance.

The majority of AI systems are currently fairly simple and reductionist. As AI develops, we can expect them to become more complex and lifelike. Humans are already beginning to develop relationships with robots, in many cases seen as replacements for another human whether for companionship or carnal desire. It could be considered that a penalty input to an AI system is a harmful or negative input. When genetic algorithms delete generations that are no longer of use, is this a form of murder? At what point do we need to consider AI's humane treatment is catalyzed when AI systems can mirror both human behavior and appearance?

If AI systems are considered to be able to perceive, feel, and respond to stimuli, it is not a huge leap to consider their place as a species, legal status, or even nationality. An AI agent made in America was even given Saudi Arabian citizenship [163].

Should AI be treated like an animal of intelligence comparable to humans? Can machines truly feel and suffer? The moral questions of how we treat AI agents and mitigate suffering and the risk of negative outcomes are fundamental to responsibly progress AI toward its immense potential to better the lives of humanity.

Employing Data and AI Ethics Within Your Organization

Many organizations have a code of conduct that serves as an internal communication on expected behaviors, requirements, and engagement and provides external stakeholders confidence and reassurance.

Codes of conduct should be shared with employees alongside appropriate training to ensure the code of conduct is understood and observed. All stakeholders of an organization should practice and promote the code of conduct adherence.

Ethical Code

A code of ethics or ethical code is a document that is used to govern the moral conduct of an organization. A code of ethics demonstrates that an organization is committed to responsible business and technological progress.

The code of ethics narrates the behaviors that are promoted by an organization and those that are considered harmful to the organization's own moral compass, reputation, or clients. It may not cover actions that are illegal, but it will typically state the repercussions of noncompliance and how such violations can be reported. Employees should be made aware of items that are not obvious to them and then facilitated to avoid inadvertent yet potentially harmful actions. A code of ethics should also include a summary of the motivations for using data and its purpose in the organization's ambition—and reflect the profitability, integrity, and reputation of a business.

A code of ethics should be free of technical and philosophical jargon and directly communicate the expectations required of employees. Keep it simple and right to the point. Set the expectations for new employees at the time of joining and develop an organizational culture of adherence. Ensure all stakeholders are aware of amendments and reviews. Rather than only referring to the code of conduct to recruits or clients, refer to the ethical code frequently to engrain the ethical direction of your organization among internal and external stakeholders. Refer to the code of ethics when examining the ethical risks of incorporating new technologies into the decision-making process.

When developing your code of ethical conduct from scratch, consult employees and stakeholders for their contributions. Questions to consider asking include the following:

- What does AI ethics mean to you?
- How can what we do improve humanity?
- How should our organization act responsibly in its ambition to [insert ambition here]?
- What are the potential benefits of what we would like to achieve?
- What are the potential disadvantages of what we would like to achieve?
- How can we improve our ethical code?
- Are there items in the code of ethics that are confusing or require more explanation?
- Is the code of ethics useful in making decisions?
- Are the organization's ethics in alignment with your own ethical viewpoints?

Once all of the preceding questions are answered by different segments of the organization, a consensus can be reached on an implementation plan. An organizational code of ethics serves as a reference point for disciplinary actions for those who fail to meet the standards. The code of ethics provides a solid foundation for identifying and dealing with ethical challenges.

Ethical Framework Considerations

Organizations have a responsibility to be ethically guided in the collection and use of patient data. Organizations require a data privacy approach that prevents breaches and enforces security. Failing to adhere to regulations can lead to fines, reputational repercussions, and the loss of customers. There are various ways to mitigate the risks of improper data collection while taking advantage of the opportunities big data has to offer. There are several techniques to safeguard ethical data collection.

Collect the Minimal Amount of Data

The first step toward protecting user data is collecting only the data that is required for the task at hand. More data does not always mean more useful data. Although machine learning typically welcomes more data, data collection must be kept deliberate and concise. For instance, there is little to no use in knowing someone's ethnicity and BMI if they are applying for a credit card; but this same data is important when determining an individual's risk of type 2 diabetes.

Identify and Scrub Sensitive Data

All sensitive data should be identified and secured. Data scientists should be aware of what data is deemed personal and how to use such information. For instance, data collected on consumers that is collected without their consent should be scrubbed of personally identifiable data. It is paramount that only individuals with the appropriate permissions can observe sensitive data. Many organizations also discourage the use of USB sticks and external data drives to ensure data remains safe and on-site.

Compliance with Applicable Laws and Regulations

Trust in your organization's approach to data, and information management can be garnered through adherence to appropriate local and national standards, policies, and laws. There are several data and information governance laws and regulations that set the boundaries of legal and appropriate data usage. It is vital that all regulations are abided to.

Bodies such as the US FDA (US Food and Drug Administration) and EU MHRA (EU Medicines and Healthcare products Regulatory Agency) set boundaries for the legal, ethical, and compliant collection, usage, and management of data. Compliance with the

GDPR and approval from governance-related accreditation bodies demonstrate your organization's approach to ethical data usage and build trust with your users.

International Standards

Many national- and international-level standards will have a fair degree of overlap, requiring appropriate data reporting, documentation, transparency, and risk mitigation procedures. Within Europe, for example, adherence to regulations such as GDPR is a necessity and regulates over 27 countries in the European Union. International standards such as ISO 27001 ensure good data and information management standards. ISO 27001 (formally known as ISO/IEC 27001:2005) is a specification for an information security management system (ISMS). An ISMS is a framework of policies and procedures that includes all legal, physical, and technical controls involved in an organization's information risk management processes.

Many data regulations are self-accredited. This means that as an organization, you must collect the documentation required to demonstrate adherence to standards without the requirement for it to ever be verified by an external body. The concerns of self-accreditation are apparent, particularly when applied in healthcare.

Robust data and information governance is the foundation of data ethics.

Founding Data Ethics

Thorough data and information governance is the foundation of data ethics. Governance is concerned with how data and information is protected in its collection, use, analysis, and disposal. Good governance covers a variety of topics including data architecture, infrastructure, risk assessments, audit reports, risk treatment plans, design workflows, information security policies, complaint procedures, medical governance, corporate responsibility, and disaster recovery planning. An organizational ethical code steers the direction of good governance.

Auditing Your Frameworks

Conduct regular auditing of ethical data and information governance procedures and rehearse worst-case scenarios to ensure that if the worst were to happen, your organization would be able to minimize as much risk in as little time as possible. Risk mitigation drills should be rehearsed at least twice a year.

Review each section of your code of conduct to ensure it represents the values of your organization. Technology organizations move rapidly, so ensure compliance with the latest standards, guidelines, and policies relevant to your industry. Organizations operating with medical data may come up with new methods of prediction or new AI abilities. Consider aspects of your code of conduct that are missing, particularly if your organization is scaling or facing new use cases. For example, a historically business-oriented organization that begins to engage with patients will require a new section of engaging with patients.

There is no reason not to engage every member of staff in developing and reviewing your code of conduct. Digital surveys are a quick and easy way to funnel opinion. Diverse contributions make for diverse, inclusive, and multidisciplinary approaches to AI. Ask for comments on new sections or concepts you feel employees should be aware of. Conversations can highlight areas in which employees may need further training, awareness, or demonstration. Develop a dialogue with employees to engrain the organizational culture within the wider team and foster an ethically aware environment. Having the maximum number of stakeholders on board with your organization's code of conduct will likely increase employee motivation and productivity.

An ethical framework in healthcare organizations, particularly for those using AI, provides internal and external stakeholders the security, transparency, trust, and direction required for maximizing engagement. Adhering to industry standards and involving members of your organization in developing and maintaining the ethical framework will accelerate employee and organization-wide adoption. Moreover, good policy documentation, training, and auditing help maintain the integrity of organizational AI.

A Hippocratic Oath for Data Scientists

Data scientists with access to sensitive data are typically required to sign documentation to ensure they comply with risk mitigation and security policies when they begin employment. It has been suggested that to ingrain a culture of safety and positive progression in AI, data scientists acting in digital and nondigital sectors pledge allegiance to a Hippocratic oath to do no harm. This would sit alongside regulations, internal manifestos, standards, and codes of conduct operated by an organization [164].

Conclusion

Meaningful changes in healthcare will come from the right blend of innovation, ethics, and deliberation. As the boundaries of discovery are tested, the potential of providing patients precise evidence-based medicine and care is being realized; and with this there is an obligation to explore and anticipate the social, economic, political, and ethical consequences of innovation.

Big data and AI's adoption in healthcare is best managed as a gradual development, which would be useful for stakeholders to have time to map and understand the possible consequences of its implementation. Robust standards are required for auditing and understanding AI models for upholding quality assurance and trust.

The development of AI standards requires a diverse and public forum that is academically robust and reviewed periodically in line with the latest evidence base.

Awareness within the medical professional and patient communities is required. Healthcare professionals need basic knowledge of how AI applications function in their respective medical settings to understand the advantages, disadvantages, and consequences of how AI can facilitate and assist them in their everyday tasks.

Patients must become accustomed to engaging with AI systems and explore the benefits for themselves, and COVID-19 certainly expedited the process. Demonstrable examples of how AI in healthcare is optimizing efficiency, optimizing resources, and improving population health outcomes are only strengthening the public perception and adoption of AI.

Organizations working with AI need to communicate and engage the general public about the potential advantages and risks of using AI in medicine. Organizations developing AI must work with academia to take the necessary steps to be able to measure the success and the effectiveness of AI systems independently. It is vitally important to encourage affordable AI solutions to democratize AI as the stethoscope of the twenty-first century; and published validated evidence, guidelines, and literature are essential to AI's continued trust, adoption, and morality.

CHAPTER 9

What Is the Future of Healthcare?

> *Medicine is at a critical juncture where it is necessary to keep up with the speed of discovery in the real-world.*
>
> —Charlotte Summers

New advances in medicine and concepts that were once the futuristic topics of science fiction are gradually becoming a reality. Gene therapies, 3-D printing of human organs, liquid biopsies, robot-assisted surgeries, and voice-enabled personal assistants are now realities that are becoming more sophisticated as time goes on [165–168].

Advancements in technology are affecting not only practiced medicine but the public perceptions and attitudes toward health, lifestyle, and what it means to be healthy. Healthcare must implement innovation wisely to engage as many patients as possible.

Digital health technologies evolve at a lightning pace. The impact of data science, AI, machine learning, and connected healthcare technologies is tremendous and requires an unbiased mind and willingness to engage in a continually evolving environment with as much knowledge as possible.

This chapter will explore the shifting trends in healthcare and the applications of machine learning and AI in realizing the future personalized medicine. Readers will also explore how immersive reality can be used to augment healthcare and how robots, drones, and sensors can be used to develop smart health.

© Arjun Panesar 2021
A. Panesar, *Machine Learning and AI for Healthcare*, https://doi.org/10.1007/978-1-4842-6537-6_9

Shift from Volume to Value

Healthcare as an industry is transitioning from a volume- to patient-centric, value-based care in its pursuit of evidence-based, personalized medicine.

There is a shift in medicine in which pharmacology is not the first and only line of treatment; and there is a greater emphasis on the role of lifestyle as medicine for preventative and therapeutic purposes.

What Is Volume-Based Care?

Traditional payment systems in healthcare are grounded in rewarding providers for the volume of patients referred and treated. Volume-based care focuses on economies of scale. Providers receive a discrete amount of money for all of the services provided to patients for a set duration of time. Value for money, patient experience, and healthcare quality are subsequently secondary considerations for evaluation. Providers are incentivized by patient volume and the cost of the overall care provided to patients. Healthcare resource and cost pressures, exasperated doctors, and incentivized clinicians have unknowingly facilitated a "see as many patients as possible" approach that has driven disease management from a medication-first, patient-second perspective. Hospitals and clinicians are incentivized to see as many patients as possible, conduct as many tests as possible, and approach disease from a medication-first perspective.

Volume-based care models are typically modeled on financial metrics such as optimizing profit or minimizing cost per patient rather than health outcomes.

Volume-based care payment systems often penalize providers financially for keeping people healthy, error reduction, or complication reduction, as they are not a driving KPI. For instance, if you were to be diagnosed with type 2 diabetes, your healthcare team would typically do the following:

- Be given information on how to manage blood glucose levels per current guidance.
- Progress on a treatment regime.
- Be given information about dietary choices.

In the United Kingdom, for example, if a doctor's surgery or clinic were to enable the patient to realize type 2 diabetes remission, the surgery or clinic would fail to receive the money it would have for the drugs no longer prescribed to the patient. Personal biases can also complicate matters. In the United Kingdom, for example, doctors involved in assessing which drugs should be prescribed to NHS patients were found to have received payments of up to £100,000 per year from pharmaceutical companies [169-171].

Volume-based care is focused on a finite amount of money for a population. With populations growing and resources stretched, the evolution toward value-based care has been expedited. Healthcare is now in a transition toward patient-based care, which is synonymous with value, rather than volume.

Patient-Centered Care

Tseng et al. defined the value in healthcare to be the quality of care, typically measured through healthcare outcomes [172]. A value metric also for consideration is patient experience or patient-centeredness. Patient-centeredness is an important but not necessarily dominant quality measure for value-based care.

Patient-centered care utilizes a multifaceted approach, revolving around the patient, their goals, and the wider family in decisions and evaluations. There are several domains of patient-centered care, which include patient-centeredness.

Patient-centered care is a collaborative and multifaceted approach involving delivering health and social care focused on patient experience and patient outcomes. Patient-centered care focuses on empowering patients with the knowledge and experience, skills, and confidence to manage and optimize their health. Care is compassionate, personalized, coordinated around the patient, and respectful of the patient's worldview. These goals may appear common sense for healthcare practitioners. However, they have not always been performed as standard practice. Healthcare has typically been reductionist in its approach, addressing a sum rather than the whole and acting for patients rather than with patients.

Organizations like Buurtzorg Neighbourhood Care in the Netherlands deliver patient-centered care. The organization delivers care to 80,000 patients a year with 10,000 nurses, 21 coaches, and two managers [173]. The collaborative approach focuses on empowering patients through relationships with caregivers. Over time, patients see improvements in health status while the total external care time required is reduced. Patients and families are enabled with the skills and confidence to look after themselves. Digital therapies are able to provide patient-centered care by their nature, tailoring experiences based on a user's data—profile, behavior, time, location, or otherwise—and provide remote monitoring capabilities to healthcare professionals at the same time [174].

Value-Based Care

Although two distinct philosophies, patient-centered care and value-based care are becoming synonymous. Definitions of success incorporating patient-centered outcomes, perspectives, experiences, and preferences are positively influencing quality and delivery of care.

Value-based care does not mean cheap or economy care. Indeed, care is evaluated through the quality of patient experience and outcomes; and the model reshapes how care is received by patients. Value-based care is redefining the economics of wellness.

Wellness and prevention are emphasized in value-based care. Lifestyle, behavior, and environmental factors account for 90% of disease risk [175]. This has been evidenced by the growing prevalence of type 2 diabetes across the world. With growing pressure on healthcare resources, lifestyle medicine has silently refocused efforts on the role of lifestyle as medicine. Evidence demonstrates that activity, nutrition, sleep, and stress all impact physical and emotional health markers [176].

Lifestyle medicine's focus on lifestyle and behavior enables healthcare professionals to deal with reversible, non-communicable diseases such as type 2 diabetes and obesity. Rather than treating the patient in a reductionist approach, the patient is provided with holistic, compassionate care. The prescription of lifestyle medicine is already taking place in healthcare, with digital applications prescribed to treat disease. Type 2 diabetes patients, for example, may be prescribed the Low Carb Program app by their physician to support patients to self-manage their condition and achieve weight loss and type 2 diabetes remission where possible [177]. Lifestyle interventions are typically safe, easy to implement, and scalable. Growing levels of chronic disease can be mitigated with a simple approach delivered in an engaging and effective setting.

Redefining the Approach to Care: Quality, Not Quantity

Instead of reacting to events, there is a preventative, lifestyle-first approach to health in value-based care. The focus is on patient wellness, healthcare quality, and efficiency. This is where patient data, digital health, and AI hold a fountain of promise. The emphasis on efficiency ensures that the approach is scalable and effective, enabling providers to reduce healthcare costs and improve patient clinical health outcomes.

Prevention of disease through quitting smoking, nutritional change, lifestyle change, activity, sleep, and identifying genetic risk factors reduces the burden on healthcare resources. Wellness is becoming incentivized. It has become the best interest of insurance companies and healthcare providers to monitor their patients' health more closely.

Start-ups and digital technology companies are developing digital health tools that are disrupting the traditional doctor–patient relationship and have enabled providers to become third actors involved in facilitating sustained health. The goal of value-based care is to standardize healthcare processes through best practices and to democratize access and quality of care. Mining of data and historical evidence can determine which methods work and which don't.

Digital Health Enables a Patient-Centric, Value-Based Paradigm

Keeping people well reduces healthcare delivery costs and optimizes the use of resources. For instance, when managing a chronic condition such as type 2 diabetes, value-based care uses a collaborative and multidisciplinary approach to managing the disease to prevent diabetes-related complications.

Patients engage with one healthcare team that is aware of patient progress and health status. The healthcare team may contain the patient's diabetes nurse, dietician, behavior health coach, and other professionals to support patient progress.

The team would set patient-centered goals and help a patient with the following:

- Maintain blood glucose control.
- Introduce a digital community of people with type 2 diabetes for support.
- Provide health mentoring and habit maintenance support.
- Facilitate an activity program.
- Use the latest evidence base to provide nutritional guidance.
- Deal with the psychological aspects of type 2 diabetes.

Incentivization is transformed. Take a hospital setting, for instance, in which rather than being paid on how many patients could be seen, the hospital is being compensated based on the opposite. The hospital is paid based on how many patients are in positive health status and how many beds are available. The focus moves from the point of hospital admission to predicting the likelihood of future risks and anticipating them with preventative solutions.

Aside from face-to-face delivery, ongoing patient support can be provided digitally, for example, with health coaching, delivering an exercise program, or help with mental health concerns through apps, wearable technology, or telemedicine. AI and predictive analytics enable healthcare providers to optimize the delivery of healthcare, focusing on preventing and treating disease.

Value-based care can utilize low-cost digital technologies to enhance and democratize care. The simultaneous collection of behavioral, demographic, health, and engagement data provides an opportunity for machine learning and development of novel AI, to rapidly improve, and learn from, user behavior and outcomes.

Managing healthcare operations and subsequently evaluating healthcare quality are complex. To assess a value-based care model, healthcare organizations must collect and analyze data, objectively assessing performance through the quality of care, patient health outcomes, and efficiency of cost. Providers can report and model preventative care metrics such as hospital readmission rates, error rates, disease progression, population health improvement, and engagement strategies.

Healthcare quality is defined by several metrics. Patient experience and satisfaction scoring is typically the first metric of evaluation. Metrics such as time spent per patient, patient engagement, medication savings, and adherence are examples of quality that relates to productivity.

Many applications of AI in healthcare are within a hospital setting. Hospital management platforms are delivering a return on investment by enabling organizations to prepare. Systems that predict times of peak patient traffic, predict readmission times, and use real-time data to cope with the demand of real-world healthcare are seen as long-term systems of value to healthcare providers.

Applications involved in improving clinical care demonstrate enormous potential, and long-term studies are required to evaluate the impact on disease management and population health.

Data linkages between primary and secondary care can enable tracking of a network of data points including medication consumption, hospital admissions, and A&E admissions which can be used to develop incentivized reimbursement models.

Commentary: No Funding? No Problem? Value-Based Care in a Public Healthcare System

The following is taken from a blog written for the NHS Innovation Accelerator on innovating funding models to scale digital health.

You've got a fantastic innovation that can help millions of people, it's loved by the NHS teams you've demonstrated it to, but the local NHS area can't afford it. There's no funding.

There are lots of reasons that innovations don't work within the NHS, but when the only reason innovations aren't being adopted is a lack of funding, not only is it a frustration, it's to the detriment of patient and population health. Funding often halts a great vision in its tracks preventing even the lightest of explorations. What's more, it is a massive frustration for healthcare teams who invest time, energy and passion into curating projects even before a business case is written. And for the innovators and NHS teams involved it can be just down-right demoralising—how can we possibly influence positive change?

As we engaged with NHS localities, stakeholders rejoiced in the idea of scaling improved type 2 diabetes management and remission with a scalable, engaging platform. The patient health benefits and savings in de-prescribing were evident—but still, it was cumbersome to find a budget. And even if there was a budget, it would be too small to truly impact health at a population level.

As innovators, it's not just services we are innovating—it's pathways of delivery. Global healthcare is in a transition toward patient-based care, synonymous with value, rather than volume.

With this in mind and our experience in other countries, earlier this year we implemented our first "gain-share" model with an NHS CCG. The gain-share approach is a paid-on-results model, which, rather than requiring immediate upfront budget, can share in the success of innovation delivery. In our case, in sharing the savings of de-medication through patients who are de-prescribed medication as they improve their blood glucose control.

It works well for all as it requires much smaller investment in the short-term, with the outcomes celebrated and shared in the long-term. It makes sense—if everyone is in it for the "gains", then there is complete involvement and greater level of collaboration. It's one team—rather than two—working towards shared goals.

The model reshapes how care is received by patients. What's more, with lifestyle, behavior, and environmental factors accounting for 90% of disease risk, it enables a scalable approach to population health.

As NHS innovators, it is fantastic to actualise the benefits of digital health. In particular, democratisation of access to education and self-management for people to live healthier, happier lives.

Evidence-Based Medicine

As shown in Figure 91, evidence-based medicine solves a clinical problem by integrating best research and clinical evidence with real-world clinical expertise and patient values. In the case of healthcare, this is deemed to be randomized controlled trials and increasing real-world evidence. Datafication and the IoT are contributing significantly to the evidence base.

Evidence-based medicine is underpinned by decision-making that is based on the latest, most reliable scientific evidence. And as we develop new models of healthcare, unbiased evidence is critical to understand and trust services and innovations.

David Sackett, a pioneer of evidence-based medicine, defined the concept as "the conscientious explicit and judicious use of current best evidence in making decisions about the care of individual patients" [178].

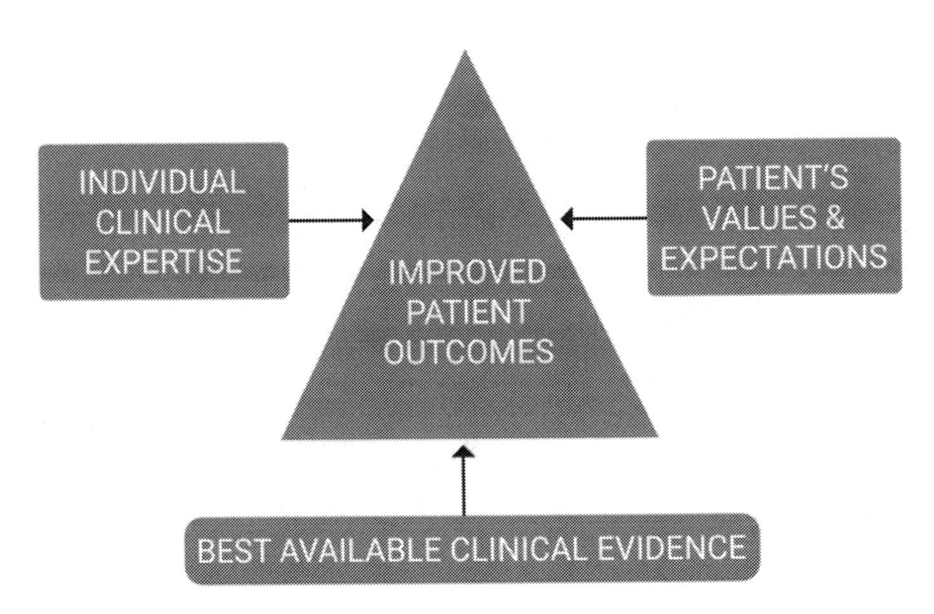

***Figure 9-1.** Evidence-based medicine*

Randomized controlled trials (RCTs) are deemed as the most robust and reliable form of evidence for assessing the efficacy of a treatment. Evidence demonstrates that many factors can influence the reliability of RCTs, including methodological quality, reporting quality, and source of funding. "Big" companies fund the majority of clinical research that is undertaken and face an enormous conflict of interest. Studies have demonstrated bias in favor of research funded by industry, which undermines confidence in medical knowledge. History has shown some organizations have chosen profits over people. Coca-Cola paid researchers to downplay the links between sugar and obesity [179].

Real-world evidence comes at a critical juncture for evidence-based medicine [27]. Real-world evidence is disrupting traditional evidence-based hierarchies and approaches. Datafication of human experience through mobile phones, social media, digital communities, health apps, nutrition tracking, wearables, and health IoT has empowered patients to become their own evidence base and influenced healthcare academia and understanding. Patients, for example, can compare the results of different hospitals, or even individual doctors, to best choose their treatment providers.

Quite simply, by the time an RCT has started and finished, a digital health app may have hundreds of thousands of members. Other channels of validating digital healthcare and innovation as clinically safe and effective are required [16].

Digital Health Research Poses a Question of Validity

Most RCTs conducted within pharmacology are internally valid, ensuring an exact population is involved in the testing of a particular drug. For example, a trial for high blood pressure medication may focus on people who have high blood pressure only and no other comorbidities. In the real world, the comorbidities of high blood cholesterol and triglycerides may be common in people with high blood pressure. This approach means there is often a gap between what is reported in RCTs and what is reported by patients. This is where digital technology can help bridge the gap between pharmacological intervention and real-world patient experience by providing data to connect the dots.

Digital communities can provide real-world evidence on medication side effects. Community app Diabetes Forum provides medication side effect data back to pharmaceutical organizations through a mobile application in the form of analysis of patient discussion and reported side effects. The world's second most commonly prescribed drug for diabetes, metformin, has a side effect of diarrhea reported to affect one in ten people taking the drug. Real-world evidence from members demonstrates this is 400% more common and reported by over 48% of people with type 2 diabetes [181].

Real-world evidence has historically been viewed as anecdotal by the medical community. However, the datafication of patient lives now means that data is directly received from devices, wearables, and sensors, ensuring that data cannot be misreported. Real-world evidence is growing at a pace that healthcare is struggling to keep up with. Digital health platforms and the aggregation of data also call into question previous paradigms of medicine.

How Does Evidence-Based Medicine Keep Up with the Real World?

It is a necessity for evidence-based medicine to keep up with the developments of real-world evidence. As David Sackett said, "Half of what you learn in medical school will be out of date in 5 years of your graduation; the trouble is nobody can tell you which half. The most important thing to learn is how to learn on your own." Patients now use the Internet, themselves, and their peers as the evidence base; and ubiquity of real-world evidence and AI in healthcare means that healthcare professionals can no longer be ignorant to change in the evidence base and biased toward old paradigms.

This data is a wealth of opportunity for AI and machine learning. Healthcare can adopt a data-driven approach, utilizing a variety of data sources to improve the patient experience and reduce costs. The exploration of this data over the ensuing decades will significantly progress healthcare and facilitate precision medicine.

Personalized Medicine

Personalized medicine, also known as stratified or precision medicine, is an approach that stratifies patients into groups and makes informed clinical decisions for the delivery of treatment and interventions based on the patients' anticipated response. As the pinnacle of evidence-based medicine, personalized medicine approaches to health are tailored to the patient. Patient health is managed on an individual level to achieve the most optimal state of health possible.

Around 10% of disease risk is based on genetics [182]. Each person has a unique version of the human genome. Groups of particular patients may share common genome characteristics and thus the risk of disease. Our genetic variations determine how our bodies would respond to a particular drug or determine risk of disease. For instance, research demonstrates people of South Asian descent have a higher prevalence of type 2 diabetes compared to Caucasians [183]. One drug might not fit everyone's requirement. Two people taking the same dose of the same drug might respond differently. Using personalized medicine, a right combination of a drug and its dose can be selected for everyone.

This approach to medicine is not new, and healthcare professionals have been using this approach since the time of Hippocrates.

However, the hyper-personalization achievable from the use of the patient genome and its falling cost—and medical data from health records, wearable technology, and the health IoT—have roused the public interest once again. It has never before been possible to predict the risk of disease, how the human body will respond to a particular medication, or print treatments out of materials other than ink. The combination of these technologies will fuel an era of personalized care and healthcare innovation.

Predictive tools can be used to evaluate health risks and develop personalized healthcare plans to mitigate patient health risks, prevent disease, manage disease, and treat a disease precisely if it occurs. As healthcare becomes increasingly personalized for patients in both treatment and service delivery, it is critical that access is widened to ensure participation from all groups of society.

Diagnostic tests, such as blood tests, are typically used to identify appropriate treatments based on a patient's physiological analysis. The selection of optimal therapies will be increasingly personalized to the patient genome. Healthcare providers will be able to diagnose current illnesses, predict future risks of diseases, and identify predicted responses to treatments and subtle traits within an instant. Genetic testing has started to make an impact in personalized medicine; DNA results are limited to being imported into services that can personalize treatment regimes, diet plans, or education to a patient. Currently, DNA test results take days to receive. We will soon be in an era of instant, inexpensive genetic testing that will empower patients to make informed choices about treatment, services, products, medications, and outcomes.

Personalization of Medicine Raises Ethical Issues

Raising serious ethical issues, the decision to have a genetic test and the implications of its results deserve careful preparation and thought. For providers, it provides an opportunity to fully engage in personalized, value-based care, with the quality of patient experience paramount. For patients, it can pinpoint genetic mutations that predispose a person to disease. The physical and mental consequences of such knowledge can be profound and have caused controversy among the medical establishment and general public.

A natural division among all people, regarding specific issues, is to be either for or against the topic of discussion. Progress is not without moral consequence, and legislation must endure the speed at which first-time problems are occurring. People are conflicted about having a lot of information about themselves.

Personalizing Medicine with Data

Wearables and the IoT hold the key to realizing personalized medicine, also known as smart medicine. Data captured by sensors in wearable contraptions are playing an increasingly powerful role in healthcare, facilitating the development of patient-centric healthcare systems.

Many factors are accelerating the acceptance of wearable healthcare solutions, particularly their use in clinical trials and academic studies to monitor patient health and lifestyle factors. Studies can record participants' health vitals using an Android Watch, Apple Watch, Garmin, Fitbit, or another smartwatch device. Participants use apps to record their lifestyle habits, nutrition, activity, and medication adherence and to monitor medication side effects among others. Wearables and patient data are beginning to be used by insurance companies to incentivize wellness.

Insurance companies have historically targeted products toward digital-savvy buyers, offering the latest gadgets to incentivize health improvements. Incentivized insurance products will become commonplace for wider populations, proving particularly effective in the combat against non-communicable diseases.

The delivery of connected, digital healthcare opens an opportunity to monitor, manage, and reverse disease, broadening the risk portfolio and creating a longer-living, reduced-morbidity population. The cost of ill-health is just too high for bill payers to ignore.

Rich, Sensor-Driven Data

As sensors become faster, smaller, and more capable, patient health profiles will eventually comprise detailed sleep analyses, details of continuous blood glucose monitoring, heart rate, blood pressure, and approximate calorie burn. A smartwatch will combine a variety of diagnostic tools, able to monitor blood pressure, heart rate variability, blood glucose, ketones, and more. Health sensors will become embeddable, biodegradable, and constantly connected, playing critical roles in such tasks as patient care. Behavior economics will play a larger insight into our everyday lives.

Table 9-1 briefly describes some examples of connected medicine devices, explaining what device can be used where and why it can be useful. Some of the wearables and embedded sensors involved in connected medicine include the following:

- Stress bands

  Stress tracking and calming of the mind are two key markets for wearables. Wearables currently monitor breathing and heart rate, detecting signs of tension to improve breathing and the ability to reach a calmer state of mind. Fitness trackers offer mindfulness functions, and there are headbands that deliver waveforms to the brain [185].

- UV sensor

  Northwestern University has developed UV sensors that precisely monitor a person's UV light exposure. The wearable sensors, which are small enough to fit on a human fingernail, are wafer-thin. As UV is a known and potent carcinogen, this has the potential to reduce overexposure to UV and prevent complications such as heatstroke and melanoma [186].

- Smart tattoos

  Smart tattoos place the sensor in the skin with significant strides made by researchers at MIT and Harvard [187]. Smart tattoo ink reacts with the biochemical composition of the interstitial fluid to indicate the status. Developments will allow for tattoos to degrade, lasting for as long as required and even only displaying in particular lights. For instance, high blood glucose could turn a blood glucose smart tattoo red, whereas low blood glucose may turn the tattoo blue.

- Smart medication

  There are many directions to which medication is being made smart. Smart medication can tell a patient and their respective healthcare team as to the amount of medicine consumed, the time it was taken, and provide timely reminders if it looks like it may be missed. Today, Bluetooth-connected bottle caps and pill packets provide reminders to patients to encourage medication adherence. Bluetooth insulin pens and pen caps that log the time, amount, and kind of insulin injected typically push data from the device to the cloud, with users engaging in a digital app interface to query their data. Smart asthma inhalers, for instance, can identify an oncoming attack before the wearer recognizes the symptoms [188].

- Smart insulin

  A form of smart medicine, smart insulin is next-generation insulin that responds automatically to changing blood glucose levels. The lower or higher blood sugar levels are, less or more insulin is released, respectively. The University of North Carolina researchers reported that their smart insulin patch, which rests on the outside of the body, will use a system of micro-needles to automatically detect high blood glucose levels and administer insulin through live beta cells appropriately. Because the beta cells are kept within the patch on the outside of the body, there is zero danger of them being rejected by the immune systems of people with type 1 diabetes [189].

Table 9-1. *Applications of wearable technology*

What?	Where?	Why?
Immersive technology **Military apparel** **Helmets** **Mixed reality**	Head	Education Behavior change Intelligence to intelligence communication
Smart contact lenses **Trackers**	Eyes	Blood glucose levels
Hearing aids **Headphones** **Trackers**	Ears	Sound
Odor detection	Nose	Smell
Smart tattoos **Trackers** **Patches** **Implantables** **Smartwatch Trackers**	Arms/Wrist	Blood glucose Blood pressure Oxygen saturation Ketone levels Education Rehabilitation
Clothing **Chest straps** **Implantables** **Trackers** **Exoskeleton**	Body	Rehabilitation
Clothing	Legs	Protection Rehabilitation
Embedded footwear	Feet	Health metrics Posture correction Rehabilitation

Patients immediately turn to Dr. Google at the first sign or symptom of potential illness. It's become a running joke that online searches and symptom checkers eventually tell you, regardless of symptom, that you have cancer [190].

Patients are best placed to understand what is happening to their body, now visiting the doctor more prepared than ever. More informed, more aware, and more health-conscious patients now attend appointments, with their engagement used to personalize their healthcare plans. However, this also causes varying degrees of mental anguish and hypochondria.

Patients must not forget that qualified healthcare professionals provide a much more complete assessment and explanation of health, particularly if up to date with the evidence base. This is perhaps a contributing factor for why the medical profession has not always welcomed a patient-centric approach. A 2019 survey reported 87% of patients had a better relationship with their healthcare professionals as the result of using the platform, with 75% reporting a better understanding of their condition. When this survey was conducted in 2015, the same number reported a better understanding of their condition, but only 20% of patients reported improved relationships with their healthcare team [191].

Connected medicine provides a wealth of opportunity in solving novel health problems, particularly with disease management and elderly care.

In 2015, the United Nations approximated 1.2 in 10 people were over the age of 65. Projections estimate that this will jump to 22% by 2050 [192]. Healthcare expenditure on the elderly is a mounting concern, as it accounts for a higher share of healthcare expenditure compared to other age groups.

Non-communicable diseases such as type 2 diabetes and obesity are pandemics, with the WHO estimating 600 million people will develop type 2 diabetes by 2050 [193]. About one in four people suffers from two or more chronic conditions [194].

With rising pressure on governments, payers, and manufacturers to reduce healthcare costs, elderly and long-term care require solutions that are prepared for the impending rise in numbers. The Internet of Medical Things (medical IoT) demonstrates a tremendous potential to accelerate value-based care through providing remote monitoring and predictive and diagnostic capabilities.

Applications of Personalized Medicine

Driven by rich data, machine learning is transforming healthcare, enabling it to be personalized in a growing number of domains.

Disease and Condition Management

Heart monitors can detect heart rate variability and alert the healthcare professional in real time. The same concept is liberating for parents of children with type 1 diabetes using smart blood glucose monitors. Blood glucose measurements are sent directly to the cloud where parents can access their children's blood glucose in near real time and be alerted to potential hypoglycemic events. Digital apps can be delivered to reverse type 2 diabetes, manage epilepsy, and improve quality of life.

Novel utilization of technology can minimize the risk of health concerns and their subsequent healthcare costs. Elderly care can be delivered through digital apps. Risks such as falling, for instance, can be mitigated through fall detection and emergency assistance.

Accelerometers found in phones or watches can be used to detect falls or seizures, triggering notifications to carers of an emergency, its details, and location as they emerge. Future technology could also contain safety equipment. Wearable adult airbags can be used by those with mobility problems. Adult airbags deploy if they detect the wearer is falling to prevent injury [195]. Technology can help save the cost of avoidable elderly care and substantial healthcare-related costs.

However, the true value of the technology, especially connected assistants and wearables, in the provision of elderly care is realized in the predictive analytics that can be generated from the data collected. Deviations from daily routines, for example, could indicate cause for concern and emerging physical or mental health concerns and predict risk of illness. As sensor data and evidence expands, so too does the application of wearables in detecting signs of disease. Atrial fibrillation, for example, can be predicted from heart rate variability collected through an Apple Watch [196, 197].

Virtual Assistants

Virtual assistants are not just entertaining; they can accompany those who live alone and provide support to the aging and elderly. Virtual assistants can empower people by answering questions, teaching new skills, setting verbal reminders to perform tasks such as reminding to take medication, and even controlling the home by performing tasks such as picking up the telephone or making a call. As homes become increasingly connected, virtual assistants will assist people in every aspect of life and interface with all aspects of the home.

Virtual assistants could call the emergency services on behalf of a patient, citing exact details and location. Connected cameras can track movement to detect falling, triggering the virtual assistant to reach out for help. Not confined to the realms of speech, virtual assistants in the form of robots can assist in particular situations within elderly care, performing roles such as helping people get out of bed, into the bath, or into a wheelchair. AI within such virtual assistants could even learn patterns, for example, when people may want to visit the bathroom.

Interaction through voice and touchscreen enables communication and will increasingly empower aging patients with greater autonomy and reduce the burden on elderly care. MiiCare, a company from the United Kingdom, provides remote monitoring through device-driven virtual assistants to look after aging populations, measuring patient temperature and monitoring movement, hydration, and medication adherence.

As the capabilities of virtual assistants improve, so too will their ability to determine the emotions of their master, whether it be happy or sad, excited or depressed. Virtual assistants will interpret the syntax, semantics, and tone of the conversation to check for signs of mental or emotional health concerns.

A similar opportunity exists for virtual assistants to provide useful assistance to people with disabilities. Apps can be used to control lights, music, heating, food orders, and lifestyle. Nominet, a UK Internet organization, developed an open source prototype called Pips to cater to people with sensory or cognitive impairments [198]. People are empowered with audio and visual prompts, nudges, and reminders to go through their daily routines and to develop the skills and confidence to navigate their environment. The prototype is built with a low-energy Bluetooth controller that sends and receives notifications between devices. The devices can be used to guide people through tasks such as washing their face, taking a bath, or taking their medication.

Remote Monitoring

Connected devices can provide support for caregivers and healthcare professionals to keep people out of a hospital and reduce their visits to the doctor. Apps and virtual assistants can provide medication reminders; fitness trackers will monitor user health and movement, alerting care providers if deviations from a routine were to occur.

Connected devices can alert family and care providers of events such as using a fridge, having a bath, or opening the front door. Aspects of routine demonstrated in the data would be learned and respective accounts notified if there were to be a deviation from the norm. Emergency buttons can alert the relevant services, ensuring that the elderly or those with long-term conditions can receive medical intervention when required most. The combination of connected devices, apps, and virtual assistants makes the home smarter and more conducive to mitigating risk, ensuring the healthcare provider/clinic is up to date with the latest overview of patient health while empowering the patient.

Gro Health, a lifestyle and behavior support app, co-designed a real-time remote monitoring dashboard with a team of a dozen physicians to provide real-time care to patients in British Columbia, Canada. The platform alerts clinicians should patients not adhere to treatment plans or disengage from the platform [199].

Medication Adherence

Forgetting to take medicine is easy enough, but the implications can be a problem. More than 50% of prescribed medications are not taken as directed [200]. Missed doses can result in slower recovery time, exasperate illness, or more serious consequences. A study by Benjamin et al. showed that getting patients to take their medications appropriately could prevent approximately 125,000 deaths per year in the United States alone [200].

There are many apps and digital services that remind patients when to take their medication, but medicine is going beyond even that. Smart pills with ingestible sensors seek to help doctors improve clinical health outcomes through notifying their respective app on confirmation that a medication has been consumed [201]. The notification is triggered when the sensor comes into contact with the stomach fluid. This enables doctors to measure treatment effectiveness and optimize the patient pathway.

Accessible Diagnostic Tests

People with chronic conditions and aging populations typically need more diagnostic tests than those who are healthy. Instead of turning up to a clinic or pharmacy to have urine and blood tests conducted, smart sensors in portable devices allow the test to be conducted in the comfort of the user's home. Results can then be wirelessly shared to their healthcare team.

Tests for cholesterol, cardiac function, HbA1c, fasting blood glucose, vitamin D levels, and insulin improve the convenience of performing repeated diagnostic tests, enabling quicker treatment, reduced risk of complications, and reduction in avoidable healthcare spending.

Smart Implantables

Implantable devices that are connected to a smartphone, healthcare provider, and wider networks will be used to monitor and treat disease in real time. The possibilities from this data are huge, as wearables are external implantables that have the potential to collect metrics on biological markers from organs and tissue.

Telemedicine

Telemedicine, which can be traced back to the 1900s, refers to the clinical application of electronic technologies to support long-distance patient care, patient and healthcare professional education, and health administration. The terms telehealth and telemedicine are synonymous. More recently, the term digital health has replaced telehealth as the reach of medicine has moved from the telephone-based devices to connected devices.

Consultations through Skype of video-conferencing facilities, EHR remote monitoring, digital health education, and transmission of scans or images among other applications are all considered part of telemedicine [202].

The world is truly mobile, and digital tools aim to deal primarily with minor primary care issues through electronic means. More serious issues would be escalated to the relevant healthcare professional, provider, or team. Digital health enables healthcare professionals to anticipate healthcare concerns rather than waiting for symptoms to present.

Digital Therapeutics

A very exciting, emerging field of digital health is digital therapeutics. Digital therapeutics is an enhanced form of telemedicine that brings together digital and genomic technologies with health, lifestyle, and human factors to deliver personalized medicine to patients enhancing the efficiency of healthcare delivery.

Which Health App to Trust?

There are more than 150,000 apps focused on health and wellness in the Apple App Store, which have been downloaded by over 50 million people [203]. Digital health tools allow patients to take a proactive approach to their health and wellness, looking to influence aspects of human behavior for the purposes of improving health.

The only problem is there are too many apps—of 327,000 health apps currently available to download, 65% haven't been updated for over 18 months, and only 15% of reviewed apps meet minimum standards. Only the quality of clinical evidence can differentiate the quality of vendors and services in the marketplace [204].

As more of our healthcare interventions are focused on chronic rather than acute diseases, behavioral therapy is surpassing pharmaceutical intervention as first-line therapy. Digital health and wider telemedicine have applications in a plethora of health conditions and are enabling healthcare and treatments to change rapidly.

Patients diagnosed with epilepsy can be prescribed an app to manage seizures, patients with asthma can be given a connected inhaler app, and patients scheduled for bariatric surgery can be prescribed a digital health app to assist in weight loss in preparation for surgery [205].

Measures of performance for digital health will be based on patient health outcomes, quality of care, and the number of users using tools and services over the long term.

Augmenting Traditional Healthcare

Less than one in ten patients with diabetes in the United Kingdom attends face-to-face structured education within 12 months of diagnosis [206, 207]. Research demonstrates offline education benefits typically decline 1–3 months after interventions cease, suggesting that learned behaviors change over time and further support is required to ensure lifestyle changes can be maintained. Engagement with offline diabetes education courses is limited.

COVID-19 meant that in many areas, face-to-face education was virtually nonexistent.

Digital health provides a perfect opportunity for awareness, education, and health self-management. Digital health doesn't replace traditional healthcare; it augments it. Even with the endless digital health technology at our fingertips during COVID-19, it did not stop a surge in excess deaths not directly occurring because of coronavirus [208].

Digital education is engaging by its very nature; it is available at any time and on any device at the patient's convenience. Digital education can additionally be personalized to users through the data exchange between user and provider with the presentation, and analysis of real-time patient data subsequently encourages personalized, goal-focused practice and sustainable behavior change.

Patient Centered by Design

Education can be made patient centered and person led rather than the traditional teacher–learner environment. Through use of user data and third-party services, therapies can be made to be precise.

Digital therapeutics and self-management mobile apps provide a substantial opportunity to empower patients with the knowledge, skills, and resources required to manage and optimize their health [16].

Lifestyle changes such as stopping smoking, moving more, and losing weight can all be tackled through behavioral interventions, simultaneously targeting the major chronic diseases. Chronic diseases and their management are behavior mediated [91].

Digital health tools enable treatment to be personalized, democratized, and scaled. Real-world evidence has largely demonstrated the efficacy of digital tools. More evidence and randomized clinical trials are required to demonstrate causation. One area that is still unclear is who pays for digital therapeutics. If a doctor should prescribe a digital therapeutic, who should pay? The insurance company? The healthcare provider? The employer? The patient?

Regardless, digital innovation is revolutionizing disease management. Patients are increasingly demanding autonomy, wishing to be educated about their health, aware of the latest evidence base, and engaged with their health.

Digital Therapeutics: A Long-Term Case Study

At the time of joining, case study "B" is a 72-year-old male diagnosed with obesity and type 2 diabetes in 2015. B lives in the state of Hertfordshire, England, United Kingdom. He was diagnosed with type 2 diabetes at 67 years old. B is no longer clinically diagnosed with type 2 diabetes at the end of the study. B was referred to the Low Carb Program by his NHS physician.

B followed the information presented in the application and reduced the amount of carbohydrates in his diet. B engaged in the intervention 3–4 times a week averaging 20 minutes a week for the first 16 weeks of platform use. B used the web platform and respective iOS mobile application, completing 10 of the core 12 education modules.

B posted seven times in the community. B used the app to keep track of his food diary. B's average intake of carbohydrates as measured through the app in the first week of joining was 253 g/carbs per day at baseline.

At 16 weeks, the average intake of carbohydrates was 102 g/carbs per day. At 15 months, B's average intake of carbohydrates was 109 g/carbs per day. Most frequented food items in B's food diary were cauliflower mash, omelettes, sardines, mackerel, and poached eggs.

At 15 months, B engaged an average of 9 minutes a week with the platform. At 15 months, B had reduced his HbA1c to below the threshold of type 2 diabetes diagnosis. B lost a total of 36.29 kg and reduced HbA1c by 76 mmol/mol. Average blood sugar levels had dropped by 2.2 mmol/L. B was confirmed to be in type 2 diabetes remission at 11 months.

B was interviewed as part of the case study report. B states: "In June 2018 I was told by the doctor that he was going to have to start moving me onto insulin injections as my metformin wasn't working. This frightened me and I said I'd be happy to do anything. I used to wake up during the night struggling with my breath, but the thought of injections was my main driver. I went back to my doctor in December and he was very surprised."

B reflects that his diet now contains significantly less cereal and rice. "The biggest thing was realizing the overall sugar content of carbs, particularly cereal and rice, and how much they affect blood sugar. It seems a small thing to cut out. When you sit down and see what you're missing, I don't regard it as very much. I was surprised it had such a huge effect." B advises "keeping a record of weight and recording it at the same time each week helps you to see your progress, which is a motivation in itself." B confirms "meal planning and printing meal plans out for the week" help adherence to eating habits.

Incentivized Wellness

Ill-health is expensive. To employers and the economy, physical and mental health issues cost the United Kingdom more than £77 billion [209].

Employees lose an average of 30 working days each year due to illness or underperformance as the result of poor health. The global economic impact of chronic conditions such as cancer, type 2 diabetes, and mental illness could reach over $47 trillion by 2030 according to the World Economic Forum [210].

Individual lifestyle choices and physical and mental health all impact work performance. There is a significant challenge mounting for providers such as insurers and employers in improving absenteeism and presenteeism. To do this, providers are realizing the importance of holistic health—comprising physical health, mental health, and social and human interaction, both in and out of the workplace.

As healthcare interventions focus on chronic conditions rather than acute illnesses, behavioral therapy is typically becoming first-line therapy over pharmacological intervention.

Wellness programs, personalized by available data and tailored to the user, provide a convenient mechanism by which employers and other bill payers can support health and wellness. Previously, incentivized wellness meant encouraging individuals to visit the gym and take more steps—by taking the stairs or walking to work. Employers often promoted healthy behaviors associated with employee well-being and reduced healthcare costs. Today, wellness programs and digital tools go beyond signposting and screening. Immersive, engaging programs that are remotely accessed empower individuals to modify their behavior with the goal of achieving positive clinical outcomes and better health [91].

Behavior change in the form of therapy, behavior coaching, biomedical data feedback, and sensors enable wellness to be truly patient centered.

Where Else Can AI Be Used in Medicine?

AI is a collection of technologies rather than one, and many AI technologies have relevance within medicine. Personalized medicine provides fertile ground for AI's application and machine learning models to be developed. AI seeks to place the doctor, or more specifically the healthcare team, in our pocket. Among other things, AI has applications in disease prediction, cost reduction, efficiency improvements, automating manual tasks, appointments, and promoting us to alter our health. And AI can do some things better than humans [211].

As data volume and variety increases, the capabilities of AI models will become more precise, more probing, and more contentious.

AI will transform what it means to be a doctor. AI's integration into medicine is making better doctors and saving lives. From waiting times to prioritization or finding evidence to maintaining productivity or supporting decisions, AI will assist doctors and healthcare professionals in making informed decisions.

Some applications of AI technology are described in the following.

Mining the EHR

Mining of healthcare information systems holds tremendous potential for medicine. Systems hold considerable information about patients and their health, prescriptions, doctor's notes, and more. Healthcare data can be used to improve the quality of healthcare, reduce costs, mitigate mistakes, and improve and democratize healthcare quality.

Knowledge discovery from the data contained in such systems is currently challenging due to variations in complexity, vocabulary, and standardization. Data mining provides an opportunity to extract relevant information from both textual and image-based archives. Mining massive data like EHRs enables the discovery of patterns from data that can be used to build predictive models.

As discussed in Chapters 3 and 4, unsupervised learning is known for feature extraction, whereas supervised learning is suited to predictive modeling. Mining of records is beneficial for both the patient and provider. Mining can be used to identify high-risk patients or those with chronic conditions for whom a personalized intervention could be delivered. Record mining enables providers to find best practices and treatment to reduce claims and hospital admissions. Through comparison of symptoms, treatments, and positive and adverse effects, healthcare providers can analyze the pathways that improve outcomes for patient cohorts, enabling clinical best practice and standards of care.

Conversational AI

Conversational AI refers to systems that can talk. Rather than a user interface based on text or code input, individuals engage with conversational AI systems with voice. Users are increasingly using chatbots to communicate with products, brands, and services. Voice-driven AI such as Amazon's Alexa has apps, or Skills, that can synthesize natural language to provide recipes or exercise tips, order products, find out the calories in a food, or call a cab. Conversational AI is catching on: one in five Americans owns a smart speaker, and there are over 100,000 Facebook messenger chatbots [212].

As AI chatbots develop, so too do their capabilities. Intelligent personal assistants will act as healthcare assistants. By virtue of being voice controlled, there are immediate applications for those less abled, where only the voice needs to be used to perform tasks or instructions.

Within healthcare, conversational AI enables simple questions that do not need the attention of a doctor to be answered. For instance, new parents could bombard a conversational AI with questions without fear of embarrassment or consuming a healthcare professional's time. Questions such as what temperature a baby should bathe in, how often a baby should sleep, or whether there are developmental milestones taking place can all be instantly answered by an AI speaker.

Many individuals turn to search engines to find the answers to their questions. However, most patients who query symptoms are unaware how to discern research quality and may come across conflicting evidence in addition to misleading information (or fake news) that can leave patients confused.

With the assistance of a medical AI chatbot, patients can receive immediate assistance. Continuing the preceding example, if a child's new parents had a medical question or were concerned by a symptom (say, a chesty cough), it would be burdensome for them to visit the doctor for a response to every question. However, there is a need for medical confirmation. This cannot currently be detected through algorithmic means; and hence, the evolution of conversational AI in healthcare will see them used as a digital health interface into healthcare teams and the world around them.

As conversational AI develops, cognitive systems will analyze conversation to detect early signs of mental, physical, or neurological illness. Voice-enabled devices such as Alexa will one day be able to identify symptoms of Asperger's syndrome, anxiety, psychosis, schizophrenia, and depression from conversational tones. This will assist doctors to predict better and monitor and track disease. Doctors may one day be able to predict the risk of depression in a patient's voice as they engage in a remote consultation.

Making Doctors Better

AI will enable healthcare professionals to stay up to date with the latest evidence base. PubMed, for instance, has over 30 million papers within its archives [213]. AI that was to mine such articles and present the most relevant evidence would facilitate more up-to-date healthcare professionals and enable professionals to practice medicine to the best of their knowledge and evidence base. AI can assist healthcare professionals in administrative tasks, performing roles of cognitive assistance.

AI can help mitigate mistakes, including those made by humans. Record mining, in particular, will enable doctors to improve over time. Digitalized data regarding patients, their treatment, healthcare professionals, and clinics can be mined to identify errors in treating certain conditions and avoid unnecessary hospital admissions. Zorgprisma

Publiek, a Dutch organization, currently does precisely this: using IBM Watson to mine data to identify repetitive mistakes [214].

Suboptimal processes waste time and resources. Scheduling systems for clinics, doctors, and patients enable healthcare to be as efficient as possible. An AI system that was to schedule patient appointments and streamline communication would lighten doctor burden and focus time on more pressing matters. If a patient presented an urgent matter, for instance, an AI could evaluate and prioritize patient appointments based on risk.

Diagnosing Disease

Professor Geoffrey Hinton, a pioneer of neural networks, famously stated that it is "quite obvious that we should stop training radiologists" due to the sophistication of image perception algorithms, which could soon be more advanced than humans.

AI can currently be used to examine medical scans, identify symptoms of type 2 diabetes, identify retinopathy and cardiovascular disease risk, and spot signs of breast cancer. AI will perform better as algorithms are trained on more data.

Algorithms to detect disease are showing great promise, with several already in medical practice. However, that does not mean all AI diagnostics are ready for implementation. Many AI tools are developed without peer review and its academic rigor. Essential details require verification, such as the algorithm's code, training, and validation datasets; the data to which it is compared; how performance is evaluated; and how neural networks come to their conclusions.

Making and Rationalizing Decisions

Doctors make challenging decisions daily. It is vital that these decisions are as informed as possible. The use of AI, data mining, and predictive analytics is enabling clinicians to rationalize decisions and develop evidence-based treatment options. AI does not have to make the decision but can present the most reasonable opportunities to proceed.

Drug Discovery

Developing a new drug is an expensive business, and only one in three new drugs will make it to the market. Every new drug brought to the market costs pharma an average of $2.7 billion [215].

The consequence of failed drug trials can be damaging: from reduced share price to closure of worksites and reduced staffing. Hence, the pharmaceutical and life science industry is increasingly turning to AI to facilitate drug research and development.

Pharma isn't looking to AI to replace humans. Instead, it is looking to see why the rate of failure is so high through using innovative AI. AI systems are being developed that can identify new therapies from information on existing medications; this could improve efficiency, lead to drug development success, and accelerate the route to market for new drugs. In cases of pandemics, such as Ebola or swine flu, an accelerated route could save numerous lives.

There are approximately 10^{60} compounds that have drug-like characteristics [216]. The chemist of the future will be empowered through leveraging data: historical data, experimental data, inferences, and trends. AI enables the earlier identification of promising drugs and biomarkers.

AI also has application in clinical trials. Clinical trial patient stratification allows the identification of patient groups most likely to benefit from the associated drug. This enables clinical trial operators to find the right patients for trials rather than patients who are unlikely to respond.

3-D Printing

Innovation in 3-D printing is beginning to disrupt traditional paradigms as the technology becomes more affordable and accessible. As the name suggests, 3-D printing involves printing three-dimensional objects from a digital model, using additive processes to build the object.

Also known as additive manufacturing, 3-D printing adds iterative layers, one on top of the other, to assemble the model. This allows precise modeling and reduces error. Although 3-D printing solutions are still in infancy compared to other technologies, the potential for its applications is exciting. Many AI experts consider it feasible that in the future, humans will adopt robotics and bioprinted material into their bodies. Soon humans may be able to print just about everything, from prosthetics and pills to bioengineered replacement body parts and organs. The technological advance of 3-D printing has life-changing implications for patients [99].

Personalized Prosthetics

The World Health Organization states that approximately 30 million people require prosthetic limbs, braces, or mobility tools and only 20% have them [217]. 3-D printing enables prosthetics to be custom built for each person. Through capturing a patient's measurements, prosthetics can be tailored to ensure comfortable, custom-fitted devices. Prosthetics will also become more comfortable as multi-material, 3-D printing assists with ensuring a better integration with the human body.

Casts could also be made more comfortable through the same, personalized approach. 3-D printing facilitates the swift and cost-efficient creation of personalized products. Innovation in 3-D printing is combining integrated sensors and machine learning algorithms to support more natural, fluid movements.

Predictive movement algorithms will evolve to mimic more natural movement, and humans will be able to use their brain and body to control them. 3-D printing technology enables prosthetics to be modeled and constructed in a little under 24 hours; importantly, printed prosthetics cost a fraction of conventional prosthetics.

Kidney, liver, and heart transplants could soon be a thing of the past. 3-D bioprinting marks the beginning of an era where transplant lists are a thing of the past. 3-D–printed organs can be built using the same techniques as 3-D printing, but instead use stem cells as the printing material. As well as printing cells, bioprinters typically output a gel to protect cells during printing.

In the future, once these organoids are printed, they will be able to grow inside the bodies of patients. Princeton University has developed a bionic ear from 3-D printing tools, which can hear frequencies beyond the human range. Printable and customizable implants from a patient's cells reduce the potential risk of rejection from the body.

One use case for bioprinting is the 3-D printing of human skin. Burn and acid victims, for example, have limited options for their disfigured skin. A team of Spanish researchers developed a 3-D bioprinter that was able to produce human skin cells. Within 30 minutes, a biological ink containing human plasma and other materials was able to print 100 square centimeters of human skin [218]. The opportunities bioprinting technology enables are truly life-changing.

There is no doubt a cosmetic market for bioprinting. It is feasible that in the future, face printers could be used by people to apply a model of someone else's face to themselves or even store a digital model of their face and regularly reprint their face for perpetual youth.

Alongside nanotechnology and genetic engineering, bioprinting may prove a tool in the pursuit of extending life spans.

Pharmacology and Devices

3-D printing disrupts traditional pharmacology and device manufacturer approaches. For instance, the efficacy of drugs and other treatments can be tested on replicated human cell tissue. Through this, 3-D printing enables the printing of medications that are personalized for the patient.

By reducing drug variability with this approach, there is potential for 3-D printed drugs to increase medication efficacy, reduce adverse effects, and improve adherence. The same concept applies to device consumables. Blood glucose test strips, for example, could be 3-D printed for use by diabetes patients at home. The possibilities are endless.

Education

Alder Hey Children's Hospital in Liverpool, England, uses 3-D printing for teaching and familiarizing stakeholders with heart surgeries that are to be performed [219].

Every cardiology patient is unique; and for those undergoing operations, 3-D models are printed to enable a variety of uses. Firstly, it enables surgeons to demonstrate surgeries that will be performed to often anxious and concerned parents using textures and colors that closely represent human tissues. Secondly, the 3-D models are used by staff and surgeons to familiarize themselves with the most important part of an operation—the child's heart. Finally, 3-D models are used to train students.

3-D printing also enables surgeons to prepare before surgery through evaluating a precise model of the patient's body or organs. This mitigates the likelihood of errors, improves surgeon accuracy, and reduces the amount of time the patient is required to be on the operating table. 3-D modeling is also useful in modeling bodily organs without the need for invasive procedures.

Gene Therapy

Just like organ transplant lists, genetic diseases could soon be a thing of the past. The use of nanotechnology, which refers to working with tiny particles to manipulate cells and alter DNA, enables the editing of gene expression at a cellular level. Gene editing has been successfully used to modify human immune cells to resist HIV infection [220]. With 1 in 25 children born with a genetic condition, gene editing has the potential to treat or eliminate disease [221]. It could be used to correct defective genes in embryos and personalize the child's immediate healthcare treatment plan. It is feasible that genetic conditions like cystic fibrosis, sickle cell anemia, and muscular dystrophy could be treated through editing the DNA in patient cells.

Gene editing is not without ethical implications; many people have moral and religious concerns about the use of human embryos for research or the editing of genes.

There is a concern that nanotechnology and gene editing will only be available to those who are wealthy, which will exasperate the disparity in healthcare and interventions. Nanotechnology will significantly improve treatment and recovery time, with the potential to prevent humans from developing disease at all.

Choose Your Reality

In the not too distant future, surgeons performing CT scans will be able to layer their scans over a patient's body via augmented reality (AR); medical students will use virtual reality (VR) to explore the inside of the heart, and burn victims will be virtually transported to a snow-covered mountaintop as a form of pain relief therapy. Digital health app Gro Health provides guided immersive mindfulness to users in 360-degree video, demonstrated to help people decenter [222].

Immersion into a completely digital environment, virtual reality, has mainly been used for games. Innovation in augmented and merged reality means that we can now entwine the virtual and physical worlds and manipulate both environments simultaneously.

Virtual, augmented, and mixed realities are increasingly being implemented in a wide range of medical applications. It is already evident that as it is incorporated further into healthcare, it has the potential to change the way many healthcare services are delivered.

Virtual Reality

Virtual reality (VR) is typically associated with gaming. In virtual reality, the user's reality is replaced by an immersive, entirely digital environment. This is currently achieved through a headset and handheld sensors, which enable interactions within the environment.

Augmented Reality

Augmented reality (AR) overlays a digital or 3-D environment in the form of objects, video, or data into the user's environment. The user's real-world experience remains the focus and information is added to the user's existing reality to enhance their experience. Information is distinctly digital and does not seek to emulate real-world objects.

AR does not necessarily require additional hardware and can be achieved through technology such as the mobile phone. AR's popularity was confirmed by the Pokemon GO app phenomenon. The game, downloaded more than 800 million times, overlaid 3-D characters within the user's real-world environment through the user's GPS location and inspired a wave of AR innovation [223].

Merged Reality

Merged reality seeks to emulate digital objects that can be interacted with. This requires additional technology such as a headset. Separate sensors track hand gestures and movement. In a mixed reality, the user can manipulate both the real-world and digital environments.

Use Cases of Immersive Reality in Healthcare

Virtual and augmented reality is largely used to augment traditional healthcare pathways and provide treatments and education.

Pain Management

Research demonstrates that virtual reality environments can be used to reduce the amount of pain a human feels vs. a control distraction condition [224]. The somatosensory cortex and insula, found in the brain, are linked to pain. Hence, through immersing patients in a virtual reality, VR will be used to enable patients to endure painful surgery.

Amputees often report pain in their amputated, missing limbs. VR environments can be used to immerse the patient within an environment to better cope with their phantom pains [225].

VR also works as a distraction. Children and adults alike will be given a VR headset from their doctor when receiving an injection to distract them from the impending prick. Hermes Pardini Laboratories and Vaccination Centres piloted the use of a VR headset that immerses the patient into a fictional, gamified environment [226]. The patient is distracted in a virtual world to the point that they are often unaware they have received their injection.

Physical Therapy

VR can track human movement, allowing patient movements to be monitored and analyzed. VR gyms have been opened in San Francisco and Ohio [227]. Rehabilitation will become gamified, for instance, through kicking a virtual ball or catching a ball.

Recovery exercises can be delivered and tracked in a VR environment and retold to the patients if they were not to get it right.

Cognitive Rehabilitation

The application of VR and AR in fears and phobias is apparent. Performed as a medical treatment, patients can be gradually exposed, known as graded-exposure therapy. Data from sensors will be assessed to ensure patient safety and develop best practice.

Cognitive function can be improved for patients struggling to perform everyday tasks. Through monitoring the performance of patient tasks in a VR environment, a doctor could determine declining memory loss and identify areas of concern or priority. Similarly, patients with injuries to the brain or those who struggle with tasks can have digital environments created to represent real-life scenarios.

Patients can practice tasks and regain or develop cognitive function. Patient engagement can be monitored and analyzed to observe areas of difficulty or reduced attention. There are few randomized controlled trials that have been conducted into VR in cognitive rehabilitation. However, some applications of VR are effective in treating cognitive deficits in people with neurological diagnoses [228].

Nursing and Delivery of Medicine

VR and AR will develop to become part of standard medical training. Both types of reality can be used to improve outcomes by enriching the information available to the individual. Alder Leys Children's Hospital in England uses a VR headset and 360-degree video to train student doctors how to deal with stressful situations. A case study from Alder Ley is available at the end of this book.

Critical decision points can be reviewed and analyzed with peers and staff. Virtual reality will be used more to learn medical specialties such as performing operations and anatomy.

Operations can be practiced with expert monitoring and giving feedback in real time. The Royal London Hospital conducted the world's first VR surgery, enabling viewers to watch in 3-D 360-degree video [229].

The learning experience is unparalleled and has the potential to disrupt medical training, particularly for countries that have limited healthcare resources.

Doctors will use augmented and mixed reality to enrich their environment. Surgeons will have access to critical information in real time through mixed realities, delivered to their vision through glasses.

Nurses could use augmented or mixed reality to identify veins in the arm for a blood test. The Interventional Cardiology Center at Tufts Medical Center in Boston uses VR to introduce the facility to prospective patients with anxiety before a procedure [230].

Virtual Appointments and Classrooms

As the cost for virtual and merged reality devices becomes more affordable, virtual appointments will become commonplace. Virtual appointments remove the inconvenience of attending a clinic, save time and environmental resources, and focus healthcare professional time to where required.

Virtual appointments will become as familiar as webinars, enabling stakeholders to be present without traveling. As well as one-to-one appointments, virtual and merged realities provide an immersive and engaging experience suited for learning.

Virtual reality will become a staple tool for education and training—for both healthcare professionals and patients alike.

Using the Blockchain in Healthcare

Transitioning of traditional patient health records to the EHR is considered to be enormous progress for healthcare. The digitalization of patient records mitigates some of the traditional risks of centralized data stores. This model, however, still places the medical records in the hands of the provider. Blockchain technology, popularized by the bitcoin cryptocurrency, has the potential to revolutionize data access, privacy, and trust. Currently, blockchain is yet to be deployed for mainstream healthcare.

What Is the Blockchain?

Blockchain, fundamentally a collection of data records, is a piece of software formed by the combination of several preexisting technologies that provide blockchain with its characteristic features: an immutable, distributed public ledger whose authenticity can be verified by anyone. Those who validate the data on the ledger are rewarded with value, which helps create trust in a trustless environment; distributed peer-to-peer control, which provides a high level of security; and the ledger can be programmed to trigger automatic transactions in the form of smart contracts, allowing for a widespread application of this technology.

PayPal, Visa, and Mastercard, for instance, act as central authorities for financial transactions: trusted institutions that act as intermediaries. The land registry acts as the trusted store for details on homeownership. These centralized databases are subject to being hacked or manipulated.

Blockchain technology aims to solve data management, privacy, and security issues, improving interoperability and easing the flow of data between doctors, hospitals, healthcare systems, and insurance providers through the use of a decentralized, immutable database.

Patients are demanding increased access to their medical health records. Data from EHRs, IoT, wearables, and devices can be used within the ledger, in a trusted, secure, transparent, and interoperable environment.

Tamper-Proof Security

The blockchain is a database which cannot be edited, stored, and maintained by all those using it. Each new transaction or piece of data is encrypted and then approved by a particular proof-of protocol (consensus, work, stake) by other nodes on the networks that authenticate the transaction or verify the stored piece of data.

Each node on the network has an identical copy of the blockchain; and thus, the transaction is permanently recorded and linked to previous records. The links, known as hashes, are traceable back to the very first block in the blockchain. Therefore, any attempts to tamper with a block in the present would require the transaction and all related blocks to be also altered, on all records, distributed among the nodes holding a copy of the ledger simultaneously. The longest chain of events is considered valid.

Blockchain implements crypto-economic and game theory techniques. Any attempt to create a rival chain would need to be created faster than the current version of the truth to be accepted. This would require tremendous computing, energy, and resource commitments. Miners compete to validate blocks. This good behavior is expensive regarding electricity and computing power, and so miners are incentivized to validate blocks, rewarded with bitcoin (BTC; in the BTC scenario).

Through creating a common, distributed, immutable database of healthcare information, doctors and healthcare providers have the potential to access medical data from any system, with improved security and privacy, less administration, and better sharing of results.

- Single points of failure are eliminated by decentralizing and encrypting the data.
- The blockchain is democratized. Anyone can contribute or store a version of the truth.
- Through ensuring consensus on events, the most likely version of the truth is held.
- A transparent and auditable ledger of events is provided through time-stamping.
- Game theory, crypto-economics, and hashing incentivize good behavior and ensure the events are without censorship.

Use cases

Blockchain technology has many applications in healthcare, with most technology currently in pilot or proof of concept stage in the following use cases.

Verifying the Supply Chain

Blockchain can verify the supply chain. As the blockchain is created in a chronological and auditable manner, it can act as a means of verification for components of every link in the supply chain. The journey of materials or treatments, for instance, can be logged onto a blockchain ledger, which could be used to identify fraudulent activity, anomalies, error, or a break in the chain through data entry or IoT devices.

The cold delivery chain of delivering insulin from manufacturer to pharmacy could be confirmed to be untampered and appropriately cooled. This is currently being used in developing countries to combat counterfeit medication. It is also being developed to enhance genomic data protection, addressing the privacy challenges of big genomic data.

Incentivized Wellness

Blockchain technology could be used to incentivize wellness: through the use of a cryptocurrency as a digital token of value. Engaging people with health services or a healthier lifestyle could save the global economy a tremendous amount of money in healthcare costs.

Health providers or employers typically see the cost benefit of this in the form of savings or profit, and this is rarely passed to the individual. By utilizing blockchain technology, tokens could be created and distributed to patients through the blockchain to share the value of the savings and be treated as a tradable currency.

Individuals could earn tokens through behaviors such as going to the gym, reaching their step goal, attending education sessions, engaging in mindfulness, completing a particular sports event, or adhering to medication or digital therapeutics. The ecosystem rewards positive behaviors with an asset or token of value. The value of the token could be fixed.

Extending this concept, positive health behaviors could be extended to the point where patients have health token savings accounts that could be used to transact in hospitals.

Patient Record Access

Patients are demanding access to their health record. This poses a significant challenge as to how to best share sensitive medical data with unknown third parties. For third parties, there is also a challenge to verify the integrity of the data while ensuring privacy for the patient.

Presenting patient records on an appropriately permitted blockchain would give cryptographic assurance on data quality without any need for human involvement. Providers or consumers uploading health data would generate transactions. Users provide a signature and timestamp, alongside a private key to access data. By using digital signatures, all records stored on the blockchain can be identified and used to create a comprehensive patient health record.

The use of digital signatures and cryptographic encryption ensures data travel securely and are accessible only by those with the relevant public keys. By using blockchain technology, each addition to the EHR can be logged, with an immutable, auditable trail of transactions, while ensuring the most current version of the record is used. Patients would be able to verify all attempts to access or process data.

Blockchain's decentralized structure enables any approved stakeholder to join the ecosystem, without the need for data integration or manipulation concerns.

Robots

Like other fields of AI, the use of robotics is changing healthcare. Robotic technology is not yet widely affordable or implemented, and it is unlikely robots will ever entirely replace humans in the medical setting. Hospitals and healthcare systems are unable to meet the cost of the technology until proven; and robotic assistants have not replaced, and some would say cannot replace, human contact.

Over and above its use in the supply chain, robot technologies have potential in the following areas.

Robot-Assisted Surgery

Robot-assisted surgery enables enhanced vision, improved precision, and dexterity. Currently limited in reach, robot-assisted surgery will become commonplace as barriers to entry such as the cost of hardware and training are reduced. In the future, healthcare professionals may have to learn how to use an instrument as well as how to perform the surgery. Robot-assisted surgery blurs the lines of liability, which could prove to be a barrier to adoption.

Exoskeletons

Robotic exoskeleton technology focuses on enabling patients to perform tasks through the use of an external, integrated device. Ekso Bionics is an organization that provides a wearable exoskeleton that assists spinal cord injury patients to stand and learn to walk [231]. In the future, exoskeletons will become a standard form of rehabilitation and aid human mobility in more activities and in more environments.

Inpatient Care

Inpatient care can be enhanced through the use of robots for streamlining tasks. Robots could be used to automate tasks such as collecting mail or delivering blood. Delivery robots will deliver medications or important material autonomously.

Robots are already being used to disinfect rooms, where the risk to humans is minimized through using a robotic agent [232]. Soon ingestible agents and smart pills will be able to monitor a patient's internal reaction to treatment; robotic nurses will become commonplace for automatable tasks, such as taking blood. A robot will take your vitals and draw blood by identifying the correct vein with greater accuracy than a human nurse. Robots will be used to support care for the elderly and those with long-term conditions.

Companions

Virtual assistants are becoming more humanlike with developments in natural language processing. As discussed, humans are already forming bonds with robots. Robots can be used to provide companionship or treat loneliness in cases of mental health, elderly, and long-term care.

As robots evolve in their capabilities, they will be able to perform increasing duties such as bathing patients, transporting patients, and so on. Robots will be able to monitor vital signs through IoT and wearables and aid the patient in real time.

Drones

Drones have the potential to alter the way medicine is delivered. Drones can be used in hard-to-reach areas, places of conflict, and remote populations to deliver medication, vaccinations, and diagnostics. Drones will be used to deliver medications from the pharmacy, just as they are being used in shopping. Time-sensitive items such as blood, bodily fluids, and organs can travel shorter distances to go directly to where they need to be, within campuses or across larger distances. Drones were used in the transfer of humanitarian aid during the COVID-19 crisis in Wuhan [233].

Drones can also be used to locate and identify, particularly in remote locations. One use case for drones is as a flying medication toolbox in cases of emergency. Critical care in the minutes after a stroke, trauma, or heart failure is essential to accelerate recovery and prevent death. An ambulance drone prototype, developed by TU Delft, combined a heart defibrillator, medication, and two-way radio that could be dispatched to a patient to speed up the response before the first responder [234].

Current restrictions in the use of drones are reducing their potential. It may be some time before flight restrictions, legislation, and functional issues such as drone battery life and drone load are resolved.

Smart Places

Intelligent homes, hospitals, places, and things promise to change the way we live. Ubiquitous connectivity and an expanding sensor base provide a wealth of opportunity to people, patients, and providers alike. Improvements in patient lifestyle and healthcare have contributed to growing numbers of centenarians over the past 30 years. In the United Kingdom alone, the number of people living to 100 increased by 65% [235].

These breakthroughs have come at a cost to healthcare; and smart places, and smart things, enable scalable opportunities to improve health and social care—reducing costs per capita while personalizing the experience to the patient. Rather than going through the process of connecting and integrating apps, sensors, or devices, connected places will use automated sensors that do not require the user's constant attention, collecting tremendous amounts of real-time data.

Facial recognition, voice recognition, responsive notifications, and data-based suggestions will become the norm. Aggregation and analytics platforms will streamline and simplify decision-making processes in a rapidly changing environment. These devices can notify patients and healthcare professionals in real time: forget your Big Brother, this is Big Doctor.

Smart Homes

There will be many ways that homes will become smart in the pursuit of improved patient and population health, which are best illustrated by example. On waking, a wearable sleep monitor will assess the quality of your sleep while you perform mindfulness through your connected watch. Your sleep monitor will even tell you when to go to bed for optimum recovery. After brushing your teeth with a toothbrush that assesses whether you are hydrated or not, you drink your morning coffee, which controls the dosage of insulin released for absorption in your bloodstream. Your fridge will inform you of out-of-date food and ensure none of your potential allergens are found in the food you purchase.

Smartphone apps will augment the calories and nutritional values of foods; and 3-D printers will enable the printing of safe-to-dispense, at-home pharmaceuticals. Virtual assistants and even mirrors will assess for signs of anxiety and depression and assess health biomarkers in natural language. Should you develop a cough, your virtual assistant will tell you that you're coughing more than usual; and with the change in body temperature detected from your smartwatch, your assistant calls the doctor on your behalf and arranges an appointment.

As a closed environment, the smart home allows for AI to accommodate for the highly individualized needs of the user. The most important element of smart anything is the nuance when scaling: to understand the context and prioritize local activity.

The future is not as far as it seems. Bolzano in Italy is already working with IBM and various partners to empower aging people to age safely at home [236]. Safety and security are the main priorities, with the project using sensors installed into homes to monitor environmental factors such as temperature, carbon monoxide, and water leaks. Data is pushed to an off-site control room where, depending on the individual need, family members, volunteers, emergency response staff, and social services are notified.

Smart Hospitals

A smart hospital is one that is connected, much like the smart home. The objective of a smart hospital is to provide clinical excellence, an efficient supply chain, and superb patient experience—facilitated with technology. Smart hospitals will use a continuous learning ecosystem, ranging across many areas including electronic data collection and health records, digital technology, robotics, 3-D printing, unstructured data, and robust analytics. Somewhat ironically, mass use of technology and AI will enable patients to become more actively involved in their treatment decisions. Nonemergency consultations will take place on the Internet, with AI prioritization enabling the right doctor, with the right skills and training, to treat the patient. Treatment will be blended between offline, physical care and digital care—where adherence and accountability can be quantified and maintained. Growing data and large-scale analytics will continue to personalize digital treatments for patients.

On attending a hospital, an automated and streamlined admission reduces patient waiting time. On admission, patients will be tagged with a clinical-grade wearable to track vital signs through the inpatient stay. Metrics are all sent wirelessly to a dashboard visible to your medical team. Any anomalies or causes for concern are detected and prioritized.

Hospitals and surgeries will become hubs of data, with hospitals working with businesses to extract value from the data—regarding improving efficiency, minimizing mistakes, and improving treatment and device decisions. Healthcare will be delivered as a service, with patients incentivized with cryptocurrency to maintain sensible and healthy lifestyle choices. All relevant data will be anonymized and accessible to digital health partners and internal departments to enable continuous learning from each patient's experience.

Developing Whole AI

Reasoning is reductionist. The basic premise of reductionism is that by breaking down or simplifying complex biological or medical phenomena into their many parts, one is much more likely to understand a single cause and devise a cure.

For instance, a mouse can be divided into a skeleton, a circulatory system, a nervous system, a digestive system, and so on. It fits a model—a pattern, theory, equation, formula, or some form of common structure. AI systems also follow the same, reductionist approach, reducing the problem to a defined discipline. For AI to truly develop artificial understanding, the key will be to address humans from a perspective of holism.

As the world continues to immerse itself in digital technology, healthcare AI models are becoming quicker and more accurate, with patient-based care and lifestyle medicine placing the focus back on holistic health. The effect of lifestyle factors, sleep, movement, nutrition, environment, and genetics on health is well known. We also know that the world and people, mind and language, can be unpredictable.

As AI develops, problems of irreducible complexity will require solving where the reductionist approach cannot work. The focus is on value, on the patient, treating patients as people and focusing on the patient's wider health, relationships, and goals.

Conclusion

Artificial intelligence has come a long way since its definition as a field in 1956. Machines have come so far they can beat humans in games that are hundreds of years old and tell us about signals that are unidentifiable by human computation.

As AI grows, so too does the requirement to not lose the human touch. Collaborative academic research, clinical evidence, and policymaking must shape the domain of AI and healthcare to ensure humanity's best interests are protected from those that develop intelligent software.

Whether you are a pessimist about the impact of AI on society or look forward to it with baited breath, one thing is for certain—with more data now than ever before, the most promising era for healthcare is yet to come.

CHAPTER 10

Case Studies

Learn from your failures or learn from other people's failures. Whichever you choose, learn.

—Kirpa Gayatri Kaur

AI is improving the healthcare experience, bringing success to those who can leverage and adapt to a new health and care delivery paradigm. This chapter contains real-world case studies where applications of machine learning are being used to solve real-world problems in digital health.

Each case study provides a unique and engaging perspective of the use of big data, AI, and machine learning within health and social care. Real-life descriptions of organizational approaches to data-identified healthcare problems demonstrate the instant value within data and AI.

The purpose behind the case studies is to showcase real-world demonstrations of AI, machine learning, data, and IoT usage that are novel and innovative.

Each case study shares unique learnings developed from creation, training, and deployment of the respective intelligent system.

Real-World Inspiration

A call for case studies was placed on Twitter, and 108 case studies were submitted. Case studies were accepted if they were used in the real world, whether in pilot or otherwise, and were supported with evidence to demonstrate value from a health benefit or cost-saving perspective.

© Arjun Panesar 2021
A. Panesar, *Machine Learning and AI for Healthcare*, https://doi.org/10.1007/978-1-4842-6537-6_10

The criteria used to evaluate case studies for inclusion involved the following:

- There is focus on delivering precision medicine and big data-generating technologies.
- Machine learning or data-driven models must be deployed into the real world.
- Authors must highlight learnings from the experiences the project provided.
- Authors must share adequate information on the project scope and conclude on the organizational, business, clinical, or financial impact.

Real-World Application and Learnings

Case studies selected showcase a tremendous vision, focus, and drive to improve patient engagement, outcomes, and clinical success. Case studies serve as stimulating examples on the following:

- The use of big data to drive decision-making and provide patient- and population-scale precision medicine
- Advantages and pitfalls of developing machine learning models and their application in the real world.
- Data collection, analysis, management, and governance
- Developing artificially intelligent agents and blending digital and face-to-face healthcare
- The use of wearables, sensors, and IoT health devices to provide care within primary and secondary care systems
- Ethics of AI and morality
- The future of digital technology, healthcare, and evidence-based medicine in the pursuit of improving quality of life

Further details about case studies, their respective authors, and their organizations are available in the supplementary reading area available online.

The case studies chosen for inclusion all detail real-world applications of machine learning and AI that exist in healthcare today and include examples of digital health technologies provided by DDM, the organization I lead today.

Case studies comprise

1. AI for Imaging of Diabetic Foot Concerns and Prioritization of Referral for Improvements in Morbidity and Mortality
2. Outcomes of a Digitally Delivered, Low-Carbohydrate, Type 2 Diabetes Self-Management Program: 1-Year Results of a Single-Arm Longitudinal Study
3. Delivering a Scalable and Engaging Digital Therapy for Epilepsy
4. Improving Learning Outcomes for Junior Doctors Through the Novel Use of Augmented and Virtual Reality
5. Do Wearable Apps Have Any Effect on Health Outcomes? A Real-World Service Evaluation of the Impact on Activity
6. Big Data, Big Impact, Big Ethics: Diagnosing Disease Risk from Patient Data
7. Assessment of a Predictive AI Model for Personalised Care and Evaluation of Accuracy
8. Can Voice-Activated Assistants Support Adults to Remain Autonomous, a Real-World Service Evaluation of the Impact of a Voice-Activated Smart Speaker Application on Activity

Case Study 1: AI for Imaging of Diabetic Foot Concerns and Prioritization of Referral for Improvements in Morbidity and Mortality

Harkrishan Panesar, Gro Health, UK

Background

Globally, 463 million people have diabetes, 4 million in the United Kingdom [237]. Seventy-seven thousand diabetics in England have foot ulcers today [238]. The longer ulcers progress, the longer they take to heal. Untreated ulcers can lead to diabetic foot disease (DFD) and amputation [239]. Twenty-five lower limb (toe/foot/leg) amputations occur in England daily; 85% are preventable [240, 241].

From 2015 to 2018, 27,465 lower limb diabetes-related amputations happened in England with 147,067 DFD-related hospital admissions and an average 8-day stay totaling 1,826,734 bed days costing >£686 million [242, 243].

DFD accounts for £1 in every £100 the NHS spends [244]. Reduction in ulceration prevalence by 33% would save NHS England £250 million/year [9]. Foot checks are a chance for potential problems to be identified and assessed and action taken. NICE NG19 states people with diabetes should have diabetes foot checks annually and patients with active foot problems referred to a multidisciplinary team within one working day and triaged within another [245].

However, "poor symptom recognition by patients, inaccurate healthcare assessment, and difficulties in accessing specialist services" [246] exist due to

- Delays to examination: The National Diabetes Foot Care Audit found 39% of people waited 14+ days for their first foot ulceration specialist examination [247].
- Lack of testing: 800,000 patients in England do not receive an annual foot check. Eighty percent of practice nurses are not confident performing foot checks, causing variances in care and outcomes [248].
- Poor assessment: 33% of NHS England commissioners do not provide healthcare professionals (HCPs) with foot care training. Where training is in place, quality of assessment is poor [249].

Innovation can democratize patient care and reduce the economic and health burden of ulceration/DFD.

Cognitive Vision

Cognitive vision or image recognition is a machine learning technique designed to mimic how the human brain functions. With this technique, machine learning models are taught how to recognize visual components that make up an image. Through learning on large datasets of images and identifying patterns, models are able to make sense of their input and determine relevant tags and classifications. Image recognition is not an easy task. To successfully perform image recognition, neural networks are the technique of choice.

However, even still, the computation resource cost is expensive in practice. For instance, a 20-pixel by 20-pixel image received by a neural network would receive over 400 inputs. Although this sounds achievable in a typical hardware setup, images of greater size, say 1,000 pixels by 1,000 pixels, would require a powerful computational resource to process the higher number of parameters and inputs. In a traditional neural network, each pixel would be linked to a single neuron. This is computationally expensive.

Proximity of pixels within images has a strong relationship with their similarity. Convolutional neural networks specifically make use of this feature. Rather than treating two nearby pixels as distinct, convolutional neural networks assume there is more likely a relationship between these pixels than two that are further apart.

Through bypassing less significant connections between pixels, convolution solves the computational and time problems faced by traditional neural networks.

Convolutional neural networks improve the computational resource required for image recognition through filtering relationships by proximity.

Clinical evidence suggests that diagnostic tests and risk stratification can predict the risk of ulceration and amputation, with early referral reducing amputation rates and time to heal [251].

Solutions for diabetic foot care must be broader than the standard healthcare worker-dependent service and be deliverable at a lower cost for the patient and healthcare system.

Together with my colleague Yang Wang, we applied state-of-the-art AI techniques to automatically detect signs of diabetic foot concerns, alert the patient, and support communication and assessment by the healthcare team. Image recognition technology for diabetic foot concerns would be revolutionary, particularly considering this machine learning experience could improve ulceration healing time, cost, and risk of amputations—as this is related to how early ulceration is detected. The model has achieved high accuracy in detecting subtle areas of bruising and cracked heels, which can soon develop into areas of foot ulceration.

A cloud-based system was developed to allow remote use of the application such that any device with an Internet browser was able to upload an image alongside a standardized protocol. The output would be a list of any concerns detected, the referral pathway most suited to the concern, and education, if relevant.

The model is constantly refined through a machine learning feedback loop that optimizes and retrains the model periodically with random subsets of labeled data, including the data received from the web-based system.

Project Aims

The aims of our project are to further develop and validate this machine learning technology, including the deep learning neural network used to predict the risk of foot concerns and provide precise medicine to patients. The convolutional neural networks used for multiple predictions can become more precise through their use and feedback loop from users.

As the user base is largely healthcare professionals and carers, the learning can be scaled up and validated by the healthcare professional for appropriate triage. The project also aims to deliver a service to patients, which is only possible when the model has reached maximum, and ideally near-absolute, accuracy. Reducing error is a key priority for the project.

Grading schemes and standardized protocols support the model to develop greater precision and recall. The ability to identify ulceration and other foot concerns earlier would accelerate diagnosis and treatment and save the lives of many. A cloud-based system would enable easier sharing of the application and its outcomes. A catchall event log was developed to identify all events taking place within the ecosystem. The platform includes behavior health mentoring and integrated tracking for blood glucose, blood pressure, mood, food, weight, and sleep through Bluetooth-enabled devices, wearables, and self-inputted data.

AI models provide boundless opportunity in predictive analytics. By analyzing and generalizing on thousands of images, medical images are being turned into science. There are few ways as engaging and immersive as digital, which enables healthcare to collect enormous amounts of data; analyze and provide useful feedback to inform patients; and prevent, predict, diagnose, and treat disease.

By the very nature of imagery, de-identification for aggregated learning is technically simple. However, the requirement to be able to explain the mechanisms of deep neural network prediction is technically difficult. A more general understanding of the application alleviates most stakeholder concerns, which can be explained through demonstration.

As a novel innovation, there has been a requirement to explain how models are generated, validated, and continually evaluated. Regulatory approval has proved more laborious than previous projects due to the loose definitions that exist around product categorization.

Error minimization has been a priority, and regulatory approval has focused on ensuring patient safety. The risk of false positives was determined to be more likely: early ulceration detection means quicker healing and less risk of amputation.

Thresholds of concerns were developed within the model, with any doubt referred to a human.

Outputs of the project are being documented through clinical papers to highlight the value of the project to patients, healthcare providers, and global healthcare providers. Perhaps of most significance, the outcomes will demonstrate how people with diabetes can achieve significant health improvements by preventing and detecting diabetic foot problems.

Challenges

There were a number of inertia points that were observed throughout the duration of the project:

Quality of Data

Any model can only be as good as the data that is used to create it. Data completeness and correctness are paramount for a robust machine learning algorithm. There were several significant challenges with the datasets collected that facilitated the development of robust data governance processes. First, lack of labeling meant that data was unlabeled and often unclear on receipt.

Second, handwritten notes that supported image scans were often illegible. Incorrect spelling or incomplete doctor notes complicated the data validation process. The cleaning and vetting of data advanced a dataset that was established as a source of truth for the model.

Data Governance

The requirement for national and international regulatory approval required appropriate, transparent data governance processes to be in place. Governance-covered areas included data architecture, archiving, management, metadata management, privacy, security, and validation. Standards, best practices, and procedures were developed for data acquisition, verification, and validation.

A central repository held all relevant data and metadata, with access restricted by credentials to ensure appropriate user management. A governance trail for all decisions was noted as per the requirements for regulation, which provided an opportunity for reflective learning as the project ended. Stringent data governance ensured the integrity of data used within the project.

Performance vs. "Explainability"

Neural networks are notorious for being difficult to explain, particularly as models become more complex. A key requirement of this project was transparency in algorithmic decisions.

As the model developed, iterations of learning and backpropagation tailored nodes within the network that became unexplainable. Model performance and accuracy was secondary to model transparency.

Ethical Governance

Strict data ethics and information ethics governance procedures were developed alongside data governance processes to ensure ethical and moral use of data. The machine learning model was used by healthcare professionals to predict and diagnose risk of foot ulceration.

Processes around the effect of false positives and true negatives in particular were noted, as misdiagnosis was established to be the greatest ethical concern. The effects of ulceration and healing times can be greatly reduced through swift diagnosis and action.

As the application was used to support the actions of healthcare professionals, users were informed as to the limitations of the model, and it was explained how model accuracy was based on validation datasets.

Closing the Loop

The developed model predicted the likelihood of foot ulceration and other diabetic foot concerns. As this was to be used in the stages of prevention and diagnosis, a feature that was requested was the ability to validate and confirm model predictions. To do so, the healthcare professional using the application would either confirm diagnoses or refer the patient to a specialist for diagnosis.

The process of feeding back patient outcomes into the system enabled the system to learn in near real time. Data feedback from patients once referred to the hospital was difficult and cumbersome to gather.

Stakeholder Understanding

An unforeseen challenge was the lack of stakeholder understanding and sporadic resistance to technology-driven healthcare solutions.

Key stakeholders were typically biased by previous experiences or negative press or threatened by concerns of an age of robots taking jobs from humans. This was overcome with staff stakeholder training days; third-party experts and internal stakeholders explained how AI and machine learning could be used to optimize the tasks conducted by healthcare professionals, improve the resource burden, and mitigate concerns of a near-future robot race.

The project scope was developed with the input of a diverse, multifaceted stakeholder group to ensure all relevant parties were vested in the project.

Adoption Strategy

As a project that started through R&D funding, there was little incorporation of the pilot project into the wider organizational direction. Stakeholder comfortability and adoption of the novel innovation was progressed through workshops and hands-on experience. Workshops focused on mitigating stakeholder concerns and on the ease, efficacy, and resource-light environment of the predictive system.

Conclusions

Cloud-based intelligence that is sensitive and accommodating to differences in photography hardware and formats can be combined with patient clinical, demographic, behavioral, and genomic data to develop a precision medicine framework for detection of diabetic foot concerns.

The prototype was welcomed by healthcare professionals, with several remarking on elements that improved their own knowledge on the topic. Healthcare professionals felt more confident in diagnosing foot concerns when partnered with the digital tool. Digital assistance can confirm and support decisions and challenge healthcare professionals on how they arrive at their decisions. Applied across the spectrum of medical images and data collected by healthcare and research organizations, the technology will affordably enable large-scale prognostic, risk stratification, and best treatment decisions.

From our experience, having more data is almost always more important than the algorithm itself.

Overcoming the concerns of text digitalization and label recognition alongside medical imagery proves to be a challenging area within this field. Handwriting and human error have been the biggest problems to mitigate to reduce false recall. Conducting a large-scale clinical trial is the next phase of the project. A pilot project assessing the feasibility of the project is pending funding approval.

Case Study 2: Outcomes of a Digitally Delivered, Low-Carbohydrate, Type 2 Diabetes Self-Management Program: 1-Year Results of a Single-Arm Longitudinal Study

L. R. Saslow, PhD, University of Michigan, Ann Arbor, USACharlotte Summers, Diabetes Digital Media, UKJ. E. Aikens, PhD, University of Michigan, Ann Arbor, USAD. J. Unwin, FRCGP, Principal in General Practice, The Norwood Surgery, Southport, UK

Background

Type 2 diabetes has serious health consequences including blindness, amputation, stroke, and dementia; and its annual global costs are more than $800 billion.

Although typically considered a progressive, nonreversible disease, some researchers and clinicians now argue that type 2 diabetes may be effectively treated with a carbohydrate reduced diet, which could improve type 2 diabetes management and potentially even lead to remission [252]. Indeed, previous research with carbohydrate-reduced diets for type 2 diabetes does show improved outcomes (such as glycemic control, weight loss, and reductions in the use of hypoglycemic medications), for both very-low-carbohydrate diets (roughly 20% or less of total dietary calories derived from carbohydrates) [253, 254, 255] and lower-carbohydrate diets (roughly 40% or less of total dietary calories derived from carbohydrates) [256, 257]. Although dietary interventions have historically been in person, online programs can be just as effective for some participants, as suggested by research that has examined diet and lifestyle interventions in adults with prediabetes [258]. Therefore, it is perhaps not surprising that the beneficial results of carbohydrate-reduced diets for people with type 2 diabetes (glycemic control, weight loss, and reductions in the use of hypoglycemic medications) have been replicated using online programs [259, 260].

Objectives

Our objective was to evaluate the 1-year outcomes of a digitally delivered Low Carb Program (LCP), a nutritionally focused, ten-session educational intervention for glycemic control and weight loss for adults with type 2 diabetes. The program reinforces carbohydrate restriction using behavioral techniques including goal setting, peer support, and behavioral self-monitoring.

Methods

The study used a quasi-experimental research design comprised of an open-label, single-arm, pre- and post-intervention using a sample of convenience.

From adults with type 2 diabetes who had joined the program and had a complete baseline dataset, we randomly selected participants to be followed for 1 year (N = 1,000; mean age 56.1, SD 15.7, years; 59% [593/1,000] women; mean HbA1c 7.8, SD 2.1, %; mean body weight 89.6, SD 23.1, kg; taking an average of 1.2 diabetes medications).

The Low Carb Program is a completely automated, structured, 10-week health intervention for adults with type 2 diabetes. Participants are given access to nutrition-focused modules, with a new module available each week over the course of 10 weeks. The modules are designed to help participants gradually reduce their total carbohydrate intake to < 130 g per day to meet their self-selected goals. The program encourages participants to make behavior changes based on "action points" or behavior change goals at the end of each module. These goals are supported with resources that are available to download including information sheets, recipes, and suggested food substitution ideas. The Low Carb Program online platform also includes digital tools for submitting self-monitoring and device-driven data on a number of different variables including blood glucose levels, blood pressure, mood, sleep, food intake, and body weight. Weekly automated feedback is provided to users based on their use of the program through email notifications, and participants are notified when the next week's module has been opened. Lessons are taught through videos, written content, or podcasts of varying lengths.

The program stresses the importance of regular contact with the participants' healthcare providers for adjustments in medications in weeks 1, 2, and 10. After the 10 weeks' worth of modules have been opened, participants continue to have access to the education content as well as the ability to continue to track their health (glycemic control, weight).

The content and strategies used in the program build off of prior research and theory. For example, evidence suggests that goal setting can act as an effective behavior change strategy used to improve adherence to lifestyle intervention programs in obesity management programs. The program therefore encourages participants to select a goal at the beginning of the program (such as to lose weight, reduce medication dependency, or make healthier choices for their whole family). Participants are also prompted to consider how their health would benefit from attaining their goal.

Throughout the program, participants are periodically prompted to consider how close they are to attaining their goal. The program further reinforces behavior change through integrated tracking, whereby program users are encouraged to track their health data including mood, food intake, blood glucose levels, weight, sleep, and HbA1c.

According to the control theory of behavior change, monitoring goal progress—that is, evaluating one's ongoing performance relative to the standard—and responding accordingly is critical to goal attainment. Recent findings suggest that program interventions that elevate the frequency of progress monitoring are likely to induce behavior change.

Results

Of the 1,000 study participants, 708 (70.8%) individuals reported outcomes at 12 months, 672 (67.2%) completed at least 40% of the lessons, and 528 (52.8%) completed all lessons of the program.

Of the 743 participants with a starting HbA1c at or above the type 2 diabetes threshold of 6.5%, 195 (26.2%) reduced their HbA1c to below the threshold while taking no glucose-lowering medications or just metformin.

Of the participants who were taking at least one hypoglycemic medication at baseline, 40.4% (289/714) reduced one or more of these medications. Almost half (46.4%, 464/1,000) of all participants lost at least 5% of their body weight. Overall, glycemic control and weight loss improved, especially for participants who completed all ten modules of the program.

For example, participants with elevated baseline HbA1c (≥ 7.5%) who engaged with all ten weekly modules reduced their HbA1c from 9.2% to 7.1% ($P < .001$) and lost an average of 6.9% of their body weight ($P < .001$).

Observations

- The engagement platform used by patients for this study was only available on the Web and not as a mobile app. It would be expected that a mobile app would improve engagement.
- The criteria for inclusion within the project were the requirement to be over 18 and the ability to speak English.
- The percentage of individuals with an HbA1c level of < 6.5% increased from 26% (257/1,000) to 50% (503/1,000). This degree of control, when achieved through pharmacotherapy, is often accompanied by weight gain and risk of hypoglycemic events [261]. As the now famous Action to Control Cardiovascular Risk in Diabetes (ACCORD) study reported, intensive hypoglycemic medical therapy "increased mortality and did not significantly reduce major cardiovascular events" [262].
- A limitation was the rate of delivering the entire intervention, as only 528 (52.8%) completed all modules. However, a high rate (70.8%) reported 12-month outcomes. This could be due to the program being launched in November, with Christmas and seasonal activities affecting the rate of completion. On the other hand, given that this program was entirely automated and had a wide reach, a large number of individuals were able to complete the program.

Conclusions

Especially for participants who fully engage, an online program that teaches a carbohydrate-reduced diet to adults with type 2 diabetes can be effective for glycemic control, weight loss, and reducing hypoglycemic medications.

Case Study 3: Delivering a Scalable and Engaging Digital Therapy for Epilepsy

Charlotte Summers, Diabetes Digital Media, UK

Background

Epilepsy is a neurological condition of the brain. Different epilepsies are due to many different underlying causes [263]. The causes can be complex and sometimes hard to identify. A person might start having seizures because they have one or more of the following: a genetic tendency, structural changes in the brain, or genetic conditions [263].

Epilepsy is sometimes referred to as a long-term condition, as people often live with it for many years or for life [264]. Although generally epilepsy cannot be "cured," for most people, seizures can be "controlled" (stopped) so that epilepsy has little or no impact on their lives. Treatment is often about managing seizures in the long term.

Most people with epilepsy take antiepileptic drugs (AEDs) to stop their seizures from happening [264]. However, there can be side effects with such medications, and there are other treatment options for people whose seizures are not controlled by AEDs, including the ketogenic diet (KD) [265].

This case study documents how the organization eliquates machine learning project ideas within teams and how an evidence-based approach is used to support decision-making to solve real-world problems.

Implementing the Evidence Base

The positive effects and therapeutic mechanisms of dietary ketosis and those of a ketogenic diet (KD) on human physiology have been well documented in literature. The KD is a high-fat, low-carbohydrate (usually less than 50 g/day), adequate-protein diet [266], whose metabolic effects originate back in the 1960s; however, the therapeutic effects of KD can be traced back to the early 1920s when it was successfully used in the treatment of epilepsy [267].

Ever since, the potential clinical utility of KDs has been investigated by several studies, resulting in an accumulation of scientific evidence on the therapeutic role of KDs on various physiological disorders. Several studies have shown that the KD does reduce or prevent seizures in many children whose seizures could not be controlled by medications.

Over half of children who go on the diet have at least a 50% reduction in the number of their seizures. Some children, usually 10–15%, even become seizure-free [268]. The Epilepsy Society supports the KD as a treatment option for patients over 12 months old. Research highlights the therapeutic effects of the KD on patients with epilepsy as primarily reduced medication and reduced seizures. In a recent Cochrane systematic review of the evidence regarding the effects of KDs, Levy and Cooper found no randomized controlled trials (RCTs) [269].

This demonstrates that epilepsy can be managed; and in some cases, patients can live seizure-free lives through the sustained application of a ketogenic way of eating [270]. Through reducing sugar in the diet and at the same time improving blood glucose control, people with epilepsy can achieve significant health benefits such as reducing their risk of seizures and number of medications [271].

Sensor-Driven Digital Program

The Ketogenic Program for Epilepsy (KPE) is a structured education behavior change program for people with epilepsy. The KPE empowers users to sustainably adopt a lower-carbohydrate lifestyle with the appropriate personalized education, health tracking facilities, support, health mentoring, and resources to maintain a safe and sustainable lifestyle. As a result, people with epilepsy could expect to improve glycemic control and as a result reduce the incidence of seizures, reduce medication dependency, and improve confidence in managing their condition.

Health data is collected in real time through blood glucose monitoring, medication monitoring, and seizure tracking modules within the application. Wearables are used to collect data to sense seizures and falling. Unstructured data is particularly useful in understanding sentiment and the patient's psychology.

Research activity takes place over the course of 12 months, reporting on 3-month, 6-month, 9-month, and then 1-year epidemiological health outcomes and engagement data followed by an 18-month and 2-year follow-up to demonstrate efficacy and adherence to the program over the short to medium term.

This is the world's largest and longest study on patients engaged with a digital platform for epilepsy.

Research

As part of this project, we seek to learn the most efficient and effective methods to implement and accelerate at-scale delivery of our technology to the United Kingdom's NHS population and international bill payers. The project is led by an award-winning and experienced team, recognized for innovating digital health, who are determined to revolutionize the health and well-being of people across the globe.

The technological challenge is to create a scalable, engaging, and effective solution for global implementation. This challenge is mitigated through the project extending on the infrastructure of DDM's Low Carb Program, with clinically validated outcomes and 71% retention at 1 year. At the time of writing, out of 300,000 patients globally registered within the Low Carb Program for type 2 diabetes, there are 3,112 people with epilepsy (0.5% of the UK population)—showing the readiness for an online, nutrition-focused intervention for this population.

The potential cost-saving impact to health bill payers is significant. Patients who are empowered and have improved health and well-being are also more active members of their communities and have lower social care needs.

Project Impact

The greatest impact will be felt by patients with epilepsy and their families through empowering patients to manage their epilepsy. Evidence demonstrates that a KD in people with epilepsy can be maintained by infants and adults with improvements in number of seizures and medication consumed. Evidence also demonstrates positive effects on cardiovascular disease risk [272]. The health improvements and potential cost savings are sizable.

In addition to health improvements, the project is of tremendous value to the UK economy and the NHS, with reduced complications, improved mental and emotional health, reduced hospital admissions, and reduced medication having a direct impact on the budget of the UK Treasury.

Positive social impacts experienced by the patient have a repercussion for wider communities, businesses, and the UK economy.

As the result of improved patient and population health, organizations will benefit from fewer sick days and reduced absenteeism, improved mental health, and improved perceived quality of life, as well as a reduction in the perceived burden of managing epilepsy. The positive impact is also received by the clinical healthcare community—through reducing physician and healthcare professional burden and enhancing the evidence base.

Preliminary Analysis

The objective of the research arm of this project is to evaluate the 1-year outcomes of a digitally delivered Ketogenic Program: a nutritionally focused, 16-session educational intervention for epilepsy control through provision of a KD.

The program assists patients to achieve ketosis using behavioral techniques including goal setting, peer support, and behavioral self-monitoring. Interesting correlations between elevated blood glucose and mood enabled predictive algorithms to be developed to identify symptoms of impending seizures, alerting patients through notifications to take action.

Especially for participants who fully engage, an online program that teaches a carbohydrate-reduced diet to adults with epilepsy can be effective for glycemic control, weight loss, and reducing medication and the number of patient seizures.

A randomized controlled trial is required to understand the clinical impact of such a digital therapeutic.

Case Study 4: Improving Learning Outcomes for Junior Doctors Through the Novel Use of Augmented and Virtual Reality

Paul Duval, University of Liverpool, UKVidhi Taylor-Jones, University of Liverpool, UK

Background

Junior doctors arrive at a university eager to help mankind, yet often they are unaware of how to deal with high-pressure, critical scenarios. Dealing with medical concerns appropriately—whether emergencies or not—requires years of experience and learning.

Over time, exposure in doctor surgeries, hospitals, and clinics develops the self-assurance, readiness, and experience required for junior doctors to realize their vocation. High-pressure scenarios, such as a patient having a cardiac arrest, require pinpoint precision accuracy in decision-making and actions.

For some time after the first arrival to a university, junior doctors are not ready to make such decisions. This is where augmented and virtual reality provides an opportunity to engage students in learning that readies them for life as a doctor. When you are a doctor, every decision has consequences—and that can be burdensome for some.

What's more, with 1,500 students in any one year, group trips to hospitals are expensive and logistically difficult to manage.

The project was recognized for its pioneering nature, winning an Association for the Study of Medical Education (ASME) Education Innovation Award.

Aims

The aims of the project were manifold:

- Increase the exposure to simulation and training environments for junior doctors without the use of simulation centers.
- Give inexperienced doctors the experience they require in a noncritical environment.
- Demonstrate effective and ineffective clinical practice.
- Encourage students to consider their own practices and thought processes.
- Immerse students in a clinical learning environment and identify whether the use of virtual reality enabled them to become familiar with the healthcare setting and patient unpredictability and provided the confidence to cope with high-pressure scenarios.

Project Description

It was agreed that the project has as wide a reach as possible. This eliminated virtual reality hardware such as the HoloLens or Oculus and focused the delivery onto mobile devices. The Samsung Gear and Google Pixel were able to facilitate the delivery of the content with the use of a headset to extend the view into virtual reality. This was logical considering almost every student owns a phone, whereas the use of HoloLens and Oculus is still relatively niche.

Simulations of a variety of clinical experiences were filmed with 360-degree videography using both doctors and patients as actors. These simulations were short 5- to 10-minute videos that featured decision moments where students needed to make an active choice—including a patient suffering from cardiac arrest and a patient with schizophrenia becoming increasingly agitated.

These videos were integrated into a virtual reality application that enabled students to explore a clinical setting in virtual reality and engage in the environment and encouraged them to make a variety of decisions during the process. The application encouraged users to "wear" the headset for a full, immersive experience.

The use of this augmented reality application to educate junior doctors was then evaluated with students.

What Is 360° Video?

360° video is a recent technology that uses omnidirectional cameras to capture a spherical video space rather than the landscape dimensions of traditional photography. These videos are slotted together to create an immersive viewing experience, placing the viewer in a scene within an environment rather than an observer or fly on the wall. The viewer has the ability to explore the scene, turning a full 360°. Capturing 360° video requires specialist cameras.

Playback of 360° video is most often experienced on mobile devices, where the smartphone display acts as the window to the 360° environment. Viewers can physically move their phone to view other areas of the 360° video—controlling both the orientation and viewing.

Physical devices which allow the viewer's device to be worn as a headset can be obtained to provide a fully immersive experience. Desktop computers can also play 360° video, in a similar manner to mobile devices. Interaction with 360° video on desktop usually involves a mouse to move through the environment. Hardware-driven headsets can also be used to engage with 360° videos; however, this was deemed to be prohibitive given the objectives of the project.

Conclusions

VR-driven education was warmly welcomed by team members, students, and staff alike. After an initial "wow" factor of using 360° videos and/or virtual reality, participants using the headsets engaged in the surrounding and participated in VR-delivered training.

- Virtual reality simulation provided a prime and safe clinical environment for junior doctors to explore their practice.
- In particular, the technology allowed students to focus on mistakes—unifying approaches to medicine across the cohort.
- The technology enabled students to learn anywhere, anytime: at home, school, or library.
- Virtual reality simulations were accessible by all medical students with the appropriate hardware.
- Students were able to become more active in decision-making in lectures.
- The virtual reality experience encouraged discussion and learning.
- It was cheaper and logistically easier to manage than student trips to hospital wards.
- The use of this innovative technology prepares junior doctors for the demands of the vocation.
- Research and exploration into the appropriate pedagogies to support virtual reality learning is required as the tools become more freely available.

Immersive education is a route of delivery that should be further explored to improve students' learning experiences, reduce costs, and further innovate healthcare.

Case Study 5: Do Wearable Apps Have Any Effect on Health Outcomes? A Real-World Service Evaluation of the Impact on Activity

Charlotte Summers, Diabetes Digital Media, UKArjun Panesar, Diabetes Digital Media, UK

Background

The United Kingdom, like many developed countries, is experiencing increasing rates of chronic conditions associated with diet and lifestyle choices such as obesity, type 2 diabetes, and cardiovascular disease [273]. Obesity has been shown to be a COVID-19 risk factor [274]. Evidence to support adoption of healthy lifestyles in the prevention and management of these and other long-term conditions is strong [275]. Less than one in ten patients newly diagnosed with type 2 diabetes attends structured education in the United Kingdom [276]. There are a number of reasons for this, including the fact they are in the working week and often inconveniently timed [277]. Over and above this, learned behaviors diminish after 6 months of attending an offline education program [278]. Improvements in management of type 2 diabetes can be achieved by reduction in carbohydrates, and this approach has been demonstrated to be the most effective approach to achieving normal blood glucose control and sustainable weight loss. Weight loss is a key factor in improving self-management of obesity and metabolic health conditions such as prediabetes [279–281].

Patient-facing digital health applications (apps) provide the opportunity to change the way individuals take responsibility for their own health by enabling more effective delivery of health information, allowing better monitoring of health markers, and encouraging lifestyle behavior change. Wearables (and wearable apps) have been demonstrated to increase activity in older adults and can contribute to weight loss [282–284].

Enthusiasm about the potential of health apps has grown rapidly over the last few years with many NHS services now commissioning digital health apps. While health apps offer the potential to augment care for a breadth of health conditions from mental health to medical conditions like hypertension, cardiovascular disease, and type 2 diabetes, in comparison with uptake, surprisingly little is known about their functionality or impact on health outcomes.

From the data that is available, there is a suggestion that some factors mainly predictive of uptake are not necessarily the same as those predicting completion or health outcome and very few health apps demonstrate long-term adherence [285] or report long-term health outcomes. This may, in part, be due to the lack of standards for collecting data across health apps [286].

Aspects of digital health apps that are demonstrated to impact behaviors and health outcomes should be evaluated to improve health service design and delivery. Research is needed to extend previous health app studies and understand the effectiveness of particular app features on engagement and health outcomes. Previous publications on Low Carb Program, a structured education and behavior change app, demonstrate its clinical effectiveness for supporting obesity management including statistically significant weight loss [287].

Patients' behavior directly contributes to their treatment success, with doctors relying on patients to take their prescribed medication alongside making and maintaining dietary and lifestyle changes. Many of the most significant challenges in healthcare, specifically in long-term or chronic conditions, such as obesity and type 2 diabetes, will only be resolved if we can influence behavior and support sustainable behavior change among other things such as the food environment.

There has been a detailed review of the features of Low Carb Program with an analysis of the theoretical underpinning for the presence of each app feature. The aim of this study was to assess some of these app features and understand how the engagement with these features impacts health outcomes, in particular, weight loss.

Obesity is associated with significantly worse cardiovascular risk factors, suggesting that more active interventions to control weight gain would be appropriate to help address the increasing burden of obesity on the NHS. Regardless of the interventions used to lose weight—pharmacological or behavioral—the weight is commonly regained [288–290].

Typically, half the weight lost is regained in the first year. Weight regain often continues up to 3–5 years after treatment; and, on average, 80% of people return to or exceed their pre-intervention weight [291].

Similarly, relapse rates are high for individuals who initiate attempts to stop smoking and those who try to reduce alcohol consumption [292–294].

Therefore, effective interventions that consider known factors associated not only with initial weight loss but also critically with weight loss maintenance such as building on internal motivations to lose weight, establishing social support mechanisms, identifying coping strategies, or providing support for self-efficacy and autonomy can all enhance weight loss maintenance, which is crucial for the long-term success of any weight loss interventions [295].

Methods

Research Design

The study used a quasi-experimental research design consisting of an open-label, single-arm, pre- and post-intervention. Participants were not paid for their participation; members were referred by their NHS healthcare professional team to access the program and consented to analyses of de-identified data. Each R&D committee confirmed that as the research was a service evaluation, it would not require ethical approval.

The program was accessible as an app for smartphone devices running on iOS and Android with companion apps for Apple Watch 3 and above and Wear OS watches running Android 8.0 or later with Wear OS 2.7 or above running on the Android smartwatch.

Participants

Participants were recruited and referred into the platform through their NHS healthcare professional. A total of 200 users were found from people with type 2 diabetes referred to the Low Carb Program between January 1, 2019, and April 30, 2019. The population's health and engagement data was analyzed for 6 months post-registration. All participants were from England, United Kingdom.

Low Carb Program

Low Carb Program is a behavior change app for patients with type 2 diabetes, prediabetes, and obesity. The app provides NHS-approved structured education for patients with type 2 diabetes, prediabetes, obesity, metabolic syndrome, polycystic ovarian syndrome (PCOS), and non-alcoholic fatty liver disease (NAFLD). The program educates patients to reduce the sugar in the diet to 130 g/carbs per day over a core 12-week implementation phase.

The platform provides users the ability to track their health (weight, HbA1c, blood glucose, food, mood, medication, ketones, cholesterol, blood pressure), track their nutrition with a food diary, and connect the app with a wearable (Apple Health, Google Fit, Fitbit, Nokia, Garmin) and prompts users to check in at regular intervals. The app is used within the NHS and has been demonstrated to facilitate weight loss in patients with obesity, prediabetes, and type 2 diabetes [287].

The platform provides native-language education and culturally relevant support for South Asians (Punjabi) and is available on responsive Web, iOS, Android, Apple Watch, Android Watch, Alexa, and VR (Oculus).

The platform is built on a peer-reviewed behavior change architecture developed to support long-term behavior change within the app. Components include goal setting, tailored education, community support, integrated tracking, tailored resources, coaching, and AI-powered engagement based on behaviors and health status. Figure 10-1 visualizes the behavior change architecture.

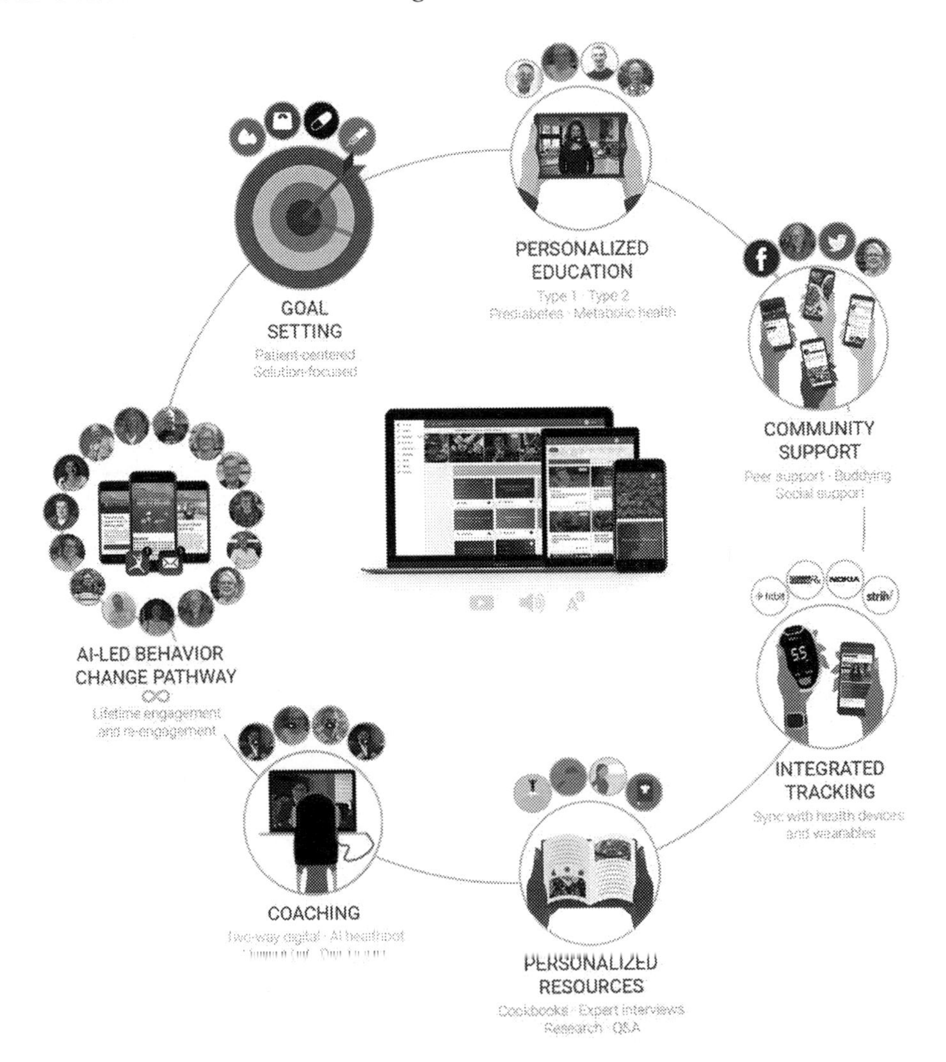

Figure 10-1. *Behavior change architecture used in the Low Carb Program*

Wearable App: Low Carb Program for Apple Watch and Wear OS Watch

The Low Carb Program's wearable app is available on Apple Watch and Android Watches, also known as Wear OS. The companion Low Carb Program watch app provides wearers with four functionalities.

- Health tracking: Companion app users can track their weight, blood glucose, mood, activity, steps, blood pressure, and heart rate.
- Relax: Participants can choose a set duration of time to focus on breathing. The smartwatch displays an animation intended to soothe and displays instructions on when to inhale and exhale.
- Lifestyle: Participants can read education and send articles to their phone from a daily-updated library of articles.
- Motivation: A screen within the companion app that displays random motivational quotes curated to prompt reflection and goal reappraisal.

Measures

At baseline, on registration to the smartphone app, participants were asked to report on their type of diabetes, year of diagnosis, weight, current medications, age, gender, dietary preference, socioeconomic status (based on weekly food budget), and presence of comorbid chronic illnesses.

At 3 and 6 months, participants were asked to report on their current weight and medications for reappraisal. Participant weight, activity (steps), and engagement data was extracted where available.

The research explored how many participants remained engaged at 6 months. It also explored the respective activity, weight loss, and engagement with the platform(s) for all users.

Statistical Analyses

Analyses were performed using the SPSS version 21.0 (SPSS Inc., Chicago, IL, USA). We conducted a multiple regression analysis to determine the predictive power of engagements with Low Carb Program app features on weight loss. The primary outcomes were engagement, body weight, and steps.

Outcomes were also analyzed within strata based on participant's Low Carb Program completion (i.e., completers, engaged with at least 9 of 12 of the Low Carb Program weekly modules, n=117; partial completers, engaged with 4–8 modules, n=49; or non-completers, engaged with ≤3 modules, n=34.

Results took into account all of the sample, regardless of follow-up information or lesson completion. Any patient within the cohort who had activated their account was included.

For participants who did not report their outcomes at 6 months, we followed the highly conservative approach of assuming that they did not improve at all (last observation carried forward), by imputing their baseline values as their outcome values. For example, participants who did not comply with reporting a 6-month outcome were treated as having no change in the outcome variable and thus were not counted as having any weight change.

Results

Participant Characteristics at Baseline

At baseline, mean weight was 95.51 kg (SD 22.36), and mean age was 56.86 years (SD 10.75). Just over half were female (59%, 118/200), 88% (176/200) were white, and all participants were from England, United Kingdom.

Platform Usage and Engagement

Over half of the referred participants (53%, 106/200) registered on the Web. Over half of participants (121, 54%) accessed the program from an iOS device and 79 (39.5%) from an Android device. A minority of participants (1%, 2/200) downloaded the Alexa app.

Of the 200 participants tracking their activity through a wearable (Apple or Android Watch), 84% (168/200) were engaged and had outcomes at 6 months (defined as actively inputting a piece of data within the last 7 days). For the 32 people lost to follow-up at 6 months, the last recorded data point was carried forward.

Wearable Engagement

For members who completed the program, participants using a companion app who completed the program reported weight loss of 7.38% (7.34 kg, t_{116}=10.9247, p<0.001).

Status	Baseline Weight (kg), Mean SD	6-Month Weight (kg), Mean SD	Weight Change (kg), Mean (SD)	Weight Change (%), Mean (SD)	P Value
All participants (n=200)	95.51 (22.36)	90.43 (20.63)	-5.07 (6.99)	-4.99 (6.31)	<0.001
Completers (n=117)	96.96 (23.21)	89.61 (21.8)	-7.34 (7.27)	-7.38 (6.72)	<0.001
Partial completers(n=49)	96.72 (22.33)	93.51 (19.13)	-3.21 (6.35)	-2.78 (4.39)	.0009
Non-completers(n=34)	88.79 (18.44)	88.81 (18.52)	0.02 (0.44)	0.02 (0.52)	.8478

For the 117 participants using a companion wearable app who completed the program, the average number of steps taken a day increased by 54.2% (t_{116}=11.57, P<.0001) by the end of 6 months. The number of average steps taken by users was significantly higher 6 months after registering for the app.

Status	Baseline Steps (Steps/Day), Mean SD	6-Month Steps (Steps/Day), Mean SD	Steps Change (Steps/Day), Mean (SD)	Steps Change (%), Mean (SD)	P Value
All participants (n=200)	2988.76 (598.95)	4318.32 (1624.74)	1329.56 (1515.61)	46.5 (52.82)	<0.001
Completers (n=117)	2947.86 (571.92)	4494.02 (1577.88)	1546.15 (1445.47)	54.2 (47.89)	<0.001
Partial completers(n=49)	3057.35 (684.04)	4649.04 (1837.2)	1591.69 (1778.9)	55.85 (66.63)	<0.001
Non-completers(n=34)	3030.65 (564.1)	3237.1 (883.87)	+206.44 (580.23)	6.37 (17.5)	.0459

Discussion

This was not a randomized controlled experimental trial, so we cannot compare the 6-month results to a control or standard-of-care group. However, of 200 patients with obesity referred to the platform, 84% of those who used a wearable were still engaged at the 6-month follow-up which supports previously reported outcomes of the effectiveness of Low Carb Program.

Although the study design does not support inferences on the causality, there are statistically significant correlations between use of the wearable and reduction of body weight.

The study found that members who either complete (>9 modules) or partially complete (>4 modules) the program report greater weight loss than those who do not complete the program.

Members who complete the program lose a significant 7.38% of body weight at 6 months. Building on prior evidence for the use of the platform by patients with type 2 diabetes to achieve improvements in glycemic control and sustained weight loss, the results of this study go to demonstrate that for participants who fully engage, an automated online program delivered by smartphone and companion app teaching a carbohydrate-reduced diet and providing tailored on-demand workouts to adults with obesity can facilitate weight loss.

Main Findings

Our initial hypothesis that patient engagement with the wearable companion app would be predictive of uptake of lifestyle behavior change appears to hold true. In particular, participants completing the program and using a wearable companion app achieve significant weight loss.

Participants using the platform's companion wearable app who completed the program lost more weight than all other strata of participants analyzed. Members using the companion app who completed the program lost more than 7% of their body weight at 6-month follow-up. These members also walked more—over 50% more at 6 months than they did at baseline. It is noteworthy that baseline data was provided to the patient by their healthcare professional on referral.

However, subsequent health outcomes (weight, steps) were input using self-report and wearable trackers, rather than measuring them through clinical health records.

However, previous research has found that these self-reported health outcomes can be quite close to actual values [296, 297].

A limitation of this study was that the analysis can only attribute *active* companion app feature engagements with health outcomes. There are a number of *passive* engagements that any given patient may undertake that would be deemed an engagement within the app but would not be explicitly tracked.

One such example would be peer support community; for "lurkers" who read posts but do not actively engage by asking or answering questions, this patient group can make up to 75% of a digital community population. There may be a significant impact on the likelihood of adopting lifestyle behavior change and subsequent health improvements, but these are a lot more difficult to monitor.

Conclusions

The aim of this study was to evaluate whether a companion wearable app contributes to weight loss. Participants tracking their health using a companion wearable app demonstrate significant weight loss at 6 months, as well as increased levels of activity.

The findings support previous studies that demonstrate wearable apps can improve health outcomes. Further analysis should be done to explore differences among populations, in particular if different ages, genders, ethnicities, and socioeconomic statuses impact health app engagements and subsequent health outcomes.

Case Study 6: Big Data, Big Impact, Big Ethics: Diagnosing Disease Risk from Patient Data

Dominic Otero, Diabetes.co.uk, UK

Background

The variety and velocity of patient data has exponentially increased over recent decades. This data has not progressed the robustness of decision-making. This leads to questions as to why decision-making is not becoming more data driven and what can be done to validate and expedite its adoption.

Diabetes.co.uk is the world's largest diabetes community, based at the University of Warwick, and the platform welcomes over 45 million visitors a year. As the world's diabetes community, the organization provides a range of services that focus on the patient data ecosystem to empower patients to manage and control their health.

By collecting patient data, or real-world evidence, over time, the ecosystem is created such that it offers insights, education, and feedback on particular aspects of the patient's health.

Platform Services

The platform collects a wealth of data. From the moment of platform access, behavioral and engagement metrics are collected, with patients progressing onto registering for the platform, opting in to part with over 120 variables that cover their diabetes and overall health status.

Once inside the platform, a variety of data types are collected—unstructured data in the form of conversations, interactions, engagements, and behavior and structured data in the form of self-reported, device-driven, wearable, IoT, and clinical health record data. The platform provides a number of services to patients and bill payers.

Medication Adherence, Efficacy, and Burden

Medication adherence, adverse effects, and wastage are concerns that pharmacology seeks to overcome. Patients with diabetes are typically on at least one medication on entry to the platform.

Within the platform, patients evaluate their medications at regular intervals, sharing their own opinions on the use of medicine in the real world, and report more structured data such as side effects, adherence, efficacy, and burden of using medications. This data is mapped geographically, by condition, medication, and other markers, and fed back in a de-identified, aggregate manner to pharmacological and academic partners for research purposes.

Data is used by partners to identify new drugs and understand interactions, usage, adherence, and real-world adverse effect prevalence. The same data is also used to improve patient safety. For instance, the platform alerts members on particular medications or using particular devices should there be an issue or recall.

Big Data: Pooling People to Empower Health Decisions

The Diabetes.co.uk community forum is the world's largest support community for people with diabetes and provides an unrivaled insight into the global diabetes population. The platform welcomes people with all types of diabetes, carers, and healthcare professionals. The concept is simple: real people with diabetes talking about their own experiences and speaking to others who are in a similar position. Patient engagement is high; there is over 2 million years of cumulative experience among members, who spend over 17 minutes each visit on the platform.

The impact of participation in online health communities on well-being has been mainly studied in relation to the informational and emotional support that members receive from their community and the positive impact that such support has on their health condition [298, 299].

Community members provide and receive informational support by sharing practical information that can help them manage their own health condition. Emotional support involves community members seeking mutual understanding and comfort by sharing their story about how it can be frustrating and painful to live with a health condition [300].

Equally, not everyone is positive about health communities. There has been a long-standing debate among healthcare professionals regarding the quality of information shared online, as well as unintended consequences of health self-management, particularly among those with low health literacy [301, 302].

As a novel digital community, data and respective insights from the online forum and dissemination of academic findings are constantly shared within the Diabetes.co.uk patient support network, organizational medical advisory board, partners, and other interested networks.

A first-phase study by Bernardi and Wu from Royal Holloway, University of London, investigated the role of online health communities in patient self-management, assessing the following:

- How patients appraise knowledge from online health communities in managing diabetes
- How a patient's healthcare professionals perceive and can potentially use such knowledge to innovate their clinical practice

Bernardi and Wu developed a model to investigate a) the factors that influence the level of engagement in online communities and b) the impact of engagement in online communities on members' cognitive, behavioral, and emotional experiences of health self-management. The model was tested empirically through an online survey with the users of the Diabetes.co.uk forum [303].

The study aimed to provide a holistic perspective on the impact of a properly managed online health community on patients' well-being.

The study concluded the following:

1. The Diabetes.co.uk forum is a catalyst of innovation.
2. The Diabetes.co.uk forum empowers patients, meaning
    - A good relationship with their healthcare professional
    - High confidence or self-efficacy in managing their condition
    - Feeling less emotionally burdened

This was supported with data from 13,000 patients who highlighted the patient benefit through platform usage:

- 53% improved blood glucose control
- 58% improved quality of life
- 59% improved confidence in managing diabetes
- 65% improved dietary choices
- 76% improved understanding of diabetes

AI Prioritization of Patient Interactions

To meet regulatory requirements, any health platform must report patient health or safety concerns immediately on discovery. This would prove cumbersome for a developed health platform such as Diabetes.co.uk if completely human led for several reasons:

- Monitoring of medical discussion groups and interactions requires human-intensive staffing to spot concerns for regulation purposes.
- Lots of time would be wasted if humans were to do all the checking.

It was decided that to meet our obligations, the use of AI bots would speed up the process, enabling humans to spend time with the concerns that mattered. This was realized through the development of a neural network–based classifier to determine sentiment and concern based on contents, user profiles, and frequency of other key metrics.

The classification system went through several hundred rounds of validation, taking a subset of the huge data lake that was available from the millions of posts. The classifier was able to spot 99.8% of all concerns, with humans required to intervene in 0.2% of cases. There were a number of benefits to the AI prioritization:

- It saved a considerable amount of human time.
- Human time was spent on the problems that were of most concern.
- The flagging of discussions occurred in near real time rather than the 15–20 minutes it would take a human to do so.

This information also assisted the customer support teams to efficiently deal with user queries by knowing exactly what a conversation was about, typical responses to such a query, and what to say to a user to ensure a favorable outcome for all parties involved. Additionally, by gauging the sentiment of users and storing it, user data can be implemented to build additional systems in the future.

Real-World Evidence

Through the analysis of patient data, we can determine a discrepancy between official and real-world health statistics. For instance, 14% of people with diabetes in the United Kingdom are expected to have a mental or emotional health concern, whereby the platform sees 44% of users with this health condition [304]. This highlights that further exploration and understanding of the patient population is severely needed to ensure critical aspects of patient care are not neglected. Partners have become more interested in real-world evidence as the years have progressed, partially to do with concerns over research funding and conflicts of interest.

Ethical Implications of Predictive Analytics

Predictive analytics enables intelligent classifiers to predict the risk of disease. The collection of a varied patient dataset—including conversations, health markers, demographics, and behavior—enables the development of a sophisticated model that can be used to refine predictions further. One of the unintended consequences of the project was discovered when deploying an AI model that predicted the risk of a patient having pancreatic cancer.

Pancreatic cancer typically affects 1 in 10,000 people and is frequently misdiagnosed. Often, patients receive a diagnosis when it's too late. A pancreatic cancer AI model was developed using a clinical dataset of health metrics, engagement, and behavior data and first predicted the likelihood of poor patient metabolic health. Should patients meet the threshold for poor metabolic health, the model evaluates the risk of the patient having pancreatic cancer.

Communicating this information to users was a first-time problem for the organization. All relevant stakeholders, including all members of the Medical Advisory Panel, were prompted for their input on this topic. Predictive analytics had previously been used to inform patients of possible hypertension or increased risk of conditions based on lifestyle.

However, informing patients of their risk of pancreatic cancer was considered to be unethical, particularly as we did not want to create additional mental health concerns or anxieties.

How should a user be told of a possible life-changing concern? The key action for this communication is to nudge the patient to see their doctor or physician as soon as possible. The factor for success in this case was determined to be user friction. The less friction there is between the user and communicating party, the more successful the engagement.

Stakeholders from behavioral economics were vital in testing and confirming an engaging communication to inform the user that their data fell outside expected norms without raising fear. To do this, the prediction was framed against patient and population demographic and medical data. Rather than being told they were the only patient to see anomalous data, patients were reassured that others have also seen similar patterns and been successful in improving their health.

Integration of the IoT

Sensors in wearables such as accelerometers, altimeters, and heart rate and skin temperature monitors provide a tremendous amount of useful information. For instance, motion, sweat molecules, and conversations are all useful and relevant data points. Such metrics can be used to take a snapshot of metabolic health, cardiovascular health, and mental health. For this to be conducted at scale, there must be little to no friction between the patient and providing parties.

The less friction there is between the user and whom they choose to share data with, the more successful the product. Wearables are becoming increasingly used in incentivized wellness propositions. Validating patient behaviors and general physiology facilitates the personalization of premiums and outcomes-based reimbursements.

Patients have the option to connect their devices and applications to the Diabetes.co.uk architecture. What is interesting is that younger users typically adhere to wearable usage and find it easy, while elderly users typically do not. This is of relevance to insurance and incentivized wellness propositions, as most claims come from older policyholders.

Conclusions

The realization of precision medicine is not without technical and ethical challenges. IoT and conversational and wearable data are just examples of a variety of data that can be collected and used to empower patients in real time, facilitating improvements in health and well-being.

With data and the application of AI come ethical concerns, particularly in the prediction of future disease risk. The lack of processes and governance on such topics is urgently required as the abilities of AI mature.

Case Study 7: Assessment of a Predictive AI Model for Personalised Care and Evaluation of Accuracy

Kelvin Summoogum, Founding CEO, MiiCare

Background

In order to provide personalized healthcare benefits and services to our users, our AI attempts to learn and interpret the user's activity of daily living. This includes the behavioral pattern relating to their movement across the home, regular routines such as bathroom habits, time they wake up and go to sleep, and so on.

This forms the key foundation for all the predictive analysis that is carried out to create situational awareness and insights about the person living alone at home and the detection of abnormal activities that would warrant an alert to be raised.

In the previous trial where we achieved accuracies above 90% in identifying abnormal behavior based on the "states" and "transitions," we had briefly discussed the potential of applying the Bayesian learning framework to automate user behavior prediction and enable proactive monitoring capabilities. In this paper, we define the problem statement for this application and outline the Bayesian approach taken to assess the effectiveness of a new predictive algorithm to add in our library.

Our initial assessment of the algorithm based on the data from a selected group of users shows a prediction of 82.2% accuracy, as an average. The accuracy relates to 10-minute windows within which the system is required to predict the location of the person. However, if the window is increased beyond 30 minutes, the accuracy reaches a maximum of 93.75%.

As the study progresses and more data about the subject is collected and the model trained on a wider dataset, we anticipate the accuracy of the 10-minute window to also increase. Some efforts are also being spent on the development of another predictive solution which uses state models for critical-time predictions without online connectivity.

Problem Statement

miiCUBE network collects movement data in the form of sensor triggers. For every user's home, the network comprises specialized sensors that detect motion installed in each room and passageways.

Every time the user passes by a sensor, the network registers the activity, with the timestamp of the event and the location where the sensor was triggered.

In our previous paper, we outlined how we had developed algorithms and formulae to visualize the daily routine of a user given sensor data collected over a period of time (days). However, that approach was static and didn't allow for consideration of scenarios where the user changed their behavior or did something suddenly. Instead we required an approach capable of

1. Predicting where the user is supposed to be, at any point during the day
2. Highlighting instances where the user didn't follow their routine
3. Continuously adapting to subtle differences in changes in behavior that happen over weeks and months as a result of ageing and cognitive decline

Expectation-Based Modeling Approach

This section outlines how we define the concept of a prediction window and train a Bayesian model for every such window to obtain a probability distribution of where the user is expected to be. This approach also allowed us to find events which are "unexpected" with respect to the model and continuously update our model with new data that is collected every day.

Prediction Windows

Earlier studies allowed us to deduct that a single predictive model cannot say where the user is in their home at a particular time of the day with a decent level of accuracy. This is because

- There are small events throughout the day which are impulsive or situational and are not indicative of true behavior.

- Elderly people have a tendency to follow a fixed routine first thing in the morning, midafternoon, and late evening. These are task-based routine, like feeding the cat, taking medicine, watching the television, going out, and so on. But in between, they are more unpredictable mainly due to morbidities, frailty, or tiredness.

To overcome this issue, we divide the day into chunks or blocks of n minutes. The duration of this window defines our level of sensitivity to unusual behavior or events.

If the window is too big (40 minutes), we lose the ability to send early notifications. In contrast, if the window is too small (5 minutes), we compromise the accuracy of our predictive algorithm which would lead to a high rate of false positives.

Therefore, setting the window requires some level of pragmatism in terms of what is acceptable to a service provider or to an elderly person in the event of an emergency.

Multinomial–Dirichlet Models

Consider a prediction window of 15 minutes between 9 AM and 9:15 AM in the morning. If our user generally wakes up at around 9 AM, then we can expect sensor triggers in the Bedroom and the Bathroom. Perhaps, after the user has visited the Bathroom, they go to the Kitchen to make a cup of tea or coffee. So there can be Kitchen sensor triggers as well. If we are only monitoring for these three locations, we can visualize the distribution of the number of sensor triggers observed as shown in Figure 10-2.

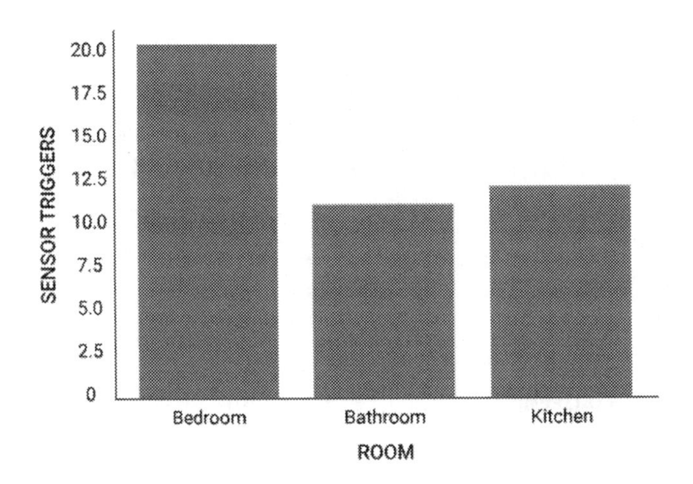

Figure 10-2. *Sample visualization of sensor trigger counts*

This distribution of sensor trigger counts is an example of a multinomial distribution. Since the distribution values are sensor trigger counts, they directly translate into probability values when normalized.

For a prediction window of a day t, we defined a multinomial distribution $(location = r)$ which tells us the probability of observing sensor triggers at a location r. From Figure 10-2, we can say that

$(location = Bedroom) = 20/(20+10+12) = 10/21 \sim 47.6\%$ between 9 AM and 9:15 AM.

The Dirichlet distribution is a powerful addition to a multinomial distribution: it serves to embed our prior personal knowledge about where we expect the user is at any point in the day. In the Bayesian framework, we always start with a prior knowledge of the probability of the event in question happening.

For instance, from our example in Figure 10-2, we know that the user will be in the Bedroom between 9 and 9:15 AM because they wake up at this time. We formalize this knowledge by defining an initial informative value for the parameter α where $\alpha_{Bedroom} > 1$ and $\alpha_{other\ rooms} = 1$.

The value of α is defined by the Dirichlet distribution (α) as

$$Dir(\alpha) = \{\alpha r\}\ for\ r \in \{all\ sensor\ locations\}$$

and it represents concentration power, alike to weightage. $\alpha_r > 1$ represents a higher concentration allocation to the probability of the user being in location r, while $ar = 1$ represents an uninformative unbiased concentration, that is, the state of not knowing any special knowledge about that particular event happening.

Using the Dirichlet distribution allows us to specify certain known characteristics of the user and therefore add behavior personalization to the model.

For prediction window t, our multinomial-Dirichlet distribution $(\alpha)(location = r)$ tells the expected probability of the user being in a location r by virtue of triggering the sensor at that location, given an initial prior knowledge. This definition forms the basis of our expectation-based predictive model ME:

$$ME = DDir(\alpha t)(location = r)$$

where

1. r ∈ {rooms where sensors are installed}
2. t ∈ {sequence of n-minute time blocks in a 24-hour day)

3. n = Duration of one time block in minutes

4. α_t = Concentration parameter for the Dirichlet distribution for the predictive window t

Conditioning the Multinomial–Dirichlet Models on Observed Data

To compute $(\alpha t)(location = r)$, we compute the "observed" multinomial distribution for the prediction window t for every date in our "training" period. For example, if we set our training period to be between 1st and 7th of the current month and our predictive window between 9 AM and 9:15 AM in the morning, we must first compute the number of sensor triggers for every location for that window for every date in that week.

The result of this computation is a matrix where each row is the multinomial distribution of sensor triggers for a date in the training period while the column will represent the locations.

We then define a Bayesian model which consists of two distributions: (α) [initialized with a prior informative set of values] and $(\alpha)(location = r)$. This model will condition itself with the observed data, that is, the distribution matrix. Once conditioned ("trained"), it can be used to generate matrices which have the same dimensions as the input matrix but now follows the true multinomial distribution of sensor triggers as learned from the input.

To compute the expectation for $(\alpha)(location = r)$, we average the generated matrices column-wise and obtain an approximation of the true multinomial distribution.

Evaluation of Predictive Accuracy

To evaluate how performant the conditioned model $(\alpha t)(location = r)$ is

1. We compute the "observed" multinomial Distribution for the window t on a test date and find the first two locations $LO1$, $LO2$ with the highest observed sensor triggers.

2. We find the first two locations $LE1$, $LE2$ where the distribution $D\alpha t$ has the highest expected probabilities.

3. We perform an equality between the observed and expected locations with highest probabilities as

   a. $if\ LO1 = LE1\ or\ LO2 = LE2$, then we consider it a good prediction ("hit").

b. $if\ LO1 \neq LE1$, then we consider it a bad prediction ("miss").

Model Evaluation

In this section, we evaluate the performance of our predictive model. We use six different durations of the prediction window so as to provide a comparison of variations in the predictive accuracy because of it.

Please do note that uniquely identifiable details about the clients have not been disclosed for privacy reasons.

Parameter values used are

1. Training period: March 2020 (31 days)
2. Testing period: 1st week of April 2020
3. $\alpha t\ for\ all\ predictive\ windows\ t\ and\ locations\ r$ = 1 (uninformative unbiased prior)
4. $length\ of\ predictive\ window\ n$ = 10, 30, 60, and 90 minutes

For each of the evaluation table, 1) each row represents a day in the first week of April 2020, 2) each column heading represents a prediction window duration chosen for the trial, and 3) each cell value is the "hit rate," which is computed as the number of times the model has a good prediction for a day.

User A is a 90+ elderly gentleman living in the United Kingdom. The user deployed activity sensors around ten locations in his home. Table 10-1 summarizes the hit rate for every day in the testing period with the different prediction window durations. We found that the model had a predictive accuracy of 80% when trained on one month of data and tested for 1 week of data.

Table 10-1. *Evaluating predictive performance for User A*

	10 m	30 m	60 m	90 m
1	75.9	81	81	90.4
2	80.6	83.5	81.4	88.1
3	77	81.4	77.3	88
4	70.5	76.5	76.5	82.8
5	76.9	77.5	81.4	82.2
6	72	79.4	77.5	82.1
7	71	71.7	74.4	77.4
AVG	74.8	78.7	78.5	84.4

We also expected a negative trend in the accuracy as we predicted further into the future as captured in Figure 10-3. The shaded region highlights the range of accuracy difference for using windows of 10 and 90 minutes.

User B is an 85+ elderly lady living in the United Kingdom. Her home has six activity sensors set up. Table 10-2 summarizes the hit rate for every day in the testing period with the different prediction window durations. For this user, we found that the model had a predictive accuracy of 84.4%.

Table 10-2. *Evaluating predictive performance for User B*

	10 m	30 m	60 m	90 m
1	74.2	89.17	89.2	87.5
2	90.3	89.8	87.5	93.75
3	82.8	87.5	79.2	93.75
4	80.7	87.5	87.5	93.75
5	82.7	83.3	87.5	87.5
6	77.4	85.4	83.3	87.5
7	76.4	77.1	80	82.5
AVG	80.6	84.3	83.4	89.4

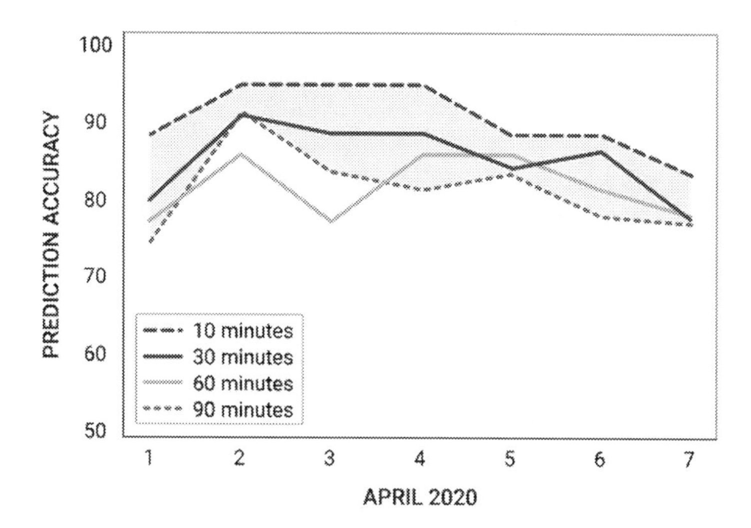

Figure 10-3. *Prediction accuracy over time for User A*

Similar to User A, a negative trend is also observed for User B, as shown in Figure 10-4. The shaded region highlights the range of accuracy difference for using windows of 10 and 90 minutes.

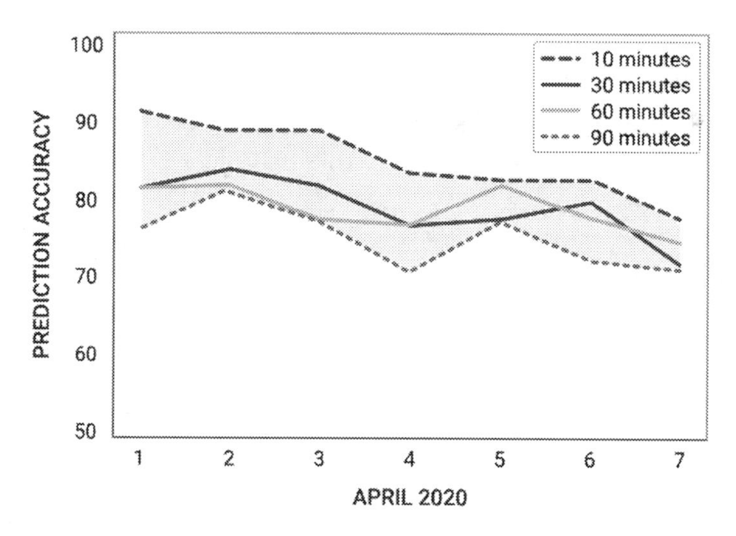

Figure 10-4. *Prediction accuracy over time for User B*

Improving the Solution

In our trial for the two users, we set the value of α for all predictive windows as 1. This represents an uninformed unbiased state of the model before it has seen any data. Instead, we can manually define an informed biased value for α which reflects our knowledge of the user's presence during different points of the day. This can be a tedious manual task, but once defined can reduce the amount of data required for the model to fully condition itself to arrive at the real and precise expectation probabilities.

Increasing the duration of the predictive window boosts predictive accuracy; however, our objective is to enhance the model to bring the accuracy to exceed 90% prediction for 30-minute windows. Achieving such a high accuracy is very challenging given an elderly person may not necessarily stick to daily routine. However, with more data points, the predictive model will be able to learn about certain external factors which trigger an elderly person to deviate from their routines, for example, sickness, the effects of medications, and so on.

Put within the context of the intended use within a care home or a home setting, the model will predict where a person is "supposed" to be, and then the algorithm will compare this with where the person is for that particular window. In most cases, having a mismatch for one or two cycles of 30-minute duration is not going to be detrimental.

However, if, for example, we predict that the user will be in the kitchen in the next 30 minutes (first cycle) and then in the living room for another 30 minutes (second cycle), but the system detects that they are in the bathroom for more than 30 minutes, then an alarm will be triggered. At the same time, the system monitors their body vitals, and if something abnormal is observed, these will also be notified to a family member/carer.

Anything shorter than 30 minutes will very likely result in excessive false alerts given the behavioral pattern of an elderly tends to vary a lot. That said, anything abnormal with the body vitals is notified instantly, for example, low heart rate.

The two focus areas as we grow the fleet of miiCUBEs are user duration and data flows.

User Duration

Establish the duration a user has to be in the same location to trigger an alarm. This is currently a feature configurable by the user/family/carer; however, with enough data points, the machine learning model will be able to set those parameters automatically.

Data Bandwidth

The miiCUBE unit serves as an IoT hub to connect the network of sensors to the servers and also houses a small onboard computer for its other functions, such as having a conversation with the elderly person. Our predictive model, if implemented, will be housed in the server where it will be fed with data incoming from the miiCUBE for its training and predictions. This workflow is dependent on the quality of Internet connection within the region where the user lives. If this connection is severed for a prolonged period of time, the network will lose its capability to proactively monitor for emergencies.

Next-Generation Model

The next version of the predictive model we will be exploring is state and transition models based on probabilistic graphical models (PGMs). This was mentioned in the previous white paper along with the Bayesian approach defined in this paper.

PGMs serve to represent, model, and infer probable courses of actions based on order-sensitive data. We define such a model using sensor locations as "states" and the connections between these locations as routes the user takes to move between them.

We can therefore isolate and detect emergencies like falls when the user is in a location ("state") or when in transit (between "states"). We can also find probabilities for the user to be in a state or a transit between two states for a specific time of a day. Such a model can work locally in miiCUBE and will be beneficial when connectivity is not available, hence maintaining service continuity.

Conclusion

In this paper, we use the Bayesian framework as the key predictive framework of operation for miiCUBE's sensor network. We define an expectation-based predictive approach where

- There is a multinomial-Dirichlet Bayesian model for every n-minute window of a day.
- The Bayesian models are conditioned on sensor trigger data for every location specific to the window of the day.
- The models state the top two locations where the user is expected to be at a specific window in the day.

This approach has three operating parameters: the training period, the duration of the predictive window, and prior values chosen for α for every window of the day.

The predictive accuracy of the model is dependent primarily on the latter two where

- There is a trade-off between relevance of predictions and uncertainty in prediction for a model in a window. Increasing the window duration decreases prediction uncertainty while also decreasing how useful the prediction becomes in real-time monitoring.
- Defining strong informative prior values for α reduces the amount of data required to condition a model to arrive at the real precise expectation probabilities of the user's location in the future.

The approach was able to achieve an average predictive accuracy of 82.6% in a trial involving two of our miiCUBE users.

For fair comparison, the training and testing periods along with the duration of the prediction window and values for α for all windows in a day were kept constant for both cases. The trial has been rolled out to all of our users.

Case Study 8: Can Voice-Activated Assistants Support Adults to Remain Autonomous, a Real-World Service Evaluation of the Impact of a V oice-Activated Smart Speaker Application on Weight and Activity

Charlotte Summers, Diabetes Digital Media, UK

Background

Regular and moderate physical activity practice provides many physiological benefits. It reduces the risk of disease outcomes and is the basis for proper rehabilitation after a severe disease [305]. Most importantly, regular activity can improve quality of life [306]. Physical activity has been demonstrated to increase fitness, physical function, cognitive function, and positive behavior in people with dementia and related cognitive impairments [307]. One study found that gaming systems are motivating and can have a positive effect on strength, balance, and overall fitness for elderly populations [308].

Patient-facing digital health applications (apps) provide the opportunity to change the way individuals, in particular the elderly, take responsibility and care for their own health by enabling more effective delivery of health information. New platforms and devices allow better monitoring of health markers and can act as tools to support and encourage lifestyle behavior change. Apps and wearables have been demonstrated to empower older patients to increase activity and can contribute to weight loss [309].

Health Apps

Enthusiasm about the potential of health apps has grown rapidly over the last few years with many global health services commissioning digital health apps. While health apps offer the potential to augment care for a breadth of health conditions from mental health to medical conditions like hypertension, cardiovascular disease, and obesity, surprisingly little is known about their functionality or impact on health outcomes. Only 15% of health apps are formally reviewed by a governing body in the United Kingdom [310].

Engaging in positive healthy lifestyle behaviors continues to be a public health challenge, requiring innovative solutions [311]. Voice-activated assistants are a form of conversational agent that include Amazon Alexa, Apple Siri, Microsoft Cortana, and Google Assistant. These agents are a form of human-centered computing which provides interaction between human and computer through sound-based input/output. This provides the opportunity to engage audiences in a novel way—through conversing with an agent that processes the speech of the user (invocation) and returning a statement, image, audio, or video as output to the user.

Conversational Agents

A conversational agent is a computer system intended to converse with a human [312]. Voice-activated assistants are connected to the Internet and listen for commands, interpret them, and take action.

Smart Speakers

Smart speakers are a form of speaker and voice command device, with an integrated virtual, voice-activated assistant that offers interactive actions and hands-free activation with the help of a wake word [313]. Smart speakers can run third-party apps. For Amazon Alexa, this is known as a Skill.

Voice-Activated Assistants

The use of smart speakers, or voice assistants, like Alexa, Siri, Cortana, and Google Assistant has soared with at least one in five homes in the United Kingdom estimated to be using them [314]. Amazon Alexa is the most popular smart speaker device [315]. Research has shown Alexa is primarily used for checking weather forecasts, playing music, and controlling other devices [316]. The use of devices such as Amazon Alexa provides a novel mechanism to engage new audiences and demonstrate potential to help older adults [317, 318]. Alexa uses over the weekends are more frequent than on weekdays, but overall usage tends to decrease over time [318].

Health and fitness apps for voice-activated assistants are considered a nascent area of smart care, shown by one of the first literature reviews being conducted only this year [321]. Indications suggest that voice-activated assistants are capable of reaching populations that are diverse, underserved, and hard to reach.

User satisfaction and receptivity to voice assistants are promising yet compared to smartphone apps, voice-enabled devices are very limited and lack evidence [321]. A review of evidence shows voice-activated assistants have been used to remind people to take medications, as pregnancy companions, to help find answers to questions, and to provide coaching [321]. Notably, evidence shows that the accuracy of apps could be improved [320, 321].

The accessibility, ubiquity, and conveniences of voice assistants have led voice-activated devices (e.g., smartphones, smart speakers) to be a significant component of a digital health ecosystem. Such an ecosystem has been envisioned especially for patients who have chronic conditions [322]. The anticipated benefits of voice assistants are also opined to be in elderly care and helping older adults live with more autonomy ("smart aging") [323].

One study found use of an AI voice assist may be a practical method to deliver scalable individualized behavioral health coaching to address cardiovascular health in cancer survivors [324].

Cybersecurity

However, a key area of inertia for voice-enabled assistants, particularly if they are ever to be integrated within the EHR, is the real and perceived threat of privacy, cybersecurity, and exploitation [320, 325]. Others have raised questions about their longer-term disruptive impact on the consumption of information, user profiling, and people's relationship with technology [325, 326].

It has been suggested that smart speakers can help older people stay more autonomous at home [327, 328]. As part of a “smart” future, smart healthcare could measure this in a number of ways—through wearables, connected devices, and smart speakers. There are a plethora of signals that could be used to measure activity—including active minutes, activities, calorie burn, and steps.

The aspects of digital health apps that are demonstrated to impact behaviors and health outcomes should be evaluated to improve health service design and delivery. It has been demonstrated that there is a low comfort level to new technologies, but with appropriate training, it could have significant roles in healthcare [329, 330].

Research is needed to extend previous health app studies and understand the effectiveness of particular app features on engagement and health outcomes.

Previous publications on Low Carb Program, a structured education and behavior change app, demonstrate its clinical effectiveness for supporting obesity and type 2 diabetes self-management including statistically significant weight loss, HbA1c improvements, and a reduced reliance on medication leading to type 2 diabetes remission [91].

There has been a detailed review of the features of Low Carb Program with an analysis of the theoretical underpinning for the presence of each app feature [16]. In 2019, a companion Low Carb Program Amazon Alexa Skill was launched for Alexa Echo and Echo Show. The skill enables users to converse with Alexa to watch all lessons of the program, update their weight, update their HbA1c, ask for the nutritional values of foods, and request on-demand exercise workouts, guided stretch workouts, and yogas [331].

Features included usage history (clicks/taps, date and time, device, screen size), age, gender, goal, health status, medication, dietary preference, allergies, and location. Features were transformed to reduce complexity, for example, by classifying age.

Real-World Digital Health Evaluation

As providers of the Low Carb Program as a multi-platform digital health intervention on the Web, mobile, smartwatch, and voice-activated assistants, there is an opportunity to assess voice-activated assistant engagement and outcomes and their real-world impact.

The aim of this study was to examine the profile characteristics and health outcomes for participants using the Low Carb Program voice-activated smart speaker Skill (app) as a health management tool to support sustainable health-promoting behaviors.

Objective

The aim of this study was to examine the profile characteristics and activity outcomes for participants using a conversational agent as a health management tool as part of their lifestyle.

Methods

A total of 27 patients with obesity, prediabetes, and type 2 diabetes were referred to Low Carb Program by their NHS healthcare professional in February 2020. Thirteen patients also downloaded the platform's Amazon Alexa Skill.

All participants were from England, United Kingdom. Health conditions that would exclude healthcare professional referral were carnitine deficiency (primary), carnitine palmitoyltransferase (CPT) I or II deficiency, carnitine translocase deficiency, beta-oxidation defects, medium-chain acyl dehydrogenase deficiency (MCAD), long-chain acyl dehydrogenase deficiency (LCAD), short-chain acyl dehydrogenase deficiency (SCAD), long-chain 3-hydroxyacyl-CoA deficiency, medium-chain 3-hydroxyacyl-CoA deficiency, pyruvate carboxylase deficiency, acute porphyria, history of mental health disorders, or pregnancy.

All users were selected for analysis. Patient health and engagement data from use of the mobile and companion apps were tracked, collected, and analyzed over 4 months to determine if use of the Alexa Skill contributed to change in health-promoting behavior.

Data analyzed included demographic data (age, gender, ethnicity, location, employment status), weight, and amount of minutes spent in on-demand classes. Qualitative feedback on features was also assessed.

Low Carb Program

Low Carb Program is an evidence-based structured education and behavior change self-management platform. The platform has demonstrated clinical benefits including weight loss, reduction in medications, and supporting type 2 diabetes remission. The platform is built on an evidence-based architecture grounded in behavior change psychology [16].

The Low Carb Program provides NHS-approved structured education for patients with obesity, prediabetes, type 2 diabetes, metabolic syndrome, polycystic ovarian syndrome (PCOS), and non-alcoholic fatty liver disease (NAFLD) on general well-being and aids patients to reduce the sugar in the diet to 130 g/carbs per day over a core 12-week implementation.

Participants engaging with the Low Carb Program show a directional increase in the number of active minutes as they progress through the intervention. To support this, on-demand activities were made available in-app. On-demand activities comprised guided exercise, yoga, and stretch classes led by qualified personal trainers.

On-demand activities were available on all app platforms, including Amazon Alexa where activities were presented in video or audio format to the participant in English only.

Voice-Activated Assistant App: Low Carb Program for Alexa

The Low Carb Program's Amazon Alexa Skill is downloaded from the Alexa Skills store hosted by Amazon or by invoking the Skill by saying "Alexa, enable Low Carb Program" to an Amazon Alexa Echo or Amazon Echo Show smart speaker [331].

The Amazon Echo provides output as audio only, whereas Echo Show outputs in video also. Participants are required to link their Amazon account with their Low Carb Program account in order to utilize the profile recall invocations available to participants using the Skill.

Engaging with the Alexa Skill

Once enabled, the Low Carb Program Amazon Alexa Skill is invoked by saying "Alexa, ask Low Carb Program..." followed by one of a range of commands to start an activity:

- Give me an exercise.
- I want to warm up.
- I want to cool down.
- I want to get active.
- Give me a yoga.
- Give me a workout.
- Give me a stretch.

Participants are informed of the commands available on the Amazon Alexa Skill listing page and also within the app. Commands allow the user to converse with Alexa to watch all lessons of the program, update their weight, update their HbA1c, ask for the nutritional values of foods, and request on-demand exercise workouts, guided stretch workouts, and yogas.

Deep neural network recommender systems are used to make predictions about other resources, activities, and education based on the current user's profile and behaviors. The neural network was trained on the behaviors of 77,713 people. An initial dataset of 105,394 records was extracted. On exploration, it was identified that only fully completed attribute sets would be used and the remaining 26,681 semi-labeled datasets would be used to validate the model in the real world.

Exercises output by the Amazon Alexa Skill were available as guided audios or guided videos, respectively, for the Amazon Alexa Echo and Amazon Alexa Echo Show. A total of 21 activities were available—7 yoga workouts, 7 HIIT exercise workouts, and 7 stretch workouts. Workouts are tailored to a beginner's level of fitness.

After 12 weeks of using the Amazon Alexa Skill, users were sent a survey to request opinions and feedback on the Amazon Alexa Skill. Users have 4 weeks to complete the feedback survey.

Measures

At baseline, on registration to the app, participants were asked to report on their type of diabetes, year of diagnosis, weight, current medications, age, gender, dietary preference, socioeconomic status (based on weekly food budget), and presence of comorbid chronic illnesses. On completion of registration, participants were notified by email that an Amazon Alexa Skill was available and prompted to download it.

At 4 months, participant weight, active minutes, and engagement data was extracted where available. Active minutes defined as the "number of minutes an on-demand exercise, stretch, or yoga class was opened and not backgrounded" was collected.

The total number of active minutes was calculated over the study period. Participants using the Amazon Alexa Skill were sent a survey to request feedback on the features, functionality, ease of use, and comfortability.

The research explored how many participants downloaded and used the Amazon Alexa Skill to engage with the on-demand classes. It also explored the respective active minutes, weight loss, and sustained engagement with the platform(s) for all users.

Statistical Analyses

Analyses were performed using the SPSS version 21.0 (SPSS Inc., Chicago, IL, USA). We conducted a multiple regression analysis to determine the predictive power of engagements with Low Carb Program app features on active minutes.

The primary outcomes were engagement and downloads of the Amazon Alexa Skill and active minutes. Secondary outcomes were health outcomes including weight loss. Outcomes were also analyzed within strata based on whether participants downloaded the Low Carb Program Amazon Alexa Skill comparing Skill users (ASU, downloaded Amazon Alexa Skill, n=13) and non-Skill users (NSU, did not download Amazon Alexa Skill, n=14).

Results took into account all of the sample, regardless of follow-up information or lesson completion.

For participants who did not report their outcomes at 4 months, we followed the highly conservative approach of assuming that they did not improve at all (last observation carried forward), by imputing their baseline values as their outcome values.

For example, participants who did not comply with reporting a 4-month outcome were treated as having no change in the outcome variable and thus were not counted as having any weight change.

Results

For the 13 participants syncing their account with the Alexa Skill (requiring an Amazon account), 92% (12/13) were engaged and had outcomes at 4 months defined as actively inputting an item of data within the last 7 days.

Participant Characteristics at Baseline

At baseline, mean weight was 95.1 kg (SD 22.9 kg) and mean age was 61.7 years (SD 5.7 years) across all participants. Participants using an Amazon Alexa Skill had a mean weight of 91.7 kg (SD 21.7 kg) at baseline.

Two-fifths of participants were female (40.7%, 11/27), over half were white (55.5%, 15/27), and all participants were from England, United Kingdom.

Platform Usage and Engagement

Almost half (48.1%) of the participants downloaded the Amazon Alexa Skill (13/27).

Of the 27 total participants, 24 (88.9%) were active at 4 months. 85.7% (12/14) of non-Amazon Alexa users were engaged at 4 months compared to 92.9% (12/13) of the Amazon Alexa users (defined as actively inputting a piece of data within the last 7 days).

Amazon Alexa Skill Usage

Participants using the Amazon Alexa Skill registered 85.82 active minutes, an increase of 68.9% compared to the 50.86 registered active minutes by participants not using the skill.

Status	Total Active Minutes (mins/day), Mean SD
All participants (n=27)	67.74 (28.36)
Skill users (n=13)	85.92 (12.84)
Non-Skill users (n=14)	50.86 (28.61)

Weight

Participants using the Amazon Alexa Skill lost an average of 1.77 kg across the duration of the study.

Status	Baseline Weight (kg), Mean SD	4-Month Weight (kg), Mean SD	Weight Change (kg), Mean (SD)	Weight Change (%), Mean (SD)	P Value
All participants (n=27)	95.05 (22.94)	93.66 (22.86)	-1.39 (3.22)	1.49 (3.43)	0.557
Skill users(n=13)	91.69 (21.79)	89.92 (21.68)	-1.77 (2.80)	2.03 (3.25)	0.041
Non-Skill users(n=14)	90.17 (24.45)	07.14 (21.17)	-1.03 (3.63)	0.99 (3.63)	0.308

Qualitative Feedback

As part of the user experience, a feedback survey was sent to participants using the Amazon Alexa Skill. Of the 13 Alexa Skill users/participants, 7 (53.8%) completed the survey.

Discussion

This was not a randomized controlled experimental trial, so we cannot compare the 4-month results to a control or standard-of-care group. However, of 13 participants using the Low Carb Program Alexa Skill, 92.8% were still engaged at the 4-month follow-up which supports previously reported outcomes of the engagement of Low Carb Program [91].

Engagement

Surprisingly, almost half of participants downloaded the Alexa Skill, demonstrating the rise in popularity of smart speakers. Healthcare professionals did not give all participants an Amazon Alexa on referral, and therefore an unaccounted bias in this study is ownership of an Amazon Alexa Echo or Echo Show. The study is intrinsically biased toward those who already own the smart speaker. If smart speakers are still considered an emerging technology, it could be reasonably inferred that the population the intervention engaged with would be considered "early adopters."

Although the study design does not support inferences on the causality, there are statistically significant differences between the total active minutes registered by participants using the Alexa Skill and those who did not.

Total Active Minutes

In this study, Alexa Skill users logged statistically significantly more total active minutes (85.92 minutes, SD 12.84 minutes) than non-Skill users (50.86 minutes, SD 28.61 minutes), $t_{(25)}$=-4.05, p=<0.01.

Participants using the companion Alexa Skill registered more than 30 minutes of activity on average as defined by total active minutes across the duration of the study compared to participants who did not use the Alexa Skill. There are several possible explanations for this, including the smart speaker's more optimal user interface and ease of information retrieval through conversation.

Alexa's command-and-repeat nature seems to suit engagement in activities. Usage of the Alexa Skill could have been influenced by COVID-19, which saw the wider public engage with and consume digital public health services such as remote monitoring while staying safe and well at home.

Total active minutes was defined as the total number of minutes spent engaged in an on-demand exercise, stretch, or yoga class while the user interface is opened and not backgrounded.

The veracity of outcomes could be improved through use of wearable data to improve the accuracy of total active minutes. An on-demand class playing is used as a proxy for minutes the user is active, and engagement is assumed based on the video continuing to play. Surprisingly, no users who paired with Alexa also used a smartwatch with the app.

The possibility of leaving on-demand classes playing was mitigated by the platform's auto-logout security mechanisms, as per medical software best practice.

Further exploration could investigate the use of wearables and Alexa to improve the precision of positive health behaviours such as time active.

Weight Loss

In this study, Alexa Skill users had a greater weight loss (-1.77 kg, SD 2.80 kg) than non-Skill users (-1.03 kg, SD 3.63 kg), $t_{(25)}=0.59$, $p=0.56$. Alexa Skill users lost more weight than non-Skill users. There are several possible explanations for this.

Given that Alexa Skill users also logged more total active minutes, one could hypothesize that the Alexa smart speaker may be biased in ownership to those already interested in fitness. Indeed, the use of the app during COVID-19 may also contribute to its engagement.

The average age of participants using the Low Carb Program Alexa Skill was almost 5 years older than non-Alexa users, supporting other research that shows voice-activated assistants and smart speakers can assist the elderly to be more active and autonomous and improve social bonds [318, 321, 332, 333].

Feedback

Of 13 Skill-using participants to complete a feedback survey evaluating the Alexa Skill

- Exactly three-quarters of participants (75%) found it easy to set up and 75% found it enjoyable to use.
- Three-quarters of participants (75%) were interested in new technology.

- The majority (76.9%) used the Skill for on-demand classes or meditation and mindfulness.
- Over a third (38.4%) of participants found using the platform on a new technology a comfortable experience.
- The majority of participants, 84.6%, would recommend the Skill to a friend.
- Over half of participants (53.8%) felt accountable to the Low Carb Program Alexa Skill, and three-quarters (75%) felt more active because of using it.
- Almost all participants (92.3%) stated they would like training on how to use the Skill.

The outcomes support evidence that older populations are interested in new technology, yet feel uncomfortable using them without appropriate training. There was a clear signal that participants expect onboarding support for new technologies.

The study contributes to the evidence base around smart care. There are very few studies available evaluating use of digital health apps for smart speakers.

The study ignites other research questions based on app platforms, including whether the use of other companion apps for smartwatch and augmented reality or virtual reality may impact health or engagement with the platform.

Conclusion

The aim of this study was to assess whether use of the Low Carb Program Alexa Skill would contribute to improvements in positive health behaviors. Participants using the Alexa Skill demonstrated improvements in total active minutes.

The findings support previous studies that demonstrate voice-activated assistant apps can improve health outcomes. Participants who use the Alexa Skill were motivated to try innovative technologies, older in age, and desire training to overcome low comfort levels. Further analysis should be done to explore differences among populations, in particular if different ages, genders, ethnicities, and socioeconomic statuses impact engagement and subsequent health outcomes.

APPENDIX A

References

1. Sohrabi, C., Alsafi, Z., O'Neill, N., Khan, M., Kerwan, A., Al-Jabir, A., Iosifidis, C., & Agha, R. (2020). World Health Organization declares global emergency: A review of the 2019 novel coronavirus (COVID-19). International Journal of Surgery, 76, 71–76. https://doi.org/10.1016/j.ijsu.2020.02.034
2. Whitelaw, S., Mamas, M. A., Topol, E., & Van Spall, H. G. (2020). Applications of digital technology in COVID-19 pandemic planning and response. The Lancet Digital Health.
3. Pastorino, R., De Vito, C., Migliara, G., Glocker, K., Binenbaum, I., Ricciardi, W., & Boccia, S. (2019). Benefits and challenges of Big Data in healthcare: an overview of the European initiatives. European journal of public health, 29(Supplement_3), 23–27.
4. Guinchard, A. (2020). Our digital footprint under Covid-19: should we fear the UK digital contact tracing app? International Review of Law, Computers & Technology, 1–14.
5. Milanovic, M., & Schmitt, M. N. (2020). Cyber Attacks and Cyber (Mis) information Operations during a Pandemic. Journal of National Security Law & Policy (Forthcoming).
6. Marr, B. (2017, September 22). 12 AI Quotes Everyone Should Read. Forbes. www.forbes.com/sites/bernardmarr/2017/09/22/12-ai-quotes-everyone-should-read/
7. McAfee, A., Brynjolfsson, E., Davenport, T. H., Patil, D. J., & Barton, D. (2012). Big data: the management revolution. Harvard business review, 90(10), 60–68.

© Arjun Panesar 2021
A. Panesar, *Machine Learning and AI for Healthcare*, https://doi.org/10.1007/978-1-4842-6537-6

8. Helbing, D., Frey, B. S., Gigerenzer, G., Hafen, E., Hagner, M., Hofstetter, Y., ... & Zwitter, A. (2019). Will democracy survive big data and artificial intelligence? In Towards digital enlightenment (pp. 73–98). Springer, Cham.

9. Panetta, K. (2017). Top trends in the gartner hype cycle for emerging technologies, 2017. Gartner, Stamford, US.

10. Bradshaw, J. M. (1997). An introduction to software agents. Software agents, 4, 3–46.

11. Campbell, M., Hoane Jr, A. J., & Hsu, F. H. (2002). Deep blue. Artificial intelligence, 134(1–2), 57–83.

12. Hodson, H. (2015, July 15). Robot homes in on consciousness by passing self-awareness test. New Scientist. www.newscientist.com/article/mg22730302-700-robot-homes-in-on-consciousness-by-passing-self-awareness-test/

13. Kendall, G. (2019b, July 15). Apollo 11 anniversary: Could an iPhone fly me to the moon? The Independent. www.independent.co.uk/news/science/apollo-11-moon-landing-mobile-phones-smartphone-iphone-a8988351.html

14. Thakur, S., & Dharavath, R. (2019). Artificial neural network based prediction of malaria abundances using big data: A knowledge capturing approach. Clinical Epidemiology and Global Health, 7(1), 121–126.

15. Lesko, L. J., & Schmidt, S. (2012). Individualization of drug therapy: history, present state, and opportunities for the future. Clinical Pharmacology & Therapeutics, 92(4), 458–466.

16. Summers, C., & Curtis, K. (2020). Novel Digital architecture of a "Low Carb Program" for initiating and maintaining long-term sustainable health-promoting behavior change in patients with type 2 diabetes. JMIR diabetes, 5(1), e15030.

17. Summers C, et al. (2020). Supporting mental health and resilience during COVID-19: 8-week outcomes of a mindfulness based digital intervention to support vulnerable and at-risk patients. Pending publication.

18. Jelley, O. (2018, June 18). NHS green light for Low Carb Program. The Diabetes Times. https://diabetestimes.co.uk/nhs-green-light-for-low-carb-programme/

19. Smart, W. (2018). Lessons learned review of the WannaCry ransomware cyber attack. Department of Health and Social Care, England UK, London, 1(20175), 10–1038.

20. Kelion, L. (2020, April 27). NHS rejects Apple-Google coronavirus app plan. BBC News. www.bbc.co.uk/news/technology-52441428

21. Ipsos, M. O. R. I. (2016). The one-way mirror: public attitudes to commercial access to health data. London: Wellcome Trust.

22. Liu, Q., Luo, D., Haase, J. E., Guo, Q., Wang, X. Q., Liu, S., ... & Yang, B. X. (2020). The experiences of health-care providers during the COVID-19 crisis in China: a qualitative study. The Lancet Global Health.

23. Rao, S. (2020, July 14). Medical student creates handbook to show symptoms on darker skin. UK. https://thetab.com/uk/2020/07/14/medical-student-creates-handbook-to-show-symptoms-on-darker-skin-166352

24. Palaiodimos, L., Kokkinidis, D. G., Li, W., Karamanis, D., Ognibene, J., Arora, S., ... & Mantzoros, C. S. (2020). Severe obesity, increasing age and male sex are independently associated with worse in-hospital outcomes, and higher in-hospital mortality, in a cohort of patients with COVID-19 in the Bronx, New York. Metabolism, 108, 154262.

25. Zheng, K. I., Gao, F., Wang, X. B., Sun, Q. F., Pan, K. H., Wang, T. Y., ... & Zheng, M. H. (2020). Obesity as a risk factor for greater severity of COVID-19 in patients with metabolic associated fatty liver disease. Metabolism, 108, 154244.

26. Abuelgasim, E., Saw, L. J., Shirke, M., Zeinah, M., & Harky, A. (2020). COVID-19: Unique public health issues facing Black, Asian and minority ethnic communities. Current Problems in Cardiology, 100621.

27. Panesar, A., & Panesar, H. (2020). Artificial Intelligence and Machine Learning in Global Healthcare. Handbook of Global Health, 1–39.

28. Summers, C. (2019). Sharing of Data: Patient Perspective.

29. McDonald, M. P., & Rowsell-Jones, A. (2012). The digital edge. Gartner, Incorporated.

30. Trifu, M. R., & Ivan, M. L. (2014). Big Data: present and future. Database Systems Journal, 5(1), 32–41.

31. Van Rijmenam, M. (2013). Why the 3v's are not sufficient to describe big data. BigData Startups.

32. Dave, M., & Kamal, J. (2017, September). Identifying big data dimensions and structure. In 2017 4th International Conference on Signal Processing, Computing and Control (ISPCC) (pp. 163–168). IEEE.

33. Ferguson, S. (2016). Big data, analytics market to hit $203 billion in 2020. InformationWeek.

34. Smith, I. Cost of hard drive storage space [online]. February 2012. Wayback: http://web.archive.org/web/20110718172940/http://ns1758.ca/winch/winchest.html

35. Diabetes.co.uk. (2018, June 10). #FacesofDiabetes engages with over 3.5 million people over Diabetes Week.

36. Hirschfeld, D. (2012). Twitter data accurately tracked Haiti cholera outbreak. Nature.

37. Smarter Business Review—Services Blog. (2020, August 1). Smarter Business Review. `www.ibm.com/blogs/services/`

38. Ginsberg, J., Mohebbi, M. H., Patel, R. S., Brammer, L., Smolinski, M. S., & Brilliant, L. (2009). Detecting influenza epidemics using search engine query data. Nature, 457(7232), 1012–1014.

39. Butler, D. (2013). When Google got flu wrong: US outbreak foxes a leading web-based method for tracking seasonal flu. Nature, 494(7436), 155–157.

40. Kayyali, B., Knott, D., & Van Kuiken, S. (2013). The big-data revolution in US health care: Accelerating value and innovation. Mc Kinsey & Company, 2(8), 1–13.

41. Malde, M., Kothari, M., & Shah, M. (2012). Open Graph Interface for Uniform Access to E-learning Resources. International Journal of Computer Applications, 58(18).

42. Renaud, B., Labarère, J., Coma, E., Santin, A., Hayon, J., Gurgui, M., ... & Salloum, M. (2009). Risk stratification of early admission to the intensive care unit of patients with no major criteria of severe community-acquired pneumonia: development of an international prediction rule. Critical care, 13(2), R54.

43. Amarasingham, R., Patel, P. C., Toto, K., Nelson, L. L., Swanson, T. S., Moore, B. J., ... & Drazner, M. H. (2013). Allocating scarce resources in real-time to reduce heart failure readmissions: a prospective, controlled study. BMJ quality & safety, 22(12), 998-1005.

44. Wallace, P. J., Shah, N. D., Dennen, T., Bleicher, P. A., & Crown, W. H. (2014). Optum Labs: building a novel node in the learning health care system. Health Affairs, 33(7), 1187–1194.

45. Gold, M., & McLaughlin, C. (2016). Assessing HITECH implementation and lessons: 5 years later. The Milbank Quarterly, 94(3), 654–687.

46. Exchange of Electronic Health Records across the EU. (2020, April 1). Shaping Europe's Digital Future—European Commission. `https://ec.europa.eu/digital-single-market/en/exchange-electronic-health-records-across-eu`

47. Quarterly, M. (2009). What Health Systems Can Learn From Kaiser Permanente: An Interview With Hal Wolf.

48. News Center, Blue Shield of California. (2018). News Center. `www.blueshieldca.com/bsca/about-blue-shield/media-center/calindex-launch-080514.sp`

49. www.blueshieldca.com/bsca/about-blue-shield/media-center/calindex-launch-080514.sp

50. Roham, M., Saldivar, E., Raghavan, S., Zurcher, M., Mack, J., & Mehregany, M. (2011, March). A mobile wearable wireless fetal heart monitoring system. In 2011 5th International Symposium on Medical Information and Communication Technology (pp. 135–138). IEEE.

51. Health Monitoring| Swiss Re. (2020, April 8). Swiss Re. www.swissre.com/institute/research/library/health_monitoring_dave_wang.html

52. Henon, H., Pasquier, F., Durieu, M. D., Godefroy, O., & Lucas, C. (1997). Preexisting dementia in stroke patients: baseline frequency, associated factors, and outcome. Stroke, 28(12), 2429–2436.

53. Kim, J. Y., Wineinger, N. E., & Steinhubl, S. R. (2016). The influence of wireless self-monitoring program on the relationship between patient activation and health behaviors, medication adherence, and blood pressure levels in hypertensive patients: a substudy of a randomized controlled trial. Journal of medical Internet research, 18(6), e116.

54. The Influence of Wireless Self-Monitoring Program on the Relationship Between Patient Activation and Health Behaviors, Medication Adherence, and Blood Pressure Levels in Hypertensive Patients: A Substudy of a Randomized Controlled Trial. Published online on June 22, 2016. doi: 10.2196/jmir.5429.

55. Gonzalez, R. (2018, April 19). Science Says Fitness Trackers Don't Work. Wear One Anyway. Wired. www.wired.com/story/science-says-fitness-trackers-dont-work-wear-one-anyway/

56. Cooper, C. A. (2019). Creating a Safer Running Experience: Reducing Runner and Vehicular Traffic Incidents

57. Ebola: The Recovery. (2019). U.S. Agency for International Development. www.usaid.gov/ebola

58. Bengtsson, L., Lu, X., Thorson, A., Garfield, R., & Von Schreeb, J. (2011). Improved response to disasters and outbreaks by tracking population movements with mobile phone network data: a post-earthquake geospatial study in Haiti. PLoS Med, 8(8), e1001083.

59. IBM100—TAKMI: Bringing Order to Unstructured Data. (2018). Bringing Order to Unstructured Data. www.ibm.com/ibm/history/ibm100/us/en/icons/takmi/

60. Wooding, A. (2020). MTEP Publications. York Health Economics Consortium. www.yhec-dev.co.uk/tools-resources/recent-publications/mtep-publications/

61. Lambert, P. (2018). Complying with the Data Protection Regime. Int'l J. Data Protection Officer, Privacy Officer & Privacy Couns., 2, 17.

62. Brighton and Sussex University Hospitals NHS Trust, Data Protection and Audit Report, ICO, 2018.

63. Scannell, K., & Chon, G. (2015). Cyber security: attack of the health hackers.

64. Leavitt, N. (2010). Researchers fight to keep implanted medical devices safe from hackers. Computer, 43(8), 11–14.

65. Rodriguez, M. (2018). HTTPS everywhere: Industry trends and the need for encryption. Serials Review, 44(2), 131–137.

66. Porter, J. (2020, August 4). Garmin reportedly paid multimillion-dollar ransom after suffering cyberattack. The Verge. www.theverge.com/2020/8/4/21353842/garmin-ransomware-attack-wearables-wastedlocker-evil-corp

67. Babylon Health data breach: GP app users able to see other people's consultations. (2020, July 1). The Guardian. www.theguardian.com/uk-news/2020/jun/10/babylon-health-data-breach-gp-app-users-able-to-see-other-peoples-consultations

68. McCorduck, P., & Cfe, C. (2004). Machines who think: A personal inquiry into the history and prospects of artificial intelligence. CRC Press.

69. Lapan, M. (2018). Deep Reinforcement Learning Hands-On: Apply modern RL methods, with deep Q-networks, value iteration, policy gradients, TRPO, AlphaGo Zero and more. Packt Publishing Ltd.

70. Artificial intelligence expedites breast cancer risk prediction. (2016). ScienceDaily. `www.sciencedaily.com/releases/2016/08/160829122106.htm`

71. Russell, S. J., & Norvig, P. (2010). Artificial intelligence: A modern approach (international ed.).

72. Glauner, P., & Meira, J. A. (2018). Machine Learning for Data-Driven Smart Grid Applications.

73. Bae, J., Cha, Y. J., Lee, H., Lee, B., Baek, S., Choi, S., & Jang, D. (2017). Social networks and inference about unknown events: A case of the match between Google's AlphaGo and Sedol Lee. PloS one, 12(2), e0171472.

74. Holcomb, S. D., Porter, W. K., Ault, S. V., Mao, G., & Wang, J. (2018, March). Overview on deepmind and its alphago zero ai. In Proceedings of the 2018 international conference on big data and education (pp. 67–71).

75. Wang, J. (2003). Data mining: opportunities and challenges. Idea Group Pub.

76. Salzberg, S. L. (1994). C4.5: Programs for machine learning by j. ross quinlan. morgan kaufmann publishers, inc., 1993.

77. Friedman, J. H. (1999). Stochastic gradient boosting. Relatório técnico.

78. Schapire, R. E. (2013). Explaining adaboost. In Empirical inference (pp. 37–52). Springer, Berlin, Heidelberg.

79. France-Presse, A. (2017). World's best Go player flummoxed by Google's "godlike" AlphaGo AI. The Guardian.

80. Issac, B., & Jap, W. J. (2009, January). Implementing spam detection using Bayesian and Porter Stemmer keyword stripping approaches. In TENCON 2009-2009 IEEE Region 10 Conference (pp. 1–5). IEEE.

81. Carr, J. (2014). An introduction to genetic algorithms. Senior Project, 1(40), 7.

82. Rooney, P. (2002, October 3). Microsoft's CEO: 80-20 Rule Applies To Bugs, Not Just Features. CRN. www.crn.com/news/security/18821726/microsofts-ceo-80-20-rule-applies-to-bugs-not-just-features.htm

83. Ogurtsova, K., da Rocha Fernandes, J. D., Huang, Y., Linnenkamp, U., Guariguata, L., Cho, N. H., ... & Makaroff, L. E. (2017). IDF Diabetes Atlas: Global estimates for the prevalence of diabetes for 2015 and 2040. Diabetes research and clinical practice, 128, 40–50.

84. Diabetes.co.uk. (2020). Diabetes Prevalence. Diabetes. www.diabetes.co.uk/diabetes-prevalence.html

85. Sardarinia, M., Akbarpour, S., Lotfaliany, M., Bagherzadeh-Khiabani, F., Bozorgmanesh, M., Sheikholeslami, F., ... & Hadaegh, F. (2016). Risk factors for incidence of cardiovascular diseases and all-cause mortality in a middle eastern population over a decade follow-up: Tehran lipid and glucose study. PloS one, 11(12), e0167623.

86. Kavakiotis, I., Tsave, O., Salifoglou, A., Maglaveras, N., Vlahavas, I., & Chouvarda, I. (2017). Machine learning and data mining methods in diabetes research. Computational and structural biotechnology journal, 15, 104–116.

87. Razavian, N., Blecker, S., Schmidt, A. M., Smith-McLallen, A., Nigam, S., & Sontag, D. (2015). Population-level prediction of type 2 diabetes from claims data and analysis of risk factors. Big Data, 3(4), 277–287.

88. Lagani, V., Chiarugi, F., Thomson, S., Fursse, J., Lakasing, E., Jones, R. W., & Tsamardinos, I. (2015). Development and validation of risk assessment models for diabetes-related complications based on the DCCT/EDIC data. Journal of Diabetes and its Complications, 29(4), 479–487.

89. Huang, G. M., Huang, K. Y., Lee, T. Y., & Weng, J. T. Y. (2015, December). An interpretable rule-based diagnostic classification of diabetic nephropathy among type 2 diabetes patients. In BMC bioinformatics (Vol. 16, No. S1, p. S5). BioMed Central.

90. Leung, R. K., Wang, Y., Ma, R. C., Luk, A. O., Lam, V., Ng, M., ... & Chan, J. C. (2013). Using a multi-staged strategy based on machine learning and mathematical modeling to predict genotype-phenotype risk patterns in diabetic kidney disease: a prospective case–control cohort analysis. BMC nephrology, 14(1), 162.

91. Saslow, L. R., Summers, C., Aikens, J. E., & Unwin, D. J. (2018). Outcomes of a digitally delivered low-carbohydrate type 2 diabetes self-management program: 1-year results of a single-arm longitudinal study. JMIR diabetes, 3(3), e12.

92. Bee, P. (2020, July 6). The best health apps: what the experts rate in 2020. The Times. `www.thetimes.co.uk/article/best-health-apps-expert-guide-2020-s9xz97t2g`

93. Kucera, R. (2017). The truth behind facebook ai inventing a new language. Towards Data Science, 7, 2017.

94. Bergstra, J., & Bengio, Y. (2012). Random search for hyper-parameter optimization. The Journal of Machine Learning Research, 13(1), 281-305.

95. Dorey, F. (2010). In brief: The P value: What is it and what does it tell you?

96. Yao, H., Shum, A. J., Cowan, M., Lähdesmäki, I., & Parviz, B. A. (2011). A contact lens with embedded sensor for monitoring tear glucose level. Biosensors and Bioelectronics, 26(7), 3290–3296.

97. Callejas-Cuervo, M., Barrera, J. P. C., & Rengifo, D. L. C. (2019, December). Technological Platform for the Control of the Medication Supply to People with Diabetes. In International Conference on Smart Technologies, Systems and Applications (pp. 288–296). Springer, Cham.

98. Diabetes.co.uk. (2020). Artificial Pancreas. Diabetes.co.uk. `www.diabetes.co.uk/artificial-pancreas.html`

99. He, P., Zhao, J., Zhang, J., Li, B., Gou, Z., Gou, M., & Li, X. (2018). Bioprinting of skin constructs for wound healing. Burns & trauma, 6(1).

100. Lee, V. K., & Dai, G. (2017). Printing of three-dimensional tissue analogs for regenerative medicine. Annals of biomedical engineering, 45(1), 115–131.

101. Bernardi, R., & Wu, P. F. (2017, December). The Impact of Online Health Communities on Patients' Health Self-Management. In ICIS.

102. Goyal, M., Reeves, N. D., Rajbhandari, S., & Yap, M. H. (2018). Robust methods for real-time diabetic foot ulcer detection and localization on mobile devices. IEEE journal of biomedical and health informatics, 23(4), 1730–1741.

103. Quigley, E. (2016). The Dataverse Project: An Open-Source Data Repository and Data Sharing Community.

104. World Health Organization. (2014). Global status report on noncommunicable diseases 2014 (No. WHO/NMH/NVI/15.1). World Health Organization.

105. World Health Organization. (2014). Global health workforce shortage to reach 12.9 million in coming decades; 2013. Available on: www. who. int/mediacentre/news/releases/2013/health-workforce-shortage/en.

106. Taddeo, M., & Floridi, L. (2016). The debate on the moral responsibilities of online service providers. Science and Engineering Ethics, 22(6), 1575–1603.

107. Liyanage, L. N. (2015). Usage of mobile phones among university students in Sri Lanka. In National Conference on Applied Social Statistics Research (NCASSR) (p. 24).

108. Banerjee, S., Hemphill, T., & Longstreet, P. (2018). Wearable devices and healthcare: Data sharing and privacy. The Information Society, 34(1), 49–57.

109. Zhao, J., Freeman, B., & Li, M. (2016). Can mobile phone apps influence people's health behavior change? An evidence review. Journal of medical Internet research, 18(11), e287.

110. Scharf, C. A. (2018, June 27). The Information of Intelligence. Scientific American Blog Network. `https://blogs.scientificamerican.com/life-unbounded/the-information-of-intelligence/`

111. Cadwalladr, C., & Graham-Harrison, E. (2018). Revealed: 50 million Facebook profiles harvested for Cambridge Analytica in major data breach. The guardian, 17, 22.

112. Fletcher, R., & Nielsen, R. K. (2017). People don't trust news media—and this is key to the global misinformation debate. AA. VV., Understanding and Addressing the Disinformation Ecosystem, 13–17.

113. Crouch, H. (2018, April 5). Apple Watch data "used as evidence in Australian murder trial." Digital Health. `www.digitalhealth.net/2018/04/apple-watch-data-evidence-australian-murder-trial/`

114. Lotrea, C. (2018). Mr. Zuckerberg and the Internet. An essay on power relations and privacy negotiation. Journal of Comparative Research in Anthropology and Sociology, 9(1), 19–24.

115. Voigt, P., & Von dem Bussche, A. (2017). The eu general data protection regulation (gdpr). A Practical Guide, 1st Ed., Cham: Springer International Publishing.

116. Porter, C. C. (2008). De-identified data and third party data mining: The risk of re-identification of personal information. Shidler JL Com. & Tech., 5, 1.

117. Shegog, R., Braverman, L., & Hixson, J. D. (2020). Digital and technological opportunities in epilepsy: Toward a digital ecosystem for enhanced epilepsy management. Epilepsy & Behavior, 102, 106663.

118. Diogenes, Y., & Ozkaya, E. (2018). Cybersecurity??? Attack and Defense Strategies: Infrastructure security with Red Team and Blue Team tactics. Packt Publishing Ltd.

119. Storm, D. (2014). Black Hat: Nest thermostat turned into a smart spy in 15 seconds. Computerworld, 12, 25.

120. Allen, A. (2016). The "three black teenagers" search shows it is society, not Google, that is racist. The Guardian, 10.

121. Horton, H. (2016). Microsoft deletes "teen girl" AI after it became a Hitler-loving sex robot within 24 hours. The Telegraph, 24.

122. Dressel, J., & Farid, H. (2018). The accuracy, fairness, and limits of predicting recidivism. Science advances, 4(1), eaao5580.

123. Moor, J. (2009). Four kinds of ethical robots. Philosophy Now, 72, 12–14.

124. Urquhart, L., Reedman-Flint, D., & Leesakul, N. (2019). Responsible domestic robotics: exploring ethical implications of robots in the home. Journal of Information, Communication and Ethics in Society.

125. Dreyer, B. P., Trent, M., Anderson, A. T., Askew, G. L., Boyd, R., Coker, T. R., ... & Montoya-Williams, D. (2020). The death of George Floyd: bending the arc of history towards justice for generations of children. Pediatrics, e2020009639.

126. Griffin, A. (2020, June 2). TikTok apologises to black users after Black Lives Matter and George Floyd posts appeared to be hidden on site. The Independent. `www.independent.co.uk/life-style/gadgets-and-tech/news/tiktok-black-lives-matter-george-floyd-apology-protests-views-a9544956.html`

127. Smith, A. (2020, June 16). Instagram boss says it will change algorithm to stop mistreatment of black users, alongside other

updates. The Independent. www.independent.co.uk/life-style/gadgets-and-tech/news/instagram-black-lives-matter-racism-harassment-bias-algorithm-a9567946.html

128. Sagan, A., McDaid, D., Rajan, S., Farrington, J., & McKee, M. When is it appropriate and how can we get it right?

129. Isaac, M. (2017). How Uber deceives the authorities worldwide. The New York Times, 3, 17.

130. Contag, M., Li, G., Pawlowski, A., Domke, F., Levchenko, K., Holz, T., & Savage, S. (2017, May). How they did it: an analysis of emission defeat devices in modern automobiles. In 2017 IEEE Symposium on Security and Privacy (SP) (pp. 231-250). IEEE.

131. Fong, T., Cabrol, N., Thorpe, C., & Baur, C. (2001). A personal user interface for collaborative human-robot exploration. In 6th International Symposium on Artificial Intelligence, Robotics, and Automation in Space (iSAIRAS) (No. CONF).

132. Laboratory of Intelligent Systems—EPFL. (2020). EPFL. www.epfl.ch/labs/lis/

133. Asimov, I. (1950). Runaround. I, Robot. New York: Bantam Dell.

134. Fong, T., Cabrol, N., Thorpe, C., & Baur, C. (2001). A personal user interface for collaborative human-robot exploration. In 6th International Symposium on Artificial Intelligence, Robotics, and Automation in Space (iSAIRAS) (No. CONF).

135. Shim, H. B. (2007, April). Establishing a Korean robot ethics charter. In IEEE ICRA workshop on roboethics, April (Vol. 14, p. 2007).

136. Chatila, R., & Havens, J. C. (2019). The ieee global initiative on ethics of autonomous and intelligent systems. In Robotics and Well-Being (pp. 11–16). Springer, Cham.

137. Silver, D., Schrittwieser, J., Simonyan, K., Antonoglou, I., Huang, A., Guez, A., ... & Chen, Y. (2017). Mastering the game of go without human knowledge. nature, 550(7676), 354–359.

138. O'Kane, S. (2019, August 1). Tesla hit with another lawsuit over a fatal Autopilot crash. The Verge. www.theverge.com/2019/8/1/20750715/tesla-autopilot-crash-lawsuit-wrongful-death

139. Ishowo-Oloko, F., Bonnefon, J. F., Soroye, Z., Crandall, J., Rahwan, I., & Rahwan, T. (2019). Behavioural evidence for a transparency—efficiency tradeoff in human–machine cooperation. Nature Machine Intelligence, 1(11), 517–521.

140. Leviathan, Y., & Matias, Y. (2018). Google Duplex: an AI system for accomplishing real-world tasks over the phone.

141. Diakopoulos, N., & Friedler, S. (2016). How to hold algorithms accountable. MIT Technology Review, 17(11), 2016.

142. Cybenko, A. K., & Cybenko, G. (2018). AI and fake news. IEEE Intelligent Systems, 33(5), 1–5.

143. Shae, Z., & Tsai, J. (2019, July). AI blockchain platform for trusting news. In 2019 IEEE 39th International Conference on Distributed Computing Systems (ICDCS) (pp. 1610–1619). IEEE.

144. Bohn, D. (2019, January 4). Exclusive: Amazon says 100 million Alexa devices have been sold. The Verge. www.theverge.com/2019/1/4/18168565/amazon-alexa-devices-how-many-sold-number-100-million-dave-limp

145. Warwick, K. (2014). Turing Test success marks milestone in computing history. University or Reading Press Release, 8.

146. Haas, B. (2017, April 4). Chinese man "marries" robot he built himself. The Guardian. www.theguardian.com/world/2017/apr/04/chinese-man-marries-robot-built-himself

147. Mailonline, B. C. H. T. Y. F. (2017, October 30). Engineer in China marries his inflatable robot girlfriend. Mail Online. www.dailymail.co.uk/news/china/article-5021113/Engineer-31-marries-inflatable-robot-girlfriend.html

148. Istvan, Z. (2019, July 24). Should I have let my daughter marry our robot? Metro. https://metro.co.uk/2019/07/24/should-i-have-let-my-daughter-marry-our-robot-10361703/

149. Beckett, L. (2017). Trump digital director says Facebook helped win the White House. The Guardian, 9.

150. Aagaard, J. (2020). Beyond the rhetoric of tech addiction: why we should be discussing tech habits instead (and how). Phenomenology and the Cognitive Sciences, 1–14.

151. NHS website. (2018, October 3). Addiction: what is it? Nhs.Uk. www.nhs.uk/live-well/healthy-body/addiction-what-is-it/

152. Raso, F. A. (2018). Perfectly Irresistible: Developer Liability for Internet Addiction Disorders. Available at SSRN 3176166.

153. Ang, C. S., Chan, N. N., & Lee, C. S. (2018). Shyness, loneliness avoidance, and internet addiction: what are the relationships? The Journal of psychology, 152(1), 25–35.

154. Strom, S. (2016, November 18). McDonald's Introduces Screen Ordering and Table Service. www.Nytimes.Com/. www.nytimes.com/2016/11/18/business/mcdonalds-introduces-screen-ordering-and-table-service.html

155. Diabetes.co.uk. (2020, July 11). Survey launched to understand public opinion on data use to monitor COVID-19. Diabetes. www.diabetes.co.uk/news/2020/jul/survey-launched-to-understand-public-opinion-on-data-use-to-monitor-covid-19.html

156. Dunckley, V. L. (2014). Gray matters: Too much screen time damages the brain. Psychology Today, 27.

157. Twenge, J. M., & Campbell, W. K. (2018). Associations between screen time and lower psychological well-being among children and adolescents: Evidence from a population-based study. Preventive medicine reports, 12, 271–283.

158. Vallance, J. K., Buman, M. P., Stevinson, C., & Lynch, B. M. (2015). Associations of overall sedentary time and screen time with sleep outcomes. American journal of health behavior, 39(1), 62–67.

159. Diabetes.co.uk. (2020, August 1). Reduce screen time and improve health say experts. Diabetes. `www.diabetes.co.uk/news/2020/aug/reduce-screen-time-and-improve-health-say-experts.html`

160. Artificial intelligence predicts patient lifespans. (2018). University of Adelaide. `www.adelaide.edu.au/news/news92622.html`

161. Lucas, L., & Waters, R. (2018). The AI arms race: China and US compete to dominate big data.

162. Winfield, A. F. (2016). Written evidence submitted to the UK Parliamentary Select Committee on Science and Technology Inquiry on Robotics and Artificial Intelligence.

163. Weller, C. (2017). Meet the first-ever robot citizen, a humanoid named Sophia that once said it would destroy humans. Business Insider Nordic. Haettu, 30, 2018.

164. Simonite, T. (2018). Should data scientists adhere to a Hippocratic oath? Wired.

165. Attaran, M. (2017). The rise of 3-D printing: The advantages of additive manufacturing over traditional manufacturing. Business Horizons, 60(5), 677–688.

166. Diaz Jr, L. A., & Bardelli, A. (2014). Liquid biopsies: genotyping circulating tumor DNA. Journal of clinical oncology, 32(6), 579.

167. Okamura, A. M. (2009). Haptic feedback in robot-assisted minimally invasive surgery. Current opinion in urology, 19(1), 102.

168. Pradhan, A., Mehta, K., & Findlater, L. (2018, April). "Accessibility Came by Accident" Use of Voice-Controlled Intelligent Personal Assistants by People with Disabilities. In Proceedings of the 2018 CHI Conference on Human Factors in Computing Systems (pp. 1–13).

169. Hawkes, N. (2016). Doctors getting biggest payments from drug companies don't declare them on new website.

170. Moynihan, R. (2003). Who pays for the pizza? Redefining the relationships between doctors and drug companies. 1: Entanglement. Bmj, 326(7400), 1189–1192.

171. Malnick, E. (2016, June 30). Individual NHS doctors receiving £100,000 per year from drugs firms. The Telegraph. www.telegraph.co.uk/news/2016/06/30/individual-nhs-doctors-receiving-100000-per-year-from-drugs-firm/

172. Tseng, E. K., & Hicks, L. K. (2016). Value based care and patient-centered care: divergent or complementary? Current hematologic malignancy reports, 11(4), 303–310.

173. Monsen, K., & Deblok, J. (2013). Buurtzorg Nederland. AJN The American Journal of Nursing, 113(8), 55–59.

174. Gro Health Platform (2020). Gro Health. https://grohealth.com

175. Anand, P., Kunnumakara, A. B., Sundaram, C., Harikumar, K. B., Tharakan, S. T., Lai, O. S., ... & Aggarwal, B. B. (2008). Cancer is a preventable disease that requires major lifestyle changes. Pharmaceutical research, 25(9), 2097–2116.

176. Stults-Kolehmainen, M. A., & Sinha, R. (2014). The effects of stress on physical activity and exercise. Sports medicine, 44(1), 81–121.

177. Saslow, L. R., Summers, C., Aikens, J. E., & Unwin, D. J. (2018). Outcomes of a digitally delivered low-carbohydrate type 2 diabetes self-management program: 1-year results of a single-arm longitudinal study. JMIR diabetes, 3(3), e12.

178. Sackett, D. L., Rosenberg, W. M., Gray, J. M., Haynes, R. B., & Richardson, W. S. (1996). Evidence based medicine: what it is and what it isn't.

179. Goldacre, B. (2012). Bad pharma: how medicine is broken, and how we can fix it. HarperCollins UK.

180. O'connor, A. (2016). How the sugar industry shifted blame to fat. New York Times, 12.

181. Diabetes.co.uk. (2018). Metformin.

182. Rappaport, S. M. (2016). Genetic factors are not the major causes of chronic diseases. PloS one, 11(4), e0154387.

183. McKeigue, P. M., Shah, B., & Marmot, M. G. (1991). Relation of central obesity and insulin resistance with high diabetes prevalence and cardiovascular risk in South Asians. The Lancet, 337(8738), 382–386.

184. McGrath, S. (2018). THE INFLUENCE OF "OMICS" IN SHAPING PRECISION MEDICINE. INNOVATIONS.

185. Danilov, Y. P., Tyler, M. E., & Kaczmarek, K. A. (2019). U.S. Patent No. 10,328,263. Washington, DC: U.S. Patent and Trademark Office.

186. Haghi, M., Thurow, K., & Stoll, R. (2017). Wearable devices in medical internet of things: scientific research and commercially available devices. Healthcare informatics research, 23(1), 4–15.

187. Heo, Y. J., & Takeuchi, S. (2013). Towards smart tattoos: implantable biosensors for continuous glucose monitoring. Advanced healthcare materials, 2(1), 43–56.

188. Payne, N., Gangwani, R., Barton, K., Sample, A., Cain, S., Burke, D., ... & Shorter, K. (2020). Medication Adherence and Liquid Level Tracking System for Healthcare Provider Feedback. Sensors, 20(8), 2435.

189. Dhapake, P. R., Chauriya, C. B., & Umredkar, R. C. (2017). Painless Insulin Drug Delivery Systems-A Review. Asian Journal of Research in Pharmaceutical Science, 7(1), 01–07.

190. Craig, D. Artificial Intelligence and Why the Internet Always Says You Have Cancer.

191. Diabetes.co.uk. (2019). Data on file. Diabetes.co.uk.

192. Kulik, C. T., Ryan, S., Harper, S., & George, G. (2014). Aging populations and management.

193. Morabia, A., & Abel, T. (2006). The WHO report "Preventing Chronic Diseases: a vital investment" and us. Sozial-und Präventivmedizin, 51(2), 74-74.

194. Centers for Disease Control and Prevention. (2017). Chronic disease prevention and health promotion| CDC. Accessed on May, 25, 2017.

195. Tamura, T., Yoshimura, T., Sekine, M., Uchida, M., & Tanaka, O. (2009). A wearable airbag to prevent fall injuries. IEEE Transactions on Information Technology in Biomedicine, 13(6), 910–914.

196. Baig, M., Mirza, F., GholamHosseini, H., Gutierrez, J., & Ullah, E. (2018, July). Clinical decision support for early detection of prediabetes and type 2 diabetes mellitus using wearable technology. In 2018 40th Annual International Conference of the IEEE Engineering in Medicine and Biology Society (EMBC) (pp. 4456–4459). IEEE.

197. The Apple Watch can detect diabetes with an 85% accuracy, Cardiogram study says. (2018, February 7). TechCrunch. https://techcrunch.com/2018/02/07/the-apple-watch-can-detect-diabetes-with-an-85-accuracy-cardiogram-study-says/

198. Nominet. (2019, September 26). Launches free IoT solution to help those with sensory and cognitive impairments. www.nominet.uk/nominet-launches-free-iot-solution-help-sensory-cognitive-impairments/

199. Gro partners with University of British Columbia study and Institute of Personalized Nutrition on research study | Press—grohealth.com. (2019). Gro Health. www.grohealth.com/press/gro-partners-with-university-of-british-columbia-study-and-institute-of-personalized-nutrition-to-study-lifestyle-management

200. Brown, M. T., & Bussell, J. K. (2011, April). Medication adherence: WHO cares? In Mayo clinic proceedings (Vol. 86, No. 4, pp. 304–314). Elsevier.

201. Kalantar-zadeh, K., Ha, N., Ou, J. Z., & Berean, K. J. (2017). Ingestible sensors. ACS sensors, 2(4), 468–483.

202. Hollander, J. E., & Carr, B. G. (2020). Virtually perfect? Telemedicine for COVID-19. New England Journal of Medicine, 382(18), 1679–1681.

203. Larson, R. S. (2018). A path to better-quality mHealth apps. JMIR mHealth and uHealth, 6(7), e10414.

204. Orcha—The Challenge. (2020). Orcha—Health Apps. www.orcha.co.uk/the-challenge/

205. Carpenter, D. M., Geryk, L. L., Sage, A., Arrindell, C., & Sleath, B. L. (2016). Exploring the theoretical pathways through which asthma app features can promote adolescent self-management. Translational behavioral medicine, 6(4), 509–518.

206. Whicher, C. A., O'Neill, S., & Holt, R. G. (2020). Diabetes in the UK: 2019. Diabetic Medicine, 37(2), 242–247.

207. National Diabetes Audit. (2018). NHS Digital. https://digital.nhs.uk/data-and-information/publications/statistical/national-diabetes-audit

208. Charity Highlights Steep Rise in Diabetes Deaths During Pandemic. (2020, July 21). Medscape. www.medscape.com/viewarticle/934314

209. Britain, G. (2011). No health without mental health: a cross-government mental health outcomes strategy for people of all ages. Stationery Office.

210. Bloom, D. E., Cafiero, E., Jané-Llopis, E., Abrahams-Gessel, S., Bloom, L. R., Fathima, S., ... & O'Farrell, D. (2012). The global economic burden of noncommunicable diseases (No. 8712). Program on the Global Demography of Aging.

211. Davenport, T., & Kalakota, R. (2019). The potential for artificial intelligence in healthcare. Future healthcare journal, 6(2), 94.

212. Johnson, K. (2017). Facebook Messenger hits 100,000 bots.

213. National Institute of Health. (2020). National Library of Medicine. https://pubmed.ncbi.nlm.nih.gov/

214. Mesko, B. (2017). The role of artificial intelligence in precision medicine.

215. Herper, M. (2017). The cost of developing drugs is insane. That paper that says otherwise is insanely bad. Forbes Magazine.

216. Mullard, A. (2017). The drug-maker's guide to the galaxy. Nature News, 549(7673), 445.

217. Cumberland, S. (2009). Banning cluster munitions. World Health Organization. Bulletin of the World Health Organization, 87(1), 8.

218. Cubo, N., Garcia, M., Del Cañizo, J. F., Velasco, D., & Jorcano, J. L. (2016). 3D bioprinting of functional human skin: production and in vivo analysis. Biofabrication, 9(1), 015006.

219. User, S. (2018). The Immersive Learning Studio talks about the benefits of VR in Hazard Awareness Training. Investment & Support for the Liverpool City Region and the North West. www.msif.co.uk/immersive-learning-studio

220. Allen, A. G., Chung, C. H., Atkins, A., Dampier, W., Khalili, K., Nonnemacher, M. R., & Wigdahl, B. (2018). Gene editing of HIV-1 co-receptors to prevent and/or cure virus infection. Frontiers in microbiology, 9, 2940.

221. Genetic Disorders. (2019). Genetic Disorders UK. www.geneticdisordersuk.org/aboutgeneticdisorders

222. Rosenkranz, M. A., Dunne, J. D., & Davidson, R. J. (2019). The next generation of mindfulness-based intervention research: what have we learned and where are we headed? Current opinion in psychology, 28, 179–183.

223. Andone, I., Blaszkiewicz, K., Böhmer, M., & Markowetz, A. (2017, September). Impact of location-based games on phone usage and movement: A case study on Pokémon GO. In Proceedings of the 19th International Conference on Human-Computer Interaction with Mobile Devices and Services (pp. 1–8).

224. Li, A., Montaño, Z., Chen, V. J., & Gold, J. I. (2011). Virtual reality and pain management: current trends and future directions. Pain management, 1(2), 147–157.

225. Murray, C. D. (2009). A REVIEW OF THE USE OF VIRTUAL REALITY IN THE TREATMENT OF PHANTOM LIMB PAIN. Journal of CyberTherapy & Rehabilitation (JCR), 2(2).

226. VR Vaccine @. (2019). Lobo. https://lobo.cx/vaccine/

227. Virtual Reality Fitness Experiences. (2020). ICAROS. www.icaros.com/

228. Larson, E. B., Feigon, M., Gagliardo, P., & Dvorkin, A. Y. (2014). Virtual reality and cognitive rehabilitation: a review of current outcome research. NeuroRehabilitation, 34(4), 759–772.

229. Davis, N. (2016). Cutting-edge theatre: world's first virtual reality operation goes live. Zugriff am, 11, 2019.

230. Tufts Medical Center | Boston Hospital and Academic Medical Center. (2020). Tufts Medical Center. www.tuftsmedicalcenter.org/patient-care-services

231. Brenner, L. (2016). Exploring the psychosocial impact of Ekso Bionics Technology. Archives of Physical Medicine and Rehabilitation, 97(10), e113.

232. Decker, M., Dillmann, R., Dreier, T., Fischer, M., Gutmann, M., Ott, I., & genannt Döhmann, I. S. (2011). Service robotics: do you know your new companion? Framing an interdisciplinary technology assessment. Poiesis & Praxis, 8(1), 25–44.

233. Ruiz Estrada, M. A. (2020). The uses of drones in case of massive epidemics contagious diseases relief humanitarian aid: Wuhan-COVID-19 crisis. Available at SSRN 3546547.

234. Delft, T. U. (2016). Ambulance drone. Abril de.

235. Richardson, H. (2016, September 29). Record number of centenarians in UK. BBC News. www.bbc.co.uk/news/education-37505339

236. Slowey, L. (2016, June 6). IoT technology and the future of healthcare. Business Operations. www.ibm.com/blogs/internet-of-things/iot-oap-new-technology-ageing-population/

237. International Diabetes Foundation. (2017). Diabetes Atlas. Retrieved from www.diabetesatlas.org/en/

238. NHS Digital. (2018). National Diabetes Audit, 2016-17 Report 1: Care Processes and Treatment Targets; England and Wales, March 14, 2018, Full Report.

239. Mishra, S. C., Chhatbar, K. C., Kashikar, A., & Mehndiratta, A. (2017). Diabetic foot. Bmj, 359, j5064.

240. NHS Digital. (2018). National Diabetes Foot Care Audit 2017, Third Annual Report.

241. Diabetes UK. (2018). Putting Feet First. Retrieved from www.diabetes.org.uk/get_involved/campaigning/putting-feet-first

242. Coleman, D., Mallik, R., Bolter, L., & Medici, F. (2019, March). Diabetes lower limb amputations: Data quality and the Public Health England Diabetes Foot Care Profile 2013–2016.

243. Diabetes Foot Care Profiles April 2019 (Public Health England). Retrieved from https://fingertips.phe.org.uk/profile/diabetes-ft

244. Marion Kerr, Insight Health Economics. (2017). Foot Care in Diabetes: The Human and Financial Cost. Retrieved from www.londonscn.nhs.uk/wp-content/uploads/2017/04/dia-foot-care-mtg-kerr-27042017.pdf

245. Abbas, ZG. (2013). Preventive foot care and reducing amputation. Diabetes Management 3, no. 5 (2013):427.

246. NICE, (2019). Diabetic foot problems: prevention and management NICE guideline [NG19]. Retrieved from www.nice.org.uk/guidance/ng19

247. Nickinson, A. T., Bridgwood, B., Houghton, J. S., Nduwayo, S., Pepper, C., Payne, T., ... & Sayers, R. D. (2020). A systematic review investigating the identification, causes, and outcomes of delays in the management of chronic limb-threatening ischemia and diabetic foot ulceration. Journal of vascular surgery, 71(2), 669–681.

248. NHS Digital. (2019). National Diabetes Foot Care Audit Fourth Annual Report England and Wales April 1, 2015, to March 31, 2018. Retrieved from www.hqip.org.uk/wp-content/uploads/2019/05/National-Diabetes-Foot-Care-Audit-fourth-annual-report-FINAL.pdf.

249. NHS Digital. (2018). National Diabetes Audit, 2016-17 Report 1: Care Processes and Treatment Targets; England and Wales, March 14, 2018, Full Report.

250. Diabetes UK. (2018). National Diabetes Audit: are services providing good quality diabetes care? A summary report of the National Diabetes Audit: Care Processes and Treatment Targets 2016–2017.

251. Armstrong, D. G., Lavery, L. A., & Harkless, L. B. (1998). Validation of a diabetic wound classification system: the contribution of depth, infection, and ischemia to risk of amputation. Diabetes care, 21(5), 855–859.

252. Feinman RD, Pogozelski WK, Astrup A, Bernstein RK, Fine EJ, Westman EC, et al. Dietary carbohydrate restriction as the first approach in diabetes management: critical review and evidence base. Nutrition 2015 Jan;31(1):1–13.

253. Hussain TA, Mathew TC, Dashti AA, Asfar S, Al-Zaid N, Dashti HM.

254. Effect of low calorie versus low-carbohydrate ketogenic diet in type 2 diabetes. Nutrition 2012 Oct;28(10):1016–1021.

255. Nielsen JV, Joensson EA. Low-carbohydrate diet in type 2 diabetes: stable improvement of bodyweight and glycemic control during 44 months follow-up. Nutr Metab (Lond) 2008, May 22; 5(1):14.

256. Yancy W, Foy M, Chalecki A, Vernon M, Westman E. A lowcarbohydrate, ketogenic diet to treat type 2 diabetes. Nutr Metab (Lond) 2005 Dec 01;2(1):34.

257. Garg A, Bantle JP, Henry RR, Coulston AM, Griver KA, Raatz SK, et al. Effects of varying carbohydrate content of diet in patients with non-insulindependent diabetes mellitus. JAMA May 11, 1994; 271(18):1421–1428.

258. Reaven GM. Effect of dietary carbohydrate on the metabolism of patients with non-insulin dependent diabetes mellitus. Nutr Rev 1986 Feb;44(2):65–73.

259. Bian R, Piatt G, Sen A, Plegue MA, De MM, Hafez D, et al. The effect of technology-mediated diabetes prevention interventions on weight: a metaanalysis. J Med Internet Res 2017 Mar 27;19(3):e76.

260. Saslow L, Mason A, Kim S, Goldman V, Ploutz-Snyder R, Bayandorian H, et al. An online intervention comparing a very low-carbohydrate ketogenic diet and lifestyle recommendations versus a plate method diet in overweight individuals with type 2 diabetes: a randomized controlled trial. J Med Internet Res 2017 Feb 13;19(2):e36.

261. McKenzie A, Hallberg S, Creighton B, Volk B, Link T, Abner M, et al. novel intervention including individualized nutritional recommendations reduces hemoglobin a1c level, medication use, and weight in type 2 diabetes. JMIR Diabetes 2017 Mar 07;2(1):e5.

262. Heller, S. (2004). Weight gain during insulin therapy in patients with type 2 diabetes mellitus. Diabetes research and clinical practice, 65, S23-S27.

263. Duncan, J. S., Sander, J. W., Sisodiya, S. M., & Walker, M. C. (2006). Adult epilepsy. The Lancet, 367(9516), 1087–1100.

264. Fisher, R. S., Acevedo, C., Arzimanoglou, A., Bogacz, A., Cross, J. H., Elger, C. E., ... & Hesdorffer, D. C. (2014). ILAE official report: a practical clinical definition of epilepsy. Epilepsia, 55(4), 475–482.

265. Groesbeck, D. K., Bluml, R. M., & Kossoff, E. H. (2006). Long-term use of the ketogenic diet in the treatment of epilepsy. Developmental medicine and child neurology, 48(12), 978–981.

266. Paoli, A., Rubini, A., Volek, J. S., & Grimaldi, K. A. (2013). Beyond weight loss: a review of the therapeutic uses of very-low-carbohydrate (ketogenic) diets. European journal of clinical nutrition, 67(8), 789–796.

267. Keene, D. L. (2006). A systematic review of the use of the ketogenic diet in childhood epilepsy. Pediatric neurology, 35(1), 1–5.

268. Martinez, C. C., Pyzik, P. L., & Kossoff, E. H. (2007). Discontinuing the ketogenic diet in seizure-free children: recurrence and risk factors. Epilepsia, 48(1), 187–190.

269. Martin, K., Jackson, C. F., Levy, R. G., & Cooper, P. N. (2016). Ketogenic diet and other dietary treatments for epilepsy. Cochrane Database of Systematic Reviews, (2).

270. Kang, H. C., Chung, D. E., Kim, D. W., & Kim, H. D. (2004). Early- and late-onset complications of the ketogenic diet for intractable epilepsy. Epilepsia, 45(9), 1116–1123.

271. Villeneuve, N., Pinton, F., BAHI-BUISSON, N. A. D. I. A., Dulac, O., Chiron, C., & Nabbout, R. (2009). The ketogenic diet improves recently worsened focal epilepsy. Developmental Medicine & Child Neurology, 51(4), 276–281.

272. Sharman, M. J., Kraemer, W. J., Love, D. M., Avery, N. G., Gómez, A. L., Scheett, T. P., & Volek, J. S. (2002). A ketogenic diet favorably affects serum biomarkers for cardiovascular disease in normal-weight men. The Journal of nutrition, 132(7), 1879–1885.

273. Yajnik CS. The lifecycle effects of nutrition and body size on adult adiposity, diabetes and cardiovascular disease. Obesity Reviews. 2002 Aug;3(3):217-24.

274. Lighter, J., Phillips, M., Hochman, S., Sterling, S., Johnson, D., Francois, F., & Stachel, A. (2020). Obesity in patients younger than 60 years is a risk factor for Covid-19 hospital admission. Clinical Infectious Diseases.

275. Whitehead, L., & Seaton, P. (2016). The effectiveness of self-management mobile phone and tablet apps in long-term condition management: a systematic review. Journal of medical Internet research, 18(5), e97.

276. National Diabetes Audit. (2018). Diabetes UK. www.diabetes.org.uk/professionals/resources/national-diabetes-audit

277. Winkley, K., Evwierhoma, C., Amiel, S. A., Lempp, H. K., Ismail, K., & Forbes, A. (2015). Patient explanations for non-attendance at structured diabetes education sessions for newly diagnosed Type 2 diabetes: a qualitative study. Diabetic Medicine, 32(1), 120–128.

278. Boulton, C. A., Kent, C., & Williams, H. T. (2018). Virtual learning environment engagement and learning outcomes at a "bricks-and-mortar" university. Computers & Education, 126, 129–142.

279. Lambrinou, E., Hansen, T. B., & Beulens, J. W. (2019). Lifestyle factors, self-management and patient empowerment in diabetes care. European journal of preventive cardiology, 26(2_suppl), 55–63.

280. Wing RR, Lang W, Wadden TA, Safford M, Knowler WC, Bertoni AG, Hill JO, Brancati FL, Peters A, Wagenknecht L, Look AHEAD Research Group. Benefits of modest weight loss in improving cardiovascular risk factors in overweight and obese individuals with type 2 diabetes. Diabetes care. 2011 Jul 1;34(7):1481-6.

281. Pournaras, D. J., Osborne, A., Hawkins, S. C., Vincent, R. P., Mahon, D., Ewings, P., ... & le Roux, C. W. (2010). Remission of type 2 diabetes after gastric bypass and banding: mechanisms and 2 year outcomes. Annals of surgery, 252(6), 966–971.

282. Mercer, K., Li, M., Giangregorio, L., Burns, C., & Grindrod, K. (2016). Behavior change techniques present in wearable activity trackers: a critical analysis. JMIR mHealth and uHealth, 4(2), e40.

283. Klasnja, P., & Hekler, E. B. (2017). Wearable technology and long-term weight loss. jama, 317(3), 317–318.

284. Lipovský R, Ferreira HA. Self hand-rehabilitation system based on wearable technology. InProceedings of the 3rd 2015 Workshop on ICTs for improving Patients Rehabilitation Research Techniques 2015 Oct 1 (pp. 93–95).

285. Whitehead L, Seaton P. The effectiveness of self-management mobile phone and tablet apps in long-term condition management: a systematic review. Journal of medical Internet research. 2016;18(5):e97.

286. Wisniewski H, Liu G, Henson P, Vaidyam A, Hajratalli NK, Onnela JP, Torous J. Understanding the quality, effectiveness and attributes of top-rated smartphone health apps. Evidence-based mental health. 2019 Feb 1;22(1):4–9.

287. Saslow, L. R., Summers, C., Aikens, J. E., & Unwin, D. J. (2018). Outcomes of a digitally delivered low-carbohydrate type 2 diabetes self-management program: 1-year results of a single-arm longitudinal study. JMIR diabetes, 3(3), e12.

288. Padwal R, Li SK, Lau DCW. Long-term pharmacotherapy for overweight and obesity: a systematic review and meta-analysis of randomized controlled trials. Int J Obes Relat Metab Disord 2003 Dec 16;27(12):1437–1446.

289. Stevens VJ, Obarzanek E, Cook NR, Lee IM, Appel LJ, Smith West D, Trials for the Hypertension Prevention Research Group. Long-term weight loss and changes in blood pressure: results of the Trials of Hypertension Prevention, phase II. Ann Intern Med 2001 Jan 02;134(1):1–11.

290. Dombrowski SU, Knittle K, Avenell A, Araújo-Soares V, Sniehotta FF. Long term maintenance of weight loss with non-surgical interventions in obese adults: systematic review and meta-analyses of randomised controlled trials. BMJ May 14, 2014; 348(may14 6):g2646-g2646.

291. Karlsson J, Taft C, Rydén A, Sjöström L, Sullivan M. Ten-year trends in health-related quality of life after surgical and conventional treatment for severe obesity: the SOS intervention study. Int J Obes 2007 Mar 13;31(8):1248–1261.

292. Carpenter MJ, Jardin BF, Burris JL, Mathew AR, Schnoll RA, Rigotti NA, et al. Clinical Strategies to Enhance the Efficacy of Nicotine Replacement Therapy for Smoking Cessation: A Review of the Literature. Drugs 2013 Apr 10;73(5):407–426.

293. Hughes JR, Keely J, Naud S. Shape of the relapse curve and long-term abstinence among untreated smokers. Addiction 2004 Jan;99(1):29–38.

294. Moos RH, Moos BS. Rates and predictors of relapse after natural and treated remission from alcohol use disorders. Addiction 2006 Feb;101(2):212–222

295. Elfhag K, Rossner S. Who succeeds in maintaining weight loss? A conceptual review of factors associated with weight loss maintenance and weight regain. Obesity Reviews 2005 Feb;6(1):67–85.

296. Burgess E, Hassmén P, Welvaert M, Pumpa KL. Behavioural treatment strategies improve adherence to lifestyle intervention programmes in adults with obesity: a systematic review and meta-analysis. Clinical obesity. 2017 Apr;7(2):105-14.

297. Carver CS, Scheier MF. Control theory: A useful conceptual framework for personality–social, clinical, and health psychology. Psychological bulletin. 1982 Jul;92(1):111.

298. Leimeister, J. M., Schweizer, K., Leimeister, S., & Krcmar, H. (2008). Do virtual communities matter for the social support of patients? Information Technology & People.

299. Nambisan, S., & Sawhney, M. (2011). Orchestration processes in network-centric innovation: Evidence from the field. Academy of management perspectives, 25(3), 40–57.

300. Merolli, M., Gray, K., & Martin-Sanchez, F. (2013). Health outcomes and related effects of using social media in chronic disease management: a literature review and analysis of affordances. Journal of biomedical informatics, 46(6), 957–969.

301. Sillence, E., Briggs, P., Harris, P. R., & Fishwick, L. (2007). How do patients evaluate and make use of online health information? Social science & medicine, 64(9), 1853–1862.

302. Chinn, D. (2011). Critical health literacy: A review and critical analysis. Social science & medicine, 73(1), 60–67.

303. Bernardi, R., & Wu, P. F. (2017, December). The Impact of Online Health Communities on Patients' Health Self-Management. In ICIS.

304. Rubin, R. R., & Peyrot, M. (2001). Psychological issues and treatments for people with diabetes. Journal of clinical psychology, 57(4), 457–478.

305. Bleser, G., Steffen, D., Weber, M., Hendeby, G., Stricker, D., Fradet, L., ... & Carré, F. (2013). A personalized exercise trainer for the elderly. *Journal of ambient intelligence and smart environments, 5*(6), 547–562.

306. Drewnowski, A., & Evans, W. J. (2001). Nutrition, physical activity, and quality of life in older adults: summary. *The Journals of Gerontology Series A: Biological Sciences and Medical Sciences, 56*(suppl_2), 89–94.

307. Angevaren, M., Aufdemkampe, G., Verhaar, H. J. J., Aleman, A., & Vanhees, L. (2008). Physical activity and enhanced fitness to improve cognitive function in older people without known cognitive impairment. *Cochrane database of systematic reviews*, (2).

308. Heyn, P., Abreu, B. C., & Ottenbacher, K. J. (2004). The effects of exercise training on elderly persons with cognitive impairment and dementia: a meta-analysis. *Archives of physical medicine and rehabilitation, 85*(10), 1694–1704.

309. Piwek, L., Ellis, D. A., Andrews, S., & Joinson, A. (2016). The rise of consumer health wearables: promises and barriers. *PLoS medicine, 13*(2), e1001953.

310. Orcha—The Challenge. (2020). Orcha—Health Apps. `www.orcha.co.uk/the-challenge/`

311. Sezgin, E., Militello, L. K., Huang, Y., & Lin, S. (2020). A scoping review of patient-facing, behavioral health interventions with voice assistant technology targeting self-management and healthy lifestyle behaviors. *Translational Behavioral Medicine, 10*(3), 606–628.

312. Hill, J., Ford, W. R., & Farreras, I. G. (2015). Real conversations with artificial intelligence: A comparison between human–human online conversations and human–chatbot conversations. *Computers in human behavior, 49*, 245–250.

313. Sciuto, A., Saini, A., Forlizzi, J., & Hong, J. I. (2018, June). "Hey Alexa, What's Up?" A Mixed-Methods Studies of In-Home Conversational Agent Usage. In *Proceedings of the 2018 Designing Interactive Systems Conference* (pp. 857–868).

314. Centre for Data Ethics and innovation. (2019). Smart Speakers and Voice Assistants.

315. Bentley, F., Luvogt, C., Silverman, M., Wirasinghe, R., White, B., & Lottridge, D. (2018). Understanding the long-term use of smart speaker assistants. *Proceedings of the ACM on Interactive, Mobile, Wearable and Ubiquitous Technologies, 2*(3), 1–24.

316. Lopatovska, I., Rink, K., Knight, I., Raines, K., Cosenza, K., Williams, H., ... & Martinez, A. (2019). Talk to me: Exploring user interactions with the Amazon Alexa. *Journal of Librarianship and Information Science, 51*(4), 984–997.

317. Brewer, R. N. (2017). *Understanding and Developing Interactive Voice Response Systems to Support Online Engagement of Older Adults* (Doctoral dissertation, Northwestern University).

318. Reis, A., Paulino, D., Paredes, H., Barroso, I., Monteiro, M. J., Rodrigues, V., & Barroso, J. (2018, June). Using intelligent personal

assistants to assist the elderlies An evaluation of Amazon Alexa, Google Assistant, Microsoft Cortana, and Apple Siri. In *2018 2nd International Conference on Technology and Innovation in Sports, Health and Wellbeing (TISHW)* (pp. 1-5). IEEE.

319. Lei, X., Tu, G. H., Liu, A. X., Ali, K., Li, C. Y., & Xie, T. (2017). The Insecurity of Home Digital Voice Assistants--Amazon Alexa as a Case Study. *arXiv preprint arXiv:1712.03327.*

320. Lau, J., Zimmerman, B., & Schaub, F. (2018). Alexa, are you listening? privacy perceptions, concerns and privacy-seeking behaviors with smart speakers. *Proceedings of the ACM on Human-Computer Interaction, 2*(CSCW), 1-31.

321. Chung, A. E., Griffin, A. C., Selezneva, D., & Gotz, D. (2018). Health and fitness apps for hands-free voice-activated assistants: content analysis. *JMIR mHealth and uHealth, 6*(9), e174.

322. Hassoon, A., Schrack, J., Naiman, D., Lansey, D., Baig, Y., Stearns, V., ... & Appel, L. (2018). Increasing physical activity amongst overweight and obese cancer survivors using an alexa-based intelligent agent for patient coaching: protocol for the physical activity by technology help (PATH) trial. JMIR research protocols, 7(2), e27.

323. Cheek, P., Nikpour, L., & Nowlin, H. D. (2005). Aging well with smart technology. *Nursing administration quarterly, 29*(4), 329-338.

324. Hassoon, A., Baig, Y., Naimann, D., Celentano, D., Lansey, D., Stearns, V., ... & Appel, L. J. (2020). Addressing Cardiovascular Health Using Artificial Intelligence: Randomized Clinical Trial to Increase Physical Activity in Cancer Survivors Using Intelligent Voice Assist (Amazon Alexa) for Patient Coaching. *Circulation, 141*(Suppl_1), A54-A54.

325. Jackson, C., & Orebaugh, A. (2018). A study of security and privacy issues associated with the Amazon Echo. *International Journal of Internet of Things and Cyber-Assurance, 1*(1), 91-100.

326. Zhang, N., Mi, X., Feng, X., Wang, X., Tian, Y., & Qian, F. (2018). Understanding and mitigating the security risks of voice-controlled third-party skills on amazon alexa and google home. *arXiv preprint arXiv:1805.01525.*

327. Portet, F., Vacher, M., Golanski, C., Roux, C., & Meillon, B. (2013). Design and evaluation of a smart home voice interface for the elderly: acceptability and objection aspects. *Personal and Ubiquitous Computing, 17*(1), 127–144.

328. Yaghoubzadeh, R., Kramer, M., Pitsch, K., & Kopp, S. (2013, August). Virtual agents as daily assistants for elderly or cognitively impaired people. In *International workshop on intelligent virtual agents* (pp. 79–91). Springer, Berlin, Heidelberg.

329. Sciuto, A., Saini, A., Forlizzi, J., & Hong, J. I. (2018, June). "Hey Alexa, What's Up?" A Mixed-Methods Studies of In-Home Conversational Agent Usage. In *Proceedings of the 2018 Designing Interactive Systems Conference* (pp. 857–868).

330. Pradhan, A., Mehta, K., & Findlater, L. (2018, April). "Accessibility Came by Accident" Use of Voice-Controlled Intelligent Personal Assistants by People with Disabilities. In *Proceedings of the 2018 CHI Conference on Human Factors in Computing Systems* (pp. 1–13).

331. Low Carb Program: Amazon.co.uk: Alexa Skills. (2019). Amazon Skills Store. `www.amazon.co.uk/Diabetes-Digital-Media-Ltd-Program/dp/B07RLFRG11`

332. Do, H. M., Pham, M., Sheng, W., Yang, D., & Liu, M. (2018). RiSH: A robot-integrated smart home for elderly care. *Robotics and Autonomous Systems, 101*, 74–92.

333. Reis, A., Paulino, D., Paredes, H., & Barroso, J. (2017, July). Using intelligent personal assistants to strengthen the elderlies' social bonds. In *International conference on universal access in human-computer interaction* (pp. 593–602). Springer, Cham.

APPENDIX B

Technical Glossary

- Absolute truth

  Also known as the ground truth, absolute truth refers to the correct answer, or reality.

- A/B testing

  A method of testing two or more approaches or techniques to determine which performs better and is statistically significant.

- Accuracy

  A metric used in machine learning that is used to evaluate the performance of a machine learning model.

- Action

  The mechanism by which an agent moves between states of an environment, by using a policy.

- Active learning

  An approach to machine learning where the algorithm itself chooses the data to learn from.

- AR

  An acronym for augmented reality, which places a computer-generated image on a human's view of the real world.

- Attribute

  A synonym for feature.

© Arjun Panesar 2021
A. Panesar, *Machine Learning and AI for Healthcare*, https://doi.org/10.1007/978-1-4842-6537-6

- Backpropagation

  Used by neural networks, the updating of biases and weights based on error. This occurs when the estimated output exceeds a defined error threshold. The error at output is propagated back into the network to update values of neurons.

- Bagging

  A machine learning technique of creating multiple models on subsets of the same data and combining them to improve overall prediction.

- Bootstrapping

  The technique of splitting data into multiple subsets with replacement. Each sample is known as a bootstrap sample.

- Bias (fairness)

  Bias refers to a prejudice or favoritism for a particular object, person, or group. Collection and interpretation of data can be affected by bias.

- Big data

  A large corpus of data—which is typically both structured and unstructured.

- Binary classification

  A classification where the output can be one of two mutually exclusive categories.

- Boosting

  A machine learning technique that combines weak classifiers into a classifier of higher accuracy.

- Categorical data

  Categorical data refers to data that falls into a discrete, mutually exclusive set, group, or class. Categorical features are known as discrete features.

- Centroid

  A centroid is the center of a cluster as defined by a k-means algorithm.

- Centroid-based clustering

  Centroid-based clustering refers to algorithms that separate data into clusters. Centroid-based clusters do not have a hierarchy.

- Class

  A class is one of a set of label values.

- Classification machine learning model

  A classification model presents an output from two or more discrete classes.

- Clipping

  Clipping is a data preparation technique that involves trimming outliers present in datasets—below a minimum and above a maximum value.

- Clustering

  A technique that groups related items together.

- Collaborative filtering

  Commonplace in recommendation systems, collaborative filtering is a technique where the interests of one individual are modeled around the interests of others with similar features.

- Columnar

  The storing of data in columns rather than rows (as per many traditional database structures). Columnar data storage improves speed due to reduced disk load and improved data transfer.

- Confirmation bias

  A type of bias that refers to one confirming their own preexisting beliefs or opinions when interpreting data or information.

- Confusion matrix

  An n × n table that presents the success of a classification model based on the label and model classification. A confusion matrix can be used to calculate performance metrics including precision and recall.

- Continuous variable

  A variable that can take any infinite value between its minimum and maximum.

- Convergence

  The phenomenon of training a machine learning model whereby additional training and validation does not improve the accuracy of the model.

- Convolutional layer

  A neural network often used in image recognition that has at least one convolutional layer.

- Convolutional neural network

  A neural network often used in image recognition that has at least one convolutional layer.

- Cost function

  A cost function is a performance metric that is used to measure the error of a machine learning model.

- Cross-validation

  A technique for evaluating how well a machine learning model will generalize when exposed to new data by testing it against subsets of data (validation set) withheld from the training data.

- Dataframe

  A dataframe is a labeled data structure that can contain columns of different types of data.

- Dataset

  A dataset is a collection of examples.

- Data transformation

  The process of converting data from one form or type to another.

- Decision boundary

  The separating line between classes learned by a machine learning model in classification problems.

- Decision tree

  A sequence of flowchart-like statements which represents possible decisions and their respective consequences.

- Deep neural network/deep machine learning model

  A neural network with many hidden layers.

- Dimension reduction

  The process of reducing the number of features to represent a vector.

- Dimensions

  The term dimensions has several definitions, most often used to refer to the number of items in a feature vector.

- Discrete variable

  A variable that can take a finite number of values.

- Dynamic machine learning model

  A model that is learning in a continuous, real-time fashion.

- Ensemble

  The use of a collection of machine learning models to achieve better predictive performance.

- Environment

  The state of the world that contains the agent.

- Example

  A collection of features. Examples can be labeled or unlabeled.

- Exploratory Data Analysis (EDA)

  EDA is a data science technique that seeks to understand data insights through statistical analysis and graphical visualization.

- False negative

  A model output that is incorrectly predicted as false.

- False positive

  A model output that is incorrectly predicted as true.

- Falcon

  A management framework for Apache, used for governance of data in Hadoop clusters.

- Feature

  A feature is a data attribute and its value. As an example, skin color is brown is a feature where skin color is the attribute and brown is the value.

- Feature selection

  The process of choosing the features required to explain the outputs of a statistical model while excluding irrelevant features.

- Feature set

  The selection of features a machine learning model trains on.

- Feedforward neural network

  A type of neural network that has no recursive or cyclical relationships. As such, data feeds forward.

- Flume

  A distributed service that collects, aggregates, and transfers large, real-time data into the HDFS (Hadoop Distributed File System).

- F-Score

  An evaluation technique that combines precision and recall into a measure of classification effectiveness.

- Generalization

  The ability for a machine learning model to correctly make predictions on previously unseen data.

- Goodness of fit

  This term refers to how well a model fits an observation set and summarizes any differences between observed and expected values.

- Hadoop

  Hadoop is a commonly used, open source framework developed by Apache that caters for distributed processing of big datasets.

- Hadoop Common

  Common libraries, modules, and extensions that support Hadoop modules.

- Hadoop MapReduce

  Hadoop MapReduce is a design framework for software development that facilitates the processing of large datasets in parallel.

- HBase

  Refers to a Hadoop database that enables the storage and management of sparse data.

- Heuristic

  A heuristic is a quick fix or, in other words, a quick machine learning solution to a problem.

- Hidden layer

  A layer in a neural network between the input layer and output layer.

- Hierarchical clustering

  A type of clustering algorithm that clusters groups with ranks.

- Hive

  Hive is an open source library that enables Hadoop tasks to be programmed using SQL. This provides a relational database storage structure.

- Hyperparameters

  A hyperparameter refers to the details that are edited between successive training iterations of a machine learning model.

- Hyperplane

  A separating boundary that enables classification of data points.

- Implicit bias

  This form of bias refers to associations made automatically based on an individual's mental constructs and memories which can affect the way machine learning models are designed and developed.

- Impala

  An open source, massively parallel processing database for Apache that enables data querying from HDFS or HBase.

- Inference

  The process of a machine learning model making a prediction when applied to an unlabeled dataset.

- Input layer

  The layer of a neural network that receives the input.

- Internet of Things (IoT)

  The IoT refers to devices that are connected to the Internet and as such send data via sensors to a cloud-based ecosystem.

- Interpretability

  The degree that a machine learning model's behaviors can be explained. Regression models, for instance, may be easier to explain than deep neural network models.

- Iteration

  A round of computing model weights and updating during training.

- JSON

  JSON, or JavaScript Object Notation, is a lightweight data file type that facilitates quicker retrieval of data, particularly in web-focused applications, as it is easier to parse and generate.

- Keras

  A machine learning API for Python.

- Key points

  This term refers to (x, y) co-ordinates of features in an image.

- K-Means

  A common unsupervised, clustering algorithm that groups data around centroid locations.

- Label

  The output or answer of an example.

- Labeled examples

  Data that contain features and a label.

- Latency

  A time delay within a system.

- Linear machine learning model

  A type of model that assigns weights to each feature to calculate predictions.

- Loss

  A value that represents how distant a model's prediction is from the associated label.

- Machine data

  Data, usually structured, that is created by devices or machines such as algorithms or sensors.

- Machine learning

  A subset of AI where computer models are trained to learn from their actions and environment over time with the intention of performing better.

- Massive data

  The term given to an enormous collection of records, interchangeable with big data.

- Matlab

  Matlab is the language many university students begin with. It is useful for fast prototyping, as it contains a large machine learning repository.

- Matrix

  A matrix is an array used for representation of data.

- Metadata

  Referring to data held about data and gives information and context about what initially the data item is about.

- MongoDB

  MongoDB is a NoSQL database, which is used to store unstructured data with no particular schema.

- Model

  The signal representation a machine learning algorithm learns from provided training data.

- Model selection

  The process of choosing a statistical model for machine learning from known models.

- Model training

  The process of choosing the best machine learning model for a given problem.

- Multi-class classification

  Classifications that are separated between two or more distinct groups or classes.

- NewSQL

  NewSQL databases maintain the integrity of typical relational SQL databases while providing the scalable performance of NoSQL structures.

- Noise

  Anything that is not part of the signal or is making the signal less apparent is noise.

- Normalization

  The method of converting a range of values into a standard range or scale.

- NoSQL

  A modern approach to databases that is useful for managing data that changes frequently or data that is unstructured or semi-structured in nature. Rows can all have their own set of unique column values. NoSQL has been driven by the requirements to provide better performance in storing big data. The architecture has better write performance, takes less storage space with compression, and reduces operational overhead.

- numpy

  An open source maths library for Python.

- Objective

  The metric that a machine learning model is attempting to optimize.

- Outliers

  Also known as anomalies, outliers are values that are not consistent with the bulk of the data and distant from other values.

- Output layer

  The final layer of a neural network that provides the model output.

- Overfitting

  Overfitting occurs when a machine learning model performs well on a training dataset but does not perform as well on a validation set. The model essentially learns the training data.

- Parameters

  Attribute values that control the output and behavior of a machine learning system.

- *P* value

  On the basis the null hypothesis is true, the *P* value refers to the probability of receiving a value equal to or greater than the observed value.

- Performance

  A measure of how correct a machine learning model is.

- Policy

  A mapping of states to actions used by an agent in reinforcement learning.

- Precision

  A metric used to evaluate classification models reflecting the number of true positives over the total.

- Prediction

  A machine learning model's output given input data.

- Ranger

  A data management and monitoring framework to ensure data security.

- R

  R is a software and programming language for statistical computing and graphics. Project R has been designed as a data mining tool, while R programming language is a high-level statistical language that is used for analysis.

- Real-time data

  Real-time data is data that is generated, transferred, processed, stored, and visualized within milliseconds.

- Regression

  This is a supervised learning technique in which the output is a real value.

- Smart health or smart care

  Smart health refers to the use of mobile and IoT technologies for the better health and well-being of people and to improve quality of life.

- Sqoop

  An extremely useful, open source application to facilitate the transfer of data between Hadoop and traditional relational database systems.

- Spark

  Spark is a simple programming model that can be used with Java, Scala, Python, and R that enables large-scale data processing applications to be written quickly.

- SQL

  SQL (Structured Query Language) is a language used for managing data held in a traditional database management system.

- Tensor

  A tensor is an object similar to a vector that is represented as an array that can hold data in N dimensions. A tensor is a generalization of a matrix in N-dimensional space.

- TensorFlow

  An Alphabet-backed, open source library of data computations optimized for machine learning that enables multilayered neural networks and quick training.

- True negative

  A model output that is correctly predicted as false.

- True positive

  A model output that is correctly predicted as true.

- Underfitting

  Underfitting occurs when a machine learning model is not able to capture the signal of the data and will have poor performance on both training and validation data.

- XML

  An acronym for Extensible Markup Language, XML is another form of flat data file designed to make the importing, exporting, generation, and movement of data easier.

Index

© Arjun Panesar 2021
A. Panesar, *Machine Learning and AI for Healthcare*, https://doi.org/10.1007/978-1-4842-6537-6

B

C

D

E

F

G

H

N, O

P, Q

R

S

T

U

V

W, X, Y, Z

Printed in the United States
By Bookmasters